FLUID MECHANICS

Stephen M. Richardson

*Department of Chemical Engineering
& Chemical Technology,
Imperial College, London, U.K.*

⊙HEMISPHERE PUBLISHING CORPORATION
A member of the Taylor & Francis Group
New York   Washington   Philadelphia   London

FLUID MECHANICS

1 2 3 4 5 6 7 8 9 0   B R B R   8 9 8 7 6 5 4 3 2 1 0 9

The editors were Brenda Brienza and Carolyn Ormes. The typesetter was Edwards Brothers, Inc. Braun-Brumfield, Inc. was the printer and binder.
Cover design by Debra Eubanks Riffe.

Library of Congress Cataloging-in-Publication Data

Richardson, Stephen M.
  Fluid mechanics.

  Includes index.
  1. Fluid mechanics. I. Title.
TA 357.R53 1989   620.1'06   88-34785

ISBN 0-89116-671-8

CONTENTS

PREFACE

This book on fluid mechanics is intended to be read by engineering and science students who wish to undertake a moderately advanced study of theoretical aspects of the subject. It is based on a course that has been taught for many years to undergraduate and postgraduate students of chemical engineering at Imperial College. The reader is not assumed to have any previous knowledge of fluid mechanics, though it is anticipated that some readers will in fact have some. Because the book concentrates on theoretical aspects, however, the reader is assumed already to have some knowledge of mathematics. No apology is made for this: very few non-trivial fluid mechanics problems of any sort can be analyzed properly without mathematics. This is not to say that arguments should only be presented mathematically and, indeed, in this book they are not: physical arguments are used wherever they are necessary or helpful. Nevertheless, the emphasis is undoubtedly mathematical. Thus the reader is presumed to have a proficiency in linear and vector algebra and partial differentiation. It is also helpful, thought it is not essential, to have an acquaintanceship with differential equations and vector calculus; some familiarity with complex algebra and tensor field theory is also an advantage.

Chapters 1 to 7 cover the fundamentals of fluid mechanics. They are intended to be read sequentially, though many details can be omitted at a first reading.

Furthermore, certain parts, notably sections 4.5 and 5.3 which deal with heat transfer, can be omitted altogether if so desired. While a deep understanding of fluid mechanics can only be obtained by mastering all of the material in these chapters, much of the material, particularly in chapters 1 to 6, can be used with only a qualitative, physical understanding of its origin. Because it is usually impossible to solve exactly the equations determining a flow, approximate solutions must be sought instead. The logical basis for such approximations is discussed in chapter 7, and it is, therefore, crucial to understand the material in it if one is to solve fluid mechanics problems.

Chapters 8 to 13 deal with applications of the fundamentals discussed in the preceding chapters. The applications are model ones, that is they are abstractions from real ones. Such abstraction is a necessary part of any theory: an attempt to deal with all the special features or peculiarities of a real problem would usually render it intractable; abstraction so that only the essential physics of the problem is dealt with often renders it tractable. Chapters 8 to 12 can be read in any order, or even omitted if so desired. Nevertheless, some material is dealt with in detail in earlier chapters and only briefly in later ones. References back to the earlier chapters should mean that this presents few difficulties, however. Chapter 13 is intended to be read last since it refers to additional problems and reading which will lead to further understanding of fluid mechanics.

Appendixes A to C are intended to be referred to whenever necessary. Appendix A summarizes the mathematics, that is the algebra and calculus, of fields, of which extensive use is made in the preceding chapters. Appendix B gives the basic equations of fluid mechanics in the three common coordinate systems used throughout the text; the equations there should also be useful when new problems are to be tackled. Appendix C and the Bibliography include the additional problems and reading, respectively, referred to in chapter 13. The Nomenclature lists the notation and gives the meanings of the most commonly used symbols.

It is a pleasure to acknowledge the encouragement and help given to me while writing this book by many of my colleagues and by three of them in particular. Alan Cornish first gave me an interest in fluid mechanics and has continued to be a ready source of good advice. Geoffrey Hewitt persuaded me to write this book and encouraged me to finish it. Anthony Pearson taught me the proper use of theory in practical problems; without him, this book could not have been written. For different reasons, this book could not have been written without the support of my wife Hilary and children Helen, Martin, and Susan and it is to them that it is dedicated.

Stephen M. Richardson

FUNDAMENTALS

Part 1 covers the fundamentals of fluid mechanics. We start in Chapter 1 by defining a continuum and a reference frame; we also consider choice of a coordinate system for a given flow situation. We then briefly introduce the theory of scalar, vector, and tensor fields. Chapter 2 considers flow kinematics, that is, the geometry of flows but not the way in which the flows might exist: thus we ignore the forces required to produce the flows. Chapter 3 considers stress and shows why the stress tensor is a fundamental quantity in continuum mechanics. Chapter 4 deals with mass, linear momentum, angular momentum, and energy conservation equations, which we develop for quite general continua, whether solid or fluid. Chapter 5 is about constitutive equations, that is, the equations of state that describe the behavior of materials. It is here that we particularize our study to fluid mechanics and specifically to the flow of incompressible (i.e., constant density) fluids. We consider a range of different types of fluid, including the Newtonian fluid, which forms the basis for most of the subsequent discussion. In Chapter 6 we discuss initial conditions and, particularly, boundary conditions. Finally, in Chapter 7, we discuss the use of dimensionless groups, in particular, the way in which simplifications can be made to the equations governing the motion of a fluid and hence how approximate solutions of the equations can be obtained. Such approximate solutions form the basis of Part 2.

ONE

PRELUDE

1.1 CONTINUUM HYPOTHESIS

When a body is viewed on a molecular, that is on a small, scale, properties of the body have an extremely nonuniform spatial distribution. Continuum mechanics in general, and fluid mechanics in particular, is normally concerned with the behavior of the body on a large scale. As a result, the molecular structure of the body need not be taken into account explicitly and we may invoke the *continuum hypothesis* and assume that the body has a continuous structure, that is, we may assume that there exist volumes, called *material points* or *particles*, in which properties of the body are constant. To be precise,

1. Let L_s denote the small length scale, which might be identified with the mean free path of the molecules, that is, the average distance traveled by molecules between collisions, or the correlation length of the molecules, which may be thought of as the typical distance over which a given molecule can influence the behavior of its neighbors, whichever is the larger.
2. Let L_l denote the large length scale, which might be identified with a characteristic geometric dimension of the flow so that, for example, for flow in a pipe of radius R, it might be identified with R.

Then we assume that there exists an intermediate length scale L_m such that

$$L_s << L_m << L_l \tag{1.1-1}$$

(see Fig. 1.1) and we identify L_m with the dimensions of a material point, so that the volume of a material point is of order L_m^3. In practice, L_s is usually (but by

3

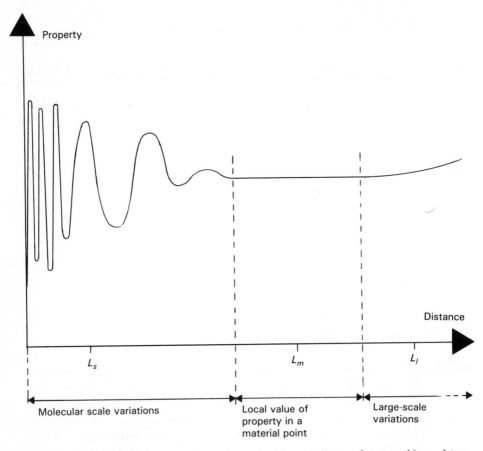

Figure 1.1. Variation of a typical property such as density with distance from an arbitrary datum; note that distance is on a nonuniform scale.

no means always) so much smaller than L_l that the existence of the intermediate length scale L_m is assured and we may, as a result, attach a definite meaning to the notion of properties at a point. Thus, for example, the density ρ at a point may be defined thus:

$$\rho = \lim_{V \to 0} \frac{M}{V} \tag{1.1-2}$$

where M denotes the mass of a volume V of a body about the point in question: the limit as V tends to zero is to be interpreted in the sense that V becomes very small compared with L_l^3 but remains very large compared with L_s^3. Provided that we only seek to describe the mechanics of bodies on a length scale comparable with or smaller than L_s in an average way—and henceforth we always will—we may regard material properties as continuous functions of position.

1.2 REFERENCE FRAMES AND COORDINATE SYSTEMS

Motions or flows are defined only relative to a reference frame, which is a structure, the component parts of which are fixed relative to one another, at least over the time scale of the motion. Clearly, choice of a reference frame is quite arbitrary: we merely choose the most convenient from an infinite set of reference frames. We assume, however, that there exists a subset of this infinite set, the subset of *inertial* reference frames, in which linear and angular momentum are conserved (see Sections 4.3 and 4.4). Given a reference frame, a three-dimensional space (which is Euclidean: we are not concerned with non-Euclidean spaces) is fixed to it, and motion is assumed to occur through it in the sense that material points or particles of the body occupy different positions **x** in the space at different times t.

Having fixed a three-dimensional space to a reference frame, we choose a three-dimensional coordinate system to span the space. A three-dimensional coordinate system is defined by an origin O and three noncoplanar directions ξ_1, ξ_2, and ξ_3. Clearly, choice of a coordinate system is, like choice of a reference frame, quite arbitrary: we merely choose the most convenient from an infinite set. We will, however, always choose an orthogonal right-handed coordinate system. An orthogonal system is one for which the directions associated with (ξ_1, ξ_2, ξ_3) are mutually perpendicular or normal. A right-handed system is one such that, if ξ_1 and ξ_2 are oriented in the plane of the page as shown in Fig. 1.2, ξ_3 is oriented out of the page (i.e., toward the viewer): the three quantities $\mathbf{i}_{\xi_1}$, $\mathbf{i}_{\xi_2}$, and $\mathbf{i}_{\xi_3}$ denote *unit vectors* (i.e., vectors of unit magnitude), aligned in the ξ_1, ξ_2, and ξ_3 directions, respectively. Note that if (ξ_1, ξ_2, ξ_3) forms a right-handed system, then so do (ξ_2, ξ_3, ξ_1) and (ξ_3, ξ_1, ξ_2), whereas (ξ_1, ξ_3, ξ_2), (ξ_3, ξ_2, ξ_1), and (ξ_2, ξ_1, ξ_3) form left-handed systems: the triplet (ξ_1, ξ_2, ξ_3) is ordered such that cyclical permutations of ξ_1, ξ_2, and ξ_3 preserve its handedness and noncyclical ones do not. The three most common orthogonal coordinate systems, and the only ones to be used here, are (see Fig. 1.3)

- Rectangular (or Cartesian) coordinates (x, y, z)
- Cylindrical polar coordinates (r, θ, z)
- Spherical polar coordinates (r, θ, α).

Note that the directions of the unit vectors $\mathbf{i}_r$ and $\mathbf{i}_\theta$ in cylindrical polar coordinates and of $\mathbf{i}_r$, $\mathbf{i}_\theta$, and $\mathbf{i}_\alpha$ in spherical polar coordinates change with position in space. Cylindrical polar coordinates are related to rectangular ones as follows:

$$x = r\cos\theta \qquad y = r\sin\theta \qquad z = z \qquad (1.2\text{-}1)$$

$$r = \sqrt{x^2 + y^2} \qquad \theta = \tan^{-1}\frac{y}{x} \qquad z = z \qquad (1.2\text{-}2)$$

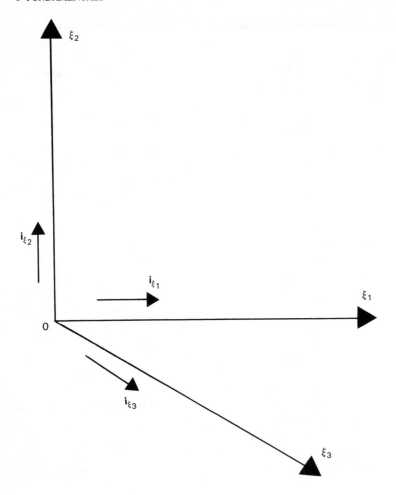

Figure 1.2. Orthogonal right-handed coordinate system.

Spherical polar coordinates are related to rectangular ones as follows:

$$x = r \sin \theta \cos \alpha \qquad y = r \sin \theta \sin \alpha \qquad z = r \cos \theta \qquad (1.2\text{-}3)$$

$$r = \sqrt{x^2 + y^2 + z^2} \qquad \theta = \tan^{-1} \frac{\sqrt{x^2 + y^2}}{z} \qquad \alpha = \tan^{-1} \frac{y}{x} \qquad (1.2\text{-}4)$$

Since choice of a coordinate system is motivated by convenience, it follows, for example, that rectangular coordinates would reasonably be used in problems involving flows past flat plates, cylindrical polar coordinates in problems involving flows through pipes of circular cross section, and spherical polar coordinates in problems involving flows about balls of circular cross section.

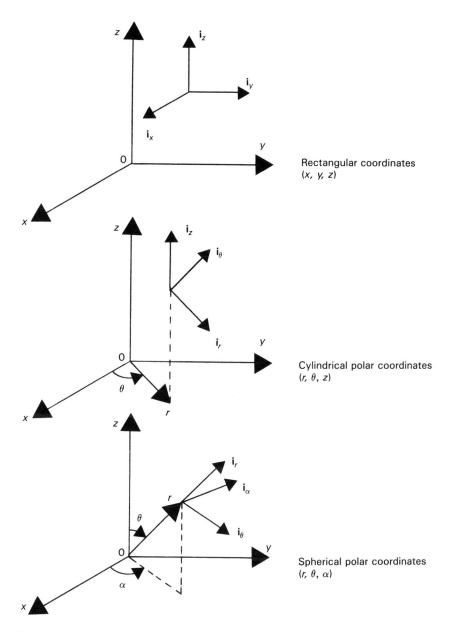

Figure 1.3. Rectangular, cylindrical polar and spherical polar coordinate systems.

1.3 SCALAR, VECTOR, AND TENSOR FIELDS

An Nth-order tensor field is a quantity that is a function of position $\mathbf{x}$ in three-dimensional space (and perhaps, but not necessarily, of time t) with size or magnitude and an ordered set of N associated directions. Thus, if $N = 0$, the quantity is a scalar field. An example is the density field ρ. Similarly, if $N = 1$, the quantity is a vector field. An example is the velocity field $\mathbf{u}$, another is the position field $\mathbf{x}$ itself. If $N = 2$, the quantity is what we will henceforth call a tensor field, since we will never here need $N > 2$. An example is the stress field $\boldsymbol{\tau}$: this has two associated directions because it is force (a vector, with one associated direction) per unit area (also a vector, with one associated direction given by a vector normal to it); the directions are ordered because the direction of the force and that of the area cannot be interchanged (unless they are both the same). Note that, here and henceforth, the symbol for a quantity with associated directions, whether a vector or a tensor, is always written in boldface.

The reason for introducing the concept of fields in general, and tensor fields in particular, is that they are an inescapable part of the physical world. Treatments of continuum mechanics and hence of fluid mechanics that do not involve tensor fields ignore, or at the very least confuse, some real physics. To be specific, they fail to recognize the tensorial nature of stress and are, as a result, severely limited in the classes of problems which they can be used to analyze. Note, however, that we do not seek here to give a complete account of scalar, vector, and tensor fields. Instead, we give in summary form properties of these fields that we will need in what follows.

We now suppose that we have selected an appropriate reference frame and an appropriate orthogonal, right-handed coordinate system (ξ_1, ξ_2, ξ_3) in three-dimensional space to span it (see Section 1.2). Then,

- A typical scalar field s has one (i.e., 3^0) component s.
- A typical vector field $\mathbf{v}$ has three (i.e., 3^1) components v_{ξ_1}, v_{ξ_2}, and v_{ξ_3} (i.e., one scalar associated with each coordinate direction), so that

$$\mathbf{v} = v_{\xi_1}\mathbf{i}_{\xi_1} + v_{\xi_2}\mathbf{i}_{\xi_2} + v_{\xi_3}\mathbf{i}_{\xi_3} \qquad (1.3\text{-}1)$$

(recall that $\mathbf{i}_{\xi_1}$, $\mathbf{i}_{\xi_2}$, and $\mathbf{i}_{\xi_3}$ are unit vectors in the ξ_1, ξ_2, and ξ_3 directions; see Fig. 1.2).

- A typical tensor field $\mathbf{T}$ has nine (i.e., 3^2) components: $T_{\xi_1\xi_1}$, $T_{\xi_1\xi_2}$, $T_{\xi_1\xi_3}$, $T_{\xi_2\xi_1}$, $T_{\xi_2\xi_2}$, $T_{\xi_2\xi_3}$, $T_{\xi_3\xi_1}$, $T_{\xi_3\xi_2}$, $T_{\xi_3\xi_3}$ (i.e., one scalar associated with each pair of coordinate directions), so that

$$\begin{aligned}
\mathbf{T} = {} & T_{\xi_1\xi_1}\,\mathbf{i}_{\xi_1}\mathbf{i}_{\xi_1} + T_{\xi_1\xi_2}\mathbf{i}_{\xi_1}\mathbf{i}_{\xi_2} + T_{\xi_1\xi_3}\,\mathbf{i}_{\xi_1}\mathbf{i}_{\xi_3} \\
& + T_{\xi_2\xi_1}\mathbf{i}_{\xi_2}\mathbf{i}_{\xi_1} + T_{\xi_2\xi_2}\mathbf{i}_{\xi_2}\mathbf{i}_{\xi_2} + T_{\xi_2\xi_3}\mathbf{i}_{\xi_2}\mathbf{i}_{\xi_3} \\
& + T_{\xi_3\xi_1}\mathbf{i}_{\xi_3}\mathbf{i}_{\xi_1} + T_{\xi_3\xi_2}\mathbf{i}_{\xi_3}\mathbf{i}_{\xi_2} + T_{\xi_3\xi_3}\mathbf{i}_{\xi_3}\mathbf{i}_{\xi_3}
\end{aligned} \qquad (1.3\text{-}2)$$

Note that the components of $\mathbf{v}$ and $\mathbf{T}$ are their physical components, that is, the components of $\mathbf{v}$ and $\mathbf{T}$ have the same dimensions as $\mathbf{v}$ and $\mathbf{T}$, respectively.

Thus the components u_{ξ_1}, u_{ξ_2}, and u_{ξ_3} of the velocity field **u** all have the same dimension as **u** (i.e., length divided by time). Such components are to be distinguished from *covariant* and *contravariant* components, which are encountered in more advanced tensor analysis and are not used here at all.

Extensive use is made in what follows of scalar fields like s, vector fields like **v**, and tensor fields like **T**. The algebra and calculus of such fields are summarized in Appendix A and frequent reference is henceforth made to it.

2.1 KINEMATICS AND DYNAMICS

Analysis of motions or flows in mechanics in general, and fluid mechanics in particular, is greatly facilitated by distinguishing between kinematics and dynamics. A kinematic description of a motion takes no consideration of the forces that cause the motion. A dynamic description, on the other hand, does consider them. It is with kinematics that we are concerned here.

2.2 EULERIAN AND LAGRANGIAN SPECIFICATIONS

Consider a motion relative to a given reference frame. Let $\mathbf{x}$ denote the position of a material point X at time t, and $\mathbf{x} + \Delta\mathbf{x}$ the position of the same material point X at time $t + \Delta t$. The velocity of X is defined as

$$\mathbf{u}(\mathbf{x}, t), \hat{\mathbf{u}}(X, t) = \lim_{\Delta t \to 0} \left(\frac{1}{\Delta t} \Delta\mathbf{x} \right) \tag{2.2-1}$$

Note that this defines two quantities: $\mathbf{u}(\mathbf{x}, t)$ is the Eulerian specification of velocity and gives the spatial distribution of velocity as a function of time; $\hat{\mathbf{u}}(X, t)$ is the Lagrangian specification of velocity and gives the velocity of material points of a deforming body as a function of time. Note that $\hat{\mathbf{u}}(X, t)$ could also be written

$\hat{\mathbf{u}}(\mathbf{x}_0(X, t_0), t - t_0)$ where $\mathbf{x}_0$ denotes the position of the material point X at time t_0. Similarly, $p(\mathbf{x}, t)$ and $\hat{p}(X, t)$ are the Eulerian and Lagrangian specifications of pressure, respectively. In general, a scalar field s, vector field $\mathbf{v}$, and tensor field $\mathbf{T}$ associated with material points X are given in an Eulerian specification by $s(\mathbf{x}, t)$, $\mathbf{v}(\mathbf{x}, t)$, and $\mathbf{T}(\mathbf{x}, t)$, respectively, and in a Lagrangian specification by $\hat{s}(X, t)$ or $\hat{s}(\mathbf{x}_0(X, t_0), t - t_0)$, $\hat{\mathbf{v}}(X, t)$, or $\hat{\mathbf{v}}(\mathbf{x}_0(X, t_0), t - t_0)$ and $\hat{\mathbf{T}}(X, t)$ or $\hat{\mathbf{T}}(\mathbf{x}_0(X, t_0), t - t_0)$, respectively.

Given a scalar, vector, or tensor field in one specification, it is in principle straightforward (though in practice often difficult) to obtain it in the other. Suppose, for example, we are given it in a Lagrangian specification and wish to obtain it in an Eulerian specification. Then we need to know the position $\mathbf{x}$ of X as a function of t. It is given by

$$\mathbf{x}(X, t) = \mathbf{x}_0(X, t_0) + \int_{t_0}^{t} \hat{\mathbf{u}}(X, t^{\#}) \, dt^{\#} \tag{2.2-2}$$

The Lagrangian specification is useful in certain contexts, for example, in flows of elasticoviscous materials in which the stress $\boldsymbol{\tau}$ in a material point X at time t depends on the deformation in the neighborhood of X at all times $t^{\#} \le t$. In general, however, use of the Lagrangian specification is cumbersome. Henceforth, therefore, we will always use an Eulerian specification unless otherwise stated.

2.3 DIFFERENTIATION FOLLOWING THE MOTION

Although a motion may be steady in an Eulerian sense, so that the velocity at each position is independent of time, a material point X of a body may accelerate. Consider, for example, a fluid of constant density flowing through a converging duct (see Fig. 2.1). Suppose that the flow is steady, so that the velocity field $\mathbf{u}$ is independent of time t at each position $\mathbf{x}$. The fluid must clearly be moving faster on average at the narrow cross section B in the duct than at the wide cross section A. It must, therefore, accelerate between A and B. Clearly, therefore, the acceleration $\mathbf{a}$ of X in an Eulerian specification is not $\partial \mathbf{u}/\partial t$ in general (though it is $d\hat{\mathbf{u}}/dt$ in a Lagrangian specification).

The correct expression for $\mathbf{a}$ in an Eulerian specification is given by differentiation following the motion. Let X have velocity $\mathbf{u}(\mathbf{x}, t)$ at position $\mathbf{x}$ and time t, and velocity $\mathbf{u}(\mathbf{x} + \Delta\mathbf{x}, t + \Delta t)$ at position $\mathbf{x} + \Delta\mathbf{x}$ and time $t + \Delta t$. Then, for small Δt, and hence small $|\Delta\mathbf{x}|$, a Taylor series expansion (in space and time, following the motion of X) leads to

$$\mathbf{u}(\mathbf{x} + \Delta\mathbf{x}, t + \Delta t) \simeq \mathbf{u}(\mathbf{x}, t) + \Delta t \frac{\partial}{\partial t} \mathbf{u}(\mathbf{x}, t) + \Delta\mathbf{x} \cdot \boldsymbol{\nabla}\mathbf{u}(\mathbf{x}, t) \tag{2.3-1}$$

Thus, in rectangular coordinates (x, y, z), with

$$\mathbf{u} = u_x \mathbf{i}_x + u_y \mathbf{i}_y + u_z \mathbf{i}_z \qquad \mathbf{x} = x\mathbf{i}_x + y\mathbf{i}_y + z\mathbf{i}_z \qquad \Delta\mathbf{x} = \Delta x \mathbf{i}_x + \Delta y \mathbf{i}_y + \Delta z \mathbf{i}_z \tag{2.3-2}$$

the expansion leads to

$$u_x((x + \Delta x)\mathbf{i}_x + (y + \Delta y)\mathbf{i}_y + (z + \Delta z)\mathbf{i}_z, (t + \Delta t))\mathbf{i}_x$$

$$+ u_y((x + \Delta x)\mathbf{i}_x + (y + \Delta y)\mathbf{i}_y + (z + \Delta z)\mathbf{i}_z, (t + \Delta t))\mathbf{i}_y$$

$$+ u_z((x + \Delta x)\mathbf{i}_x + (y + \Delta y)\mathbf{i}_y + (z + \Delta z)\mathbf{i}_z, (t + \Delta t))\mathbf{i}_z$$

$$\simeq u_x(x\mathbf{i}_x + y\mathbf{i}_y + z\mathbf{i}_z, t)\mathbf{i}_x + u_y(x\mathbf{i}_x + y\mathbf{i}_y + z\mathbf{i}_z, t)\mathbf{i}_y$$

$$+ u_z(x\mathbf{i}_x + y\mathbf{i}_y + z\mathbf{i}_z, t)\mathbf{i}_z$$

$$+ \Delta x \frac{\partial}{\partial x} [u_x(x\mathbf{i}_x + y\mathbf{i}_y + z\mathbf{i}_z, t)\mathbf{i}_x + u_y(x\mathbf{i}_x + y\mathbf{i}_y + z\mathbf{i}_z, t)\mathbf{i}_y$$

$$+ u_z(x\mathbf{i}_x + y\mathbf{i}_y + z\mathbf{i}_z, t)\,\mathbf{i}_z]$$

$$+ \Delta y \frac{\partial}{\partial y} [u_x(x\mathbf{i}_x + y\mathbf{i}_y + z\mathbf{i}_z, t)\mathbf{i}_x + u_y(x\mathbf{i}_x + y\mathbf{i}_y + z\mathbf{i}_z, t)\mathbf{i}_y$$

$$+ u_z(x\mathbf{i}_x + y\mathbf{i}_y + z\mathbf{i}_z, t)\mathbf{i}_z]$$

$$+ \Delta z \frac{\partial}{\partial z} [u_x(x\mathbf{i}_x + y\mathbf{i}_y + z\mathbf{i}_z, t)\mathbf{i}_x + u_y(x\mathbf{i}_x + y\mathbf{i}_y + z\mathbf{i}_z, t)\mathbf{i}_y$$

$$+ u_z(x\mathbf{i}_x + y\mathbf{i}_y + z\mathbf{i}_z, t)\mathbf{i}_z]$$

$$+ \Delta t \frac{\partial}{\partial t} [u_x(x\mathbf{i}_x + y\mathbf{i}_y + z\mathbf{i}_z, t)\mathbf{i}_x + u_y(x\mathbf{i}_x + y\mathbf{i}_y + z\mathbf{i}_z, t)\mathbf{i}_y$$

$$+ u_z(x\mathbf{i}_x + y\mathbf{i}_y + z\mathbf{i}_z, t)\mathbf{i}_z] \tag{2.3-3}$$

The acceleration of a fluid particle X is the rate of change of the velocity of that particle with respect to time. Thus,

$$\mathbf{a} = \lim_{\Delta t \to 0} \left\{ \frac{1}{\Delta t} [\mathbf{u}(\mathbf{x} + \Delta\mathbf{x}, t + \Delta t) - \mathbf{u}(\mathbf{x}, t)] \right\} = \frac{\partial \mathbf{u}}{\partial t} + \lim_{\Delta t \to 0} \left(\frac{1}{\Delta t} \Delta\mathbf{x} \cdot \nabla\mathbf{u} \right) \tag{2.3-4}$$

and so,

$$\mathbf{a} = \frac{\partial \mathbf{u}}{\partial t} + \mathbf{u} \cdot \nabla\mathbf{u} \tag{2.3-5}$$

We define the *substantial derivative*, *material derivative*, or *derivative following the motion* (or, though in continuum mechanics it sometimes has a different meaning: the *convected derivative*):

$$\frac{D}{Dt} = \frac{\partial}{\partial t} + \mathbf{u} \cdot \nabla \tag{2.3-6}$$

so that, in particular, $\mathbf{a} = D\mathbf{u}/Dt$. In general, the time rates of change of a scalar field s, vector field $\mathbf{v}$, and tensor field $\mathbf{T}$ associated with material points X are given by $\partial s/\partial t + \mathbf{u} \cdot \nabla s$, $\partial \mathbf{v}/\partial t + \mathbf{u} \cdot \nabla \mathbf{v}$, and $\partial \mathbf{T}/\partial t + \mathbf{u} \cdot \nabla \mathbf{T}$, respectively, in an Eulerian specification; in a Lagrangian specification they are, of course, given by $d\hat{s}/dt$, $d\hat{\mathbf{v}}/dt$, and $d\hat{\mathbf{T}}/dt$.

2.4 DECOMPOSITION OF MOTION

Let $\mathbf{u}$ denote the velocity at a position $\mathbf{x}$ in a moving body. Then, a small distance $\Delta\mathbf{x}$ from $\mathbf{x}$ (i.e., for small $|\Delta\mathbf{x}|$) at $\mathbf{x} + \Delta\mathbf{x}$, a Taylor series expansion leads to

$$\mathbf{u}(\mathbf{x} + \Delta\mathbf{x}) \simeq \mathbf{u}(\mathbf{x}) + \Delta\mathbf{x} \cdot \boldsymbol{\nabla}\mathbf{u}(\mathbf{x}) \tag{2.4-1}$$

Thus, in rectangular coordinates (x, y, z), with

$$\mathbf{u} = u_x\mathbf{i}_x + u_y\mathbf{i}_y + u_z\mathbf{i}_z \qquad \mathbf{x} = x\mathbf{i}_x + y\mathbf{i}_y + z\mathbf{i}_z \qquad \Delta\mathbf{x} = \Delta x\mathbf{i}_x + \Delta y\mathbf{i}_y + \Delta z\mathbf{i}_z \tag{2.4-2}$$

the expansion leads to

$$\begin{aligned}
&u_x[(x + \Delta x)\mathbf{i}_x + (y + \Delta y)\mathbf{i}_y + (z + \Delta z)\mathbf{i}_z]\mathbf{i}_x \\
&\quad + u_y[(x + \Delta x)\mathbf{i}_x + (y + \Delta y)\mathbf{i}_y + (z + \Delta z)\mathbf{i}_z]\mathbf{i}_y \\
&\quad + u_z[(x + \Delta x)\mathbf{i}_x + (y + \Delta y)\mathbf{i}_y + (z + \Delta z)\mathbf{i}_z]\mathbf{i}_z \\
&\simeq u_x(x\mathbf{i}_x + y\mathbf{i}_y + z\mathbf{i}_z)\mathbf{i}_x + u_y(x\mathbf{i}_x + y\mathbf{i}_y + z\mathbf{i}_z)\mathbf{i}_y \\
&\quad + u_z(x\mathbf{i}_x + y\mathbf{i}_y + z\mathbf{i}_z)\mathbf{i}_z \\
&\quad + \Delta x\frac{\partial}{\partial x}[u_x(x\mathbf{i}_x + y\mathbf{i}_y + z\mathbf{i}_z)\mathbf{i}_x \\
&\quad + u_y(x\mathbf{i}_x + y\mathbf{i}_y + z\mathbf{i}_z)\mathbf{i}_y + u_z(x\mathbf{i}_x + y\mathbf{i}_y + z\mathbf{i}_z)\mathbf{i}_z] \\
&\quad + \Delta y\frac{\partial}{\partial y}[u_x(x\mathbf{i}_x + y\mathbf{i}_y + z\mathbf{i}_z)\mathbf{i}_x \\
&\quad + u_y(x\mathbf{i}_x + y\mathbf{i}_y + z\mathbf{i}_z)\mathbf{i}_y + u_z(x\mathbf{i}_x + y\mathbf{i}_y + z\mathbf{i}_z)\mathbf{i}_z] \\
&\quad + \Delta z\frac{\partial}{\partial z}[u_x(x\mathbf{i}_x + y\mathbf{i}_y + z\mathbf{i}_z)\mathbf{i}_x \\
&\quad + u_y(x\mathbf{i}_x + y\mathbf{i}_y + z\mathbf{i}_z)\mathbf{i}_y + u_z(x\mathbf{i}_x + y\mathbf{i}_y + z\mathbf{i}_z)\mathbf{i}_z]
\end{aligned} \tag{2.4-3}$$

Note that this is *not* an expansion in time t, unlike the expansion in Section 2.3; hence it is not an expansion following the motion of a material point.

We now define the *rate of strain* (or strain rate, or rate of deformation) tensor field $\mathbf{e}$:

$$\mathbf{e} = \boldsymbol{\nabla}\mathbf{u} + (\boldsymbol{\nabla}\mathbf{u})^T \tag{2.4-4}$$

and the *vorticity* (or spin) tensor field $\mathbf{w}$:

$$\mathbf{w} = \boldsymbol{\nabla}\mathbf{u} - (\boldsymbol{\nabla}\mathbf{u})^T \tag{2.4-5}$$

Clearly, $\mathbf{e}$ is symmetric ($\mathbf{e} = \mathbf{e}^T$) and $\mathbf{w}$ is antisymmetric ($\mathbf{w} = -\mathbf{w}^T$). Indeed,

$$\boldsymbol{\nabla}\mathbf{u} = \tfrac{1}{2}(\mathbf{e} + \mathbf{w}) \tag{2.4-6}$$

which shows that, to within a factor of two, $\mathbf{e}$ is the symmetric part, and $\mathbf{w}$ the

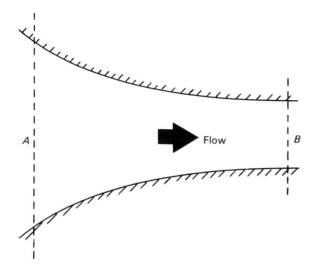

Figure 2.1. Flow in a converging duct: The fluid is moving faster on average at cross section B than at cross section A.

antisymmetric part, of $\nabla \mathbf{u}$. Let $\mathbf{v}$ denote an arbitrary three-dimensional vector. Then it can be shown that there exists a vector field called the vorticity $\boldsymbol{\omega}$, which is not to be confused with the tensor field $\mathbf{w}$ also called the vorticity, though the two are related, as we now see, such that

$$\boldsymbol{\omega}_{\wedge}\mathbf{v} = -\mathbf{v}_{\wedge}\boldsymbol{\omega} = \mathbf{v} \cdot \mathbf{w} = \mathbf{v} \cdot (\nabla \mathbf{u} - (\nabla \mathbf{u})^T) \tag{2.4-7}$$

and, entirely equivalently,

$$\boldsymbol{\omega} = \nabla_{\wedge}\mathbf{u} \tag{2.4-8}$$

Note that it is possible to define the vector (strictly, the pseudovector; see Sections A1.2 and A1.3 of Appendix A) field $\boldsymbol{\omega}$ because space is three dimensional: in n-dimensional space, an antisymmetric tensor has at most $n (n - 1)$ nonzero components, half of which are independent; we can form an n-dimensional vector from this tensor only if $n = \frac{1}{2} n (n - 1)$, that is, nontrivially, only when $n = 3$. Thus, in rectangular coordinates (x, y, z), with $\mathbf{u}$ given by Eq. (2.4-2):

$$\begin{aligned}
\mathbf{e} = {} & 2 \frac{\partial u_x}{\partial x} \mathbf{i}_x \mathbf{i}_x + \left(\frac{\partial u_y}{\partial x} + \frac{\partial u_x}{\partial y} \right) \mathbf{i}_x \mathbf{i}_y + \left(\frac{\partial u_z}{\partial x} + \frac{\partial u_x}{\partial z} \right) \mathbf{i}_x \mathbf{i}_z \\
& + \left(\frac{\partial u_y}{\partial x} + \frac{\partial u_x}{\partial y} \right) \mathbf{i}_y \mathbf{i}_x + 2 \frac{\partial u_y}{\partial y} \mathbf{i}_y \mathbf{i}_y + \left(\frac{\partial u_z}{\partial y} + \frac{\partial u_y}{\partial z} \right) \mathbf{i}_y \mathbf{i}_z \\
& + \left(\frac{\partial u_z}{\partial x} + \frac{\partial u_x}{\partial z} \right) \mathbf{i}_z \mathbf{i}_x + \left(\frac{\partial u_z}{\partial y} + \frac{\partial u_y}{\partial z} \right) \mathbf{i}_z \mathbf{i}_y + 2 \frac{\partial u_z}{\partial z} \mathbf{i}_z \mathbf{i}_z
\end{aligned} \tag{2.4-9}$$

(clearly, $\mathbf{e} = \mathbf{e}^T$) and

$$\boldsymbol{\omega} = \left(\frac{\partial u_z}{\partial y} - \frac{\partial u_y}{\partial z}\right)\mathbf{i}_x + \left(\frac{\partial u_x}{\partial z} - \frac{\partial u_z}{\partial x}\right)\mathbf{i}_y + \left(\frac{\partial u_y}{\partial x} - \frac{\partial u_x}{\partial y}\right)\mathbf{i}_z \qquad (2.4\text{-}10)$$

Note, incidentally, that other definitions of $\mathbf{e}$ and $\boldsymbol{\omega}$ exist, differing by factors of two and by signs. The components of $\mathbf{e}$ and $\boldsymbol{\omega}$ in cylindrical polar and spherical polar coordinates, as well as in rectangular coordinates, are given in Section A2.1 of Appendix B.

It follows from Eqs. (2.4-1), (2.4-6), and (2.4-7) that, in the limit as $|\Delta\mathbf{x}| \to 0$:

$$\mathbf{u}(\mathbf{x} + \Delta\mathbf{x}) = \mathbf{u}(\mathbf{x}) + \tfrac{1}{2}\,\Delta\mathbf{x}\cdot\mathbf{e}(\mathbf{x}) - \tfrac{1}{2}\,\Delta\mathbf{x}\wedge\boldsymbol{\omega}(\mathbf{x}) \qquad (2.4\text{-}11)$$

$$\uparrow \qquad\qquad \uparrow \qquad\qquad \uparrow$$
$$(a) \qquad\qquad (b) \qquad\qquad (c)$$

Thus we have decomposed the velocity field $\mathbf{u}$ into three components (a), (b), and (c), which we identify as *translation*, *straining*, and *rotation*, respectively. In order to see why we identify them in this way, we consider first two material points X_1 and X_2. At time t_1, the positions of X_1 and X_2 are $\mathbf{x}(X_1, t_1)$ and $\mathbf{x}(X_2, t_1)$, respectively, while at time $t_2 > t_1$ their positions are $\mathbf{x}(X_1, t_2)$ and $\mathbf{x}(X_2, t_2)$. It then follows that translation, straining, and rotation cause X_1 and X_2 to move as shown in Fig. 2.2. More generally, translation, straining, and rotation cause an elementary volume of material to move as shown in Fig. 2.3. In fact, a combination of translation, straining, and rotation constitutes the most general possible motion. If no straining occurs, so that the distances between all pairs of material points are constant with respect to time, then the motion, called a *rigid body motion*, comprises a combination of translation and rotation alone. We now consider three particular velocity fields in order to elucidate the nature of the terms (a), (b), and (c) in Eq. (2.4-11), and hence see why we have identified them in the way that we have.

If the velocity field $\mathbf{u}$ is uniform in the vicinity of some position $\mathbf{x}$, it follows that all the material in the vicinity of $\mathbf{x}$ moves with the fluid at $\mathbf{x}$, that is, the material in the vicinity of $\mathbf{x}$ translates with uniform rectilinear velocity $\mathbf{u}(\mathbf{x})$. Because the velocity is uniform in the vicinity of $\mathbf{x}$, it also follows that $\nabla\mathbf{u}(\mathbf{x})$ vanishes and hence that $\mathbf{e}(\mathbf{x})$ and $\boldsymbol{\omega}(\mathbf{x})$ vanish too. Thus Eq. (2.4-11) yields

$$\mathbf{u}(\mathbf{x} + \Delta\mathbf{x}) = \mathbf{u}(\mathbf{x}) \qquad (2.4\text{-}12)$$

As a result, we may identify the term (a) in it with translation.

Consider the velocity field $\mathbf{u}$ given in rectangular coordinates (x, y, z) by

$$\mathbf{u} = \epsilon\, x\, \mathbf{i}_x \qquad (2.4\text{-}13)$$

where ϵ may be a function of time t. (Note that, for kinematic purposes, it is unnecessary to specify how such a velocity field arises; it is sufficient merely to require it to exist.) For $\epsilon > 0$, this corresponds to a flow extending in the x direction (see Fig. 2.4). Clearly, for the velocity field given by Eq. (2.4-13), there

(a) Translation

(b) Straining

(c) Rotation

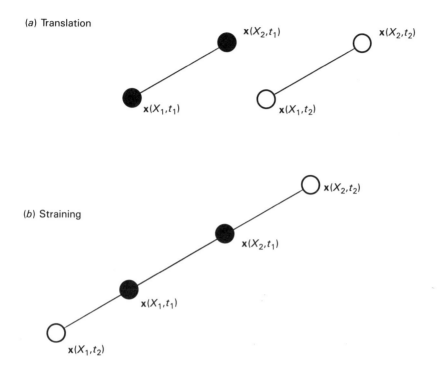

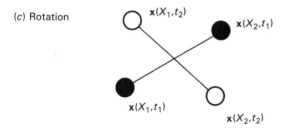

Figure 2.2. The three components of motion are (a) translation. (b) straining and (c) rotation.

is no motion in the y or z directions and no rotation. Consider now two material points X_1 and X_2, the x coordinates of which at time t are x_1 [$= x(X_1, t)$] and x_2 [$= x(X_2, t)$], respectively, while at time $t + \Delta t$ their x coordinates are $x_1 + \Delta x_1$ [$= x(X_1, t + \Delta t)$] and $x_2 + \Delta x_2$ [$= x(X_2, t + \Delta t)$]. The distance l between the points at time t is clearly given by $|x_2 - x_1|$ while at time $t + \Delta t$ the distance $l + \Delta l$ between them is given by $|(x_2 + \Delta x_2) - (x_1 + \Delta x_1)|$ (see Fig. 2.5). In the limit as $\Delta t \rightarrow 0$, $\Delta x_1 = \epsilon\, x_1\, \Delta t$ and $\Delta x_2 = \epsilon\, x_2\, \Delta t$, so that $l + \Delta l$ is given by

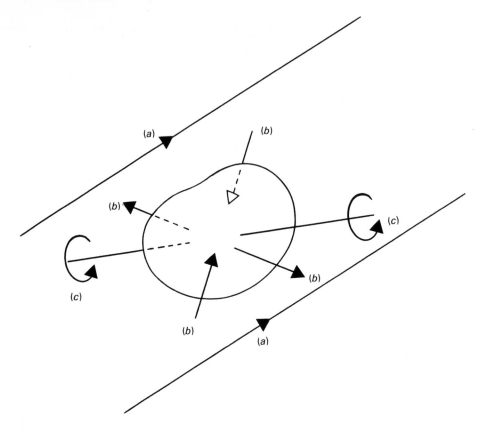

Figure 2.3. Decomposition of a general motion into (*a*) translation, (*b*) straining and (*c*) rotation.

$|x_2 - x_1|$ $(1 + \epsilon \, \Delta t)$. Thus the relative extension $\Delta l/l$ between the two material points is given by $\epsilon \, \Delta t$ and so the extension rate

$$\frac{1}{\Delta t} \frac{\Delta l}{l}$$

is given by ϵ. (Note that ϵ is independent of position, so the extension rate is also independent of position.) We now note that it follows from Eqs. (2.4-9), (2.4-10), and (2.4-13) that

$$\mathbf{e} = 2 \, \epsilon \, \mathbf{i}_x \mathbf{i}_x \qquad \boldsymbol{\omega} = \mathbf{0} \tag{2.4-14}$$

We note that

1. The only nonvanishing component of $\mathbf{e}$ is precisely twice the extension rate and is moreover aligned with $\mathbf{i}_x \mathbf{i}_x$, which is to be expected since the extension is in the x direction of elements aligned in the x direction.
2. $\boldsymbol{\omega}$ vanishes because (as we will see) the motion does not involve rotation.

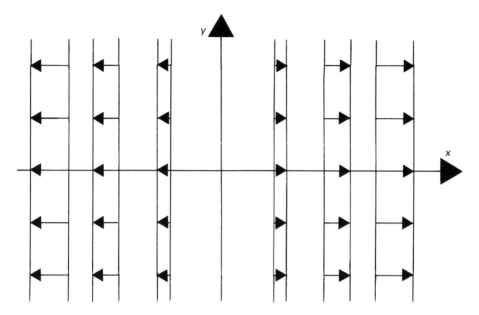

Figure 2.4. Flow that is extending in one direction.

3. **u** given by Eq. (2.4-13) and **e** given by Eq. (2.4-14) are related [using Eq. (A1.2-9) of Appendix A] by

$$\mathbf{u} = \tfrac{1}{2}\,\mathbf{x} \cdot \mathbf{e} \qquad (2.4\text{-}15)$$

where **x** denotes position [see Eqs. (2.4-2)].

As a result, we may identify the term (*b*) in Eq. (2.4-11) with straining, that is, with the expansion or contraction of elements. The factor of one-half appearing in it [and also in Eq. (2.4-15)] arises because the components of **e** are twice the corresponding extension rates. Note, incidentally, that it can be shown, though it

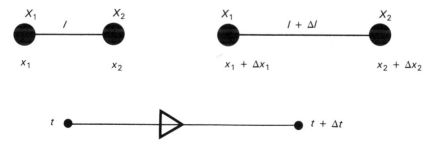

Figure 2.5. Change in distance between material points with time in a flow that is extending in one direction.

is too lengthy to do so here, that a local rectangular coordinate system (x, y, z) can always be chosen such that e becomes diagonal, that is,

$$e = 2\frac{\partial u_x}{\partial x}\mathbf{i}_x\mathbf{i}_x + 2\frac{\partial u_y}{\partial y}\mathbf{i}_y\mathbf{i}_y + 2\frac{\partial u_z}{\partial z}\mathbf{i}_z\mathbf{i}_z \qquad (2.4\text{-}16)$$

where $\mathbf{u} = u_x\mathbf{i}_x + u_y\mathbf{i}_y + u_z\mathbf{i}_z$. In fact, in this local coordinate system, the vectors $\mathbf{i}_x$, $\mathbf{i}_y$, and $\mathbf{i}_z$ are *eigenvectors* of the tensor e and the components $e_{xx} = 2\,\partial u_x/\partial x$, $e_{yy} = 2\,\partial u_y/\partial y$, and $e_{zz} = 2\,\partial u_z/\partial z$ are its *eigenvalues*.

Consider the velocity field $\mathbf{u}$ given in rectangular coordinates (x, y, z) by

$$\mathbf{u} = -\Omega y\mathbf{i}_x + \Omega x\mathbf{i}_y \qquad (2.4\text{-}17)$$

or in equivalent cylindrical polar coordinates (r, θ, z) by

$$\mathbf{u} = \Omega r\mathbf{i}_\theta \qquad (2.4\text{-}18)$$

where Ω may be a function of time t. (How such a velocity field exists is irrelevant for kinematic purposes.) For $\Omega > 0$, this corresponds to a flow which, in a right-handed coordinate system viewed in the direction of increasing z (see Fig. 2.6), is rotating clockwise about the z axis $(r = 0)$ with angular velocity given by

$$\Omega = \Omega\mathbf{i}_z \qquad (2.4\text{-}19)$$

Note that the direction associated with the angular velocity Ω is the axis of rotation (i.e., the z axis). Clearly, for the velocity field given by Eqs. (2.4-17) and (2.4-18), there is no motion in the z direction and no straining: the distances between all pairs of material points are constant with respect to time. We now note that it follows from Eqs. (2.4-9), (2.4-10), and (2.4-17) [or Eq. (2.4-18), using the expression for ω in cylindrical polar coordinates given in Eq. (A2.1-8) of Appendix B] that

$$e = 0 \qquad \omega = 2\Omega\mathbf{i}_z \qquad (2.4\text{-}20)$$

We note that

1. e vanishes because (as we have already seen) the motion does not involve straining.
2. The vorticity ω is precisely twice the angular velocity Ω.
3. $\mathbf{u}$ given by Eq. (2.4-17) and Ω given by Eq. (2.4-19) are related [using Eq. (A1.2-19) of Appendix A] by

$$\mathbf{u} = \Omega_\wedge\mathbf{x} = -\mathbf{x}_\wedge\Omega \qquad (2.4\text{-}21)$$

where $\mathbf{x}$ denotes position [see Eq. (2.4-2)].

As a result, we may identify the term (c) in Eq. (2.4-11) with rotation: the factor of one-half appearing in it arises because ω is twice Ω. Thus, to be explicit, the term $-\frac{1}{2}\,\Delta\mathbf{x}_\wedge\omega$ in Eq. (2.4-11) represents a rigid body rotation (a particular form of rigid body motion in which there is no translation) with angular velocity

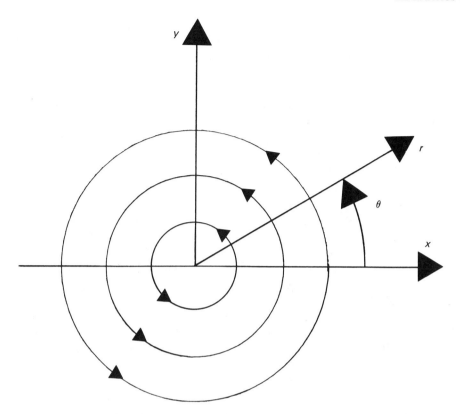

Figure 2.6. Flow that is rotating about the z axis.

$\frac{1}{2}$ $\boldsymbol{\omega}$. Note that the direction associated with the pseudovector $\boldsymbol{\omega}$ is the local axis of rotation, which is, of course, normal to the local velocity $\mathbf{u}$.

2.5 IRROTATIONAL AND SOLENOIDAL FLOWS

It was shown in Section 2.4 that the vorticity field $\boldsymbol{\omega}$ is a measure of the local rotation in a flow; an irrotational flow (strictly, an irrotational velocity field $\mathbf{u}$) is one for which $\boldsymbol{\omega}$ vanishes, that is,

$$\boldsymbol{\omega} = \boldsymbol{\nabla}_{\!\wedge}\mathbf{u} = \mathbf{0} \qquad (2.5\text{-}1)$$

We note from Eq. (A1.3-23) of Appendix A that $\boldsymbol{\nabla}_{\!\wedge}\boldsymbol{\nabla}s = \mathbf{0}$ for any scalar field s. Suppose that

$$\mathbf{u} = \boldsymbol{\nabla}\phi \qquad (2.5\text{-}2)$$

Then, clearly $\boldsymbol{\omega} = \boldsymbol{\nabla}_{\!\wedge}\mathbf{u} = \boldsymbol{\nabla}_{\!\wedge}\boldsymbol{\nabla}\phi = \mathbf{0}$, that is, $\mathbf{u}$ is irrotational. Conversely, we can show that, if $\mathbf{u}$ is irrotational (i.e., $\boldsymbol{\omega} = \mathbf{0}$), there exists a scalar field ϕ, called a *scalar potential* field, such that Eq. (2.5-2) holds. Note that adding a constant ϕ_0 to ϕ does not alter $\nabla\phi$, so ϕ is determinate only to within a constant.

A solenoidal flow (strictly, a solenoidal velocity field **u**) is one for which

$$\nabla \cdot \mathbf{u} = 0 \qquad (2.5\text{-}3)$$

We note from Eq. (A1.3-24) of Appendix A that $\nabla \cdot \nabla_{\wedge} \mathbf{v} = 0$ for any three-dimensional vector field **v**. Suppose that

$$\mathbf{u} = \nabla_{\wedge}\mathbf{\Psi} \qquad (2.5\text{-}4)$$

Then, clearly, $\nabla \cdot \mathbf{u} = \nabla \cdot \nabla_{\wedge}\mathbf{\Psi} = 0$ (i.e., **u** is solenoidal). Conversely, we can show that, if **u** is solenoidal (i.e., $\nabla \cdot \mathbf{u} = 0$), there exists a vector field $\mathbf{\Psi}$, called a *vector potential* field, such that Eq. (2.5-4) holds. Note that adding a field $\nabla\Psi_0$ to $\mathbf{\Psi}$ does not alter $\nabla_{\wedge}\mathbf{\Psi}$ (recall that $\nabla_{\wedge}\nabla\Psi_0 = \mathbf{0}$), so $\mathbf{\Psi}$ is determinate only to within a field of the form $\nabla\Psi_0$. It can be shown, though it is too lengthy to do so here, that a suitable Ψ_0 can always be chosen so that $\nabla \cdot \mathbf{\Psi} = 0$.

We see, therefore, that if a velocity field is irrotational—and we will see in Section 9.1 that this is generally the case for the flow of an inviscid fluid (i.e., a fluid of negligible viscosity)—we can define the flow in terms of the single variable ϕ rather than the three variables that comprise the components of **u**. This reduction in the number of variables leads to a substantial saving in the labor involved in solving the equations of motion of the flow and is the reason why we have introduced the concept of the scalar potential.

Similarly, if a velocity field is solenoidal, we can define the flow in terms of $\mathbf{\Psi}$ rather than **u** and automatically ensure that mass is conserved. (We will show in Section 4.2 that conservation of mass means that any flow of an incompressible material, that is, a material of constant density, is solenoidal.) This again leads to a substantial saving in the labor required to solve the equations of motion of the flow and is the reason for introducing the concept of the vector potential. Still further savings can be made if a velocity field is not only solenoidal but also two rather than three dimensional, since we can then define the flow in terms of a single variable instead of the three that comprise the components of $\mathbf{\Psi}$ or the two that comprise the components of **u**.

Thus, if **u** is a two-dimensional planar field, that is, in an appropriate rectangular coordinate system, it only has x and y components and no z component (i.e., its x and y components do not vanish but its z component does) or, in an appropriate cylindrical polar coordinate system, only r and θ components and no z component, the vector potential [see Eq. (2.5-4)] has only a single nonvanishing component:

$$\mathbf{u} = \nabla_{\wedge}\mathbf{\Psi} \qquad \mathbf{\Psi} = \psi \mathbf{i}_z \qquad (2.5\text{-}5)$$

so that

$$\mathbf{u} = \frac{\partial\psi}{\partial y}\mathbf{i}_x - \frac{\partial\psi}{\partial x}\mathbf{i}_y = \frac{1}{r}\frac{\partial\psi}{\partial\theta}\mathbf{i}_r - \frac{\partial\psi}{\partial r}\mathbf{i}_\theta \qquad (2.5\text{-}6)$$

The proof of the validity of this representation in (x, y) coordinates is as follows [that in (r, θ) coordinates is entirely analogous]. If u_x and u_y denote the x and y components of $\mathbf{u}$, respectively, then, since $\nabla \cdot \mathbf{u} = 0$, it follows from Eq. (A1.3-7) of Appendix A that

$$\frac{\partial u_x}{\partial x} + \frac{\partial u_y}{\partial y} = 0 \qquad (2.5\text{-}7)$$

We now define a quantity χ thus:

$$\chi = \int u_x \, dy + f(x) \qquad (2.5\text{-}8)$$

where $f(x)$ denotes an arbitrary function of x. It then follows that

$$u_x = \frac{\partial \chi}{\partial y} \qquad (2.5\text{-}9)$$

so that Eq. (2.5-7) becomes

$$\frac{\partial^2 \chi}{\partial x \, \partial y} + \frac{\partial u_y}{\partial y} = 0 \qquad (2.5\text{-}10)$$

whence

$$u_y = -\frac{\partial \chi}{\partial x} + \frac{d}{dx} g(x) \qquad (2.5\text{-}11)$$

where $g(x)$ denotes a function of x. Since $f(x)$ is arbitrary, we now choose it such that

$$f(x) = g(x) \qquad (2.5\text{-}12)$$

and then we choose to define a quantity ψ such that

$$\psi = \chi - f(x) = \chi - g(x) \qquad (2.5\text{-}13)$$

from which it follows that

$$u_x = \frac{\partial \psi}{\partial y} \qquad u_y = -\frac{\partial \psi}{\partial x} \qquad (2.5\text{-}14)$$

which is what we had to prove. The scalar field ψ is called a *stream function*; it is constant along a *streamline*, which is a line which is instantaneously at every point aligned with the local flow direction (i.e., with $\mathbf{u}$). The proof of this in (x, y) coordinates is as follows [that in (r, θ) coordinates is again entirely analogous]. If u_x and u_y again denote the x and y components of $\mathbf{u}$, respectively, then (see Fig. 2.7) the slope of the streamline is $dy/dx = u_y/u_x$ and so

$$u_x \, dy - u_y \, dx = 0 \qquad (2.5\text{-}15)$$

or, since $u_x = \partial \psi/\partial y$ and $u_y = -\partial \psi/\partial x$ from Eq. (2.5-6),

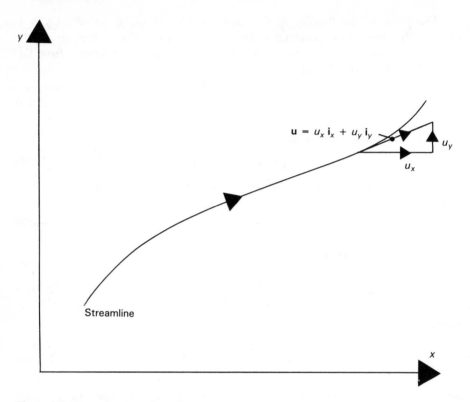

Figure 2.7. Streamline in a planar flow.

$$\frac{\partial \psi}{\partial y} \, dy + \frac{\partial \psi}{\partial x} \, dx = 0 \qquad (2.5\text{-}16)$$

But the total differential $d\psi$ of ψ is given by

$$d\psi = \frac{\partial \psi}{\partial x} \, dx + \frac{\partial \psi}{\partial y} \, dy \qquad (2.5\text{-}17)$$

so that $d\psi = 0$, from which it follows immediately that ψ is constant along a streamline. As a result, it follows that ψ must vary at most in the direction normal to a streamline. Because its gradient in that direction must increase with the speed $|\mathbf{u}|$ of a flow, it immediately follows that streamlines are close together in the parts of a flow that are moving fast and are far apart in parts that are moving slowly. Note that the sign of ψ is essentially irrelevant: we can equally validly put $\mathbf{u} = \mathbf{\nabla.\Psi}$, $\mathbf{\Psi} = -\psi \mathbf{i}_z$ instead of Eq. (2.5-5).

Similarly, if $\mathbf{u}$ is an axisymmetric field (i.e., it is independent of angular position about the axis of symmetry) and swirl-free (i.e., its component about the axis of symmetry vanishes), we can also simplify Eq. (2.5-4) to yield, in an appropriate cylindrical polar coordinate system,

$$\mathbf{u} = \nabla_\wedge \Psi \qquad \Psi = \frac{1}{r} \psi_s \mathbf{i}_\theta \qquad \text{(2.5-18)}$$

so that

$$\mathbf{u} = -\frac{1}{r} \frac{\partial \psi_s}{\partial z} \mathbf{i}_r + \frac{1}{r} \frac{\partial \psi_s}{\partial r} \mathbf{i}_z \qquad \text{(2.5-19)}$$

that is, $\mathbf{u}$ has no θ component (so that it is swirl-free) and its r and z components are independent of θ (so that it is axisymmetric); or, in an appropriate spherical polar coordinate system,

$$\mathbf{u} = \nabla_\wedge \Psi \qquad \Psi = \frac{1}{r \sin \theta} \psi_s \mathbf{i}_\alpha \qquad \text{(2.5-20)}$$

so that

$$\mathbf{u} = \frac{1}{r^2 \sin \theta} \frac{\partial \psi_s}{\partial \theta} \mathbf{i}_r - \frac{1}{r \sin \theta} \frac{\partial \psi_s}{\partial r} \mathbf{i}_\theta \qquad \text{(2.5-21)}$$

that is, $\mathbf{u}$ has no α component and its r and θ components are independent of α. The field ψ_s is called a *Stokes stream function*; the subscript s is often omitted. Like ψ, ψ_s is constant along a streamline and, again like ψ, the sign of ψ_s is essentially irrelevant.

So far we have distinguished between irrotational and solenoidal vector fields. It can in fact be shown that, subject to certain minor restrictions, any velocity field (indeed, any three-dimensional vector field) $\mathbf{u}$ can be decomposed, in what is called a *Helmholtz decomposition*, into an irrotational component and a solenoidal component:

$$\mathbf{u} = \nabla \phi + \nabla_\wedge \Psi \qquad \text{(2.5-22)}$$

The components $\nabla \phi$ and $\nabla_\wedge \Psi$ are not, however, unique in general since some fields are both irrotational and solenoidal. A trivial example is a constant vector field; a more important example is the velocity field of an incompressible material in irrotational flow.

We now consider three particular planar velocity fields $\mathbf{u}$ in order to see whether, and if so how, they might be represented in terms of a scalar potential ϕ and/or a stream function ψ. The examples are all members of a class of flows which we might call linear in the sense that the velocity $\mathbf{u}$ is a linear function of position $\mathbf{x}$. As a result, the velocity gradient $\nabla \mathbf{u}$ is independent of position and hence the rate-of-strain $\mathbf{e}$ and vorticity $\boldsymbol{\omega}$ are independent of position. If the origin ($\mathbf{x} = \mathbf{0}$) of the reference frame being used is such that $\mathbf{u} = \mathbf{0}$ there, then a Taylor series expansion (in space, and not in time, just as in Section 2.4) leads to

$$\mathbf{u}(\mathbf{x}) = \mathbf{x} \cdot \nabla \mathbf{u} = \tfrac{1}{2} \mathbf{x} \cdot \mathbf{e} - \tfrac{1}{2} \mathbf{x}_\wedge \boldsymbol{\omega} \qquad \text{(2.5-23)}$$

where $\nabla \mathbf{u}$, $\mathbf{e}$, and $\boldsymbol{\omega}$ are independent of $\mathbf{x}$, as already noted. [Note that Eq. (2.5-23) is exact whereas the clearly analogous Eqs. (2.4-1) and (2.4-11) are only approximate. This is because, whereas for general flows second and higher spatial derivatives of velocity do not vanish, for linear flows they do.]

We consider first the velocity field given in rectangular coordinates (x, y, z) by

$$\mathbf{u} = \epsilon x \mathbf{i}_x - \epsilon y \mathbf{i}_y \qquad (2.5\text{-}24)$$

[which, we note, is a slight generalization of Eq. (2.4-13)], where ϵ may be a function of time t but not of position $\mathbf{x} = x\mathbf{i}_x + y\mathbf{i}_y + z\mathbf{i}_z$. For $\epsilon > 0$ (which we henceforth assume), it corresponds to a flow extending in the x direction and contracting in the y direction. It follows from Eqs. (2.4-17) and (2.4-18) that

$$\mathbf{e} = 2\epsilon \mathbf{i}_x \mathbf{i}_x - 2\epsilon \mathbf{i}_y \mathbf{i}_y \qquad \boldsymbol{\omega} = \mathbf{0} \qquad (2.5\text{-}25)$$

from which we see that $\mathbf{e}$ and $\boldsymbol{\omega}$ are independent of position, so that the flow is linear [in fact, Eq. (2.5-23) can be simplified in this case to yield $\mathbf{u} = \frac{1}{2}\,\mathbf{x}\cdot\mathbf{e}$]. It is also irrotational, and there exists a scalar potential ϕ such that

$$\mathbf{u} = \nabla\phi \qquad \phi = \tfrac{1}{2}\epsilon\,(x^2 - y^2) \qquad (2.5\text{-}26)$$

It follows from Equation (A1.3-7) of Appendix A that

$$\nabla\cdot\mathbf{u} = 0 \qquad (2.5\text{-}27)$$

Thus the flow is also solenoidal and there exists a stream function ψ such that

$$\mathbf{u} = \nabla_\wedge\boldsymbol{\Psi} \qquad \boldsymbol{\Psi} = \psi\mathbf{i}_z \qquad \psi = \epsilon xy \qquad (2.5\text{-}28)$$

The potential lines (i.e., lines of constant scalar potential ϕ) and streamlines (i.e., lines of constant stream function ψ) are shown in Fig. 2.8; c_1 and c_2 denote arbitrary positive constants. Note that the potential lines and streamlines are mutually orthogonal. Note also that this flow involves only straining and no rotation: it is called a *simple extensional* or *simple elongational flow*; ϵ is called the *extension rate*. Its generalization to an axisymmetric velocity field is either

$$\mathbf{u} = \epsilon x\mathbf{i}_x - \tfrac{1}{2}\epsilon y\mathbf{i}_y - \tfrac{1}{2}\epsilon z\mathbf{i}_z \qquad (2.5\text{-}29)$$

which, since $\epsilon > 0$, corresponds to a flow extending in the x direction and contracting similarly in the y and z directions and is called a *simple uniaxial extensional flow* or

$$\mathbf{u} = -2\epsilon x\mathbf{i}_x + \epsilon y\mathbf{i}_y + \epsilon z\mathbf{i}_z \qquad (2.5\text{-}30)$$

which, since $\epsilon > 0$, corresponds to a flow contracting in the x direction and extending similarly in the y and z directions and is called a *simple biaxial extensional flow*.

We consider next the velocity field given by

$$\mathbf{u} = -\Omega y\mathbf{i}_x + \Omega x\mathbf{i}_y \qquad (2.5\text{-}31)$$

[which, we note, is just Eq. (2.4-17)], where Ω may be a function of t but not of $\mathbf{x}$. For $\Omega > 0$ (which we henceforth assume), it corresponds to a flow that, in

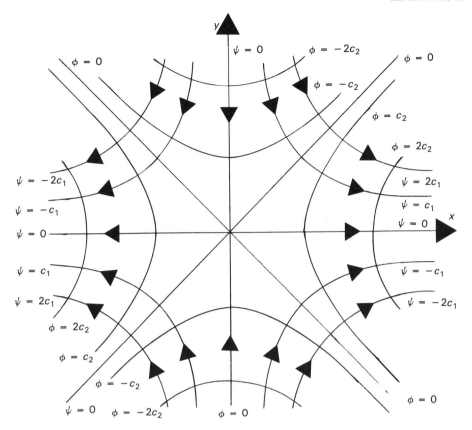

Figure 2.8. Potential lines and streamlines for simple extensional flow.

a right-handed coordinate system viewed in the direction of increasing z, is rotating clockwise about the z axis. It follows from Eqs. (2.4-17) and (2.4-18) that

$$\mathbf{e} = 0 \qquad \boldsymbol{\omega} = 2\Omega\mathbf{i}_z \qquad (2.5\text{-}32)$$

[see Eqs. (2.4-20)], from which we see that the flow is linear [in fact, Eq. (2.5-23) can be simplified in this case to yield $\mathbf{u} = -\tfrac{1}{2}\,\mathbf{x}{\scriptstyle\wedge}\boldsymbol{\omega}$]. It is also rotational so no scalar potential exists for it. It follows from Eq. (A1.3-7) of Appendix A that

$$\nabla \cdot \mathbf{u} = 0 \qquad (2.5\text{-}33)$$

Thus the flow is solenoidal and there exists a stream function ψ such that

$$\mathbf{u} = \nabla{\scriptstyle\wedge}\boldsymbol{\Psi} \qquad \boldsymbol{\Psi} = \psi\mathbf{i}_z \qquad \psi = -\tfrac{1}{2}\Omega\,(x^2 + y^2) \qquad (2.5\text{-}34)$$

The streamlines are shown in Fig. 2.9; c_3 denotes an arbitrary positive constant. Note that this flow involves no straining and only rotation. It is called a *rigid body rotation* (and is thus a rigid body motion with no translation; see Section 2.4); Ω is the angular speed (and $\boldsymbol{\Omega} = \Omega\mathbf{i}_z$ the angular velocity) of the motion.

③ We consider finally the velocity field given by

$$\mathbf{u} = \gamma y \mathbf{i}_x \qquad (2.5\text{-}35)$$

where γ may be a function of t but not of $\mathbf{x}$. It follows from Eqs. (2.4-17) and (2.4-18) that

$$\mathbf{e} = \gamma \mathbf{i}_x \mathbf{i}_y + \gamma \mathbf{i}_y \mathbf{i}_x \qquad \boldsymbol{\omega} = -\gamma \mathbf{i}_z \qquad (2.5\text{-}36)$$

from which we see that the flow is again linear. It is also rotational: for $\gamma > 0$ (which we henceforth assume), in a right-handed coordinate system viewed in the direction of increasing z, the local rotation is anticlockwise about the z axis (recall from Section 2.4 that the direction of $\boldsymbol{\omega}$ is the local axis of rotation). It also follows that the flow is solenoidal and there exists a stream function ψ such that

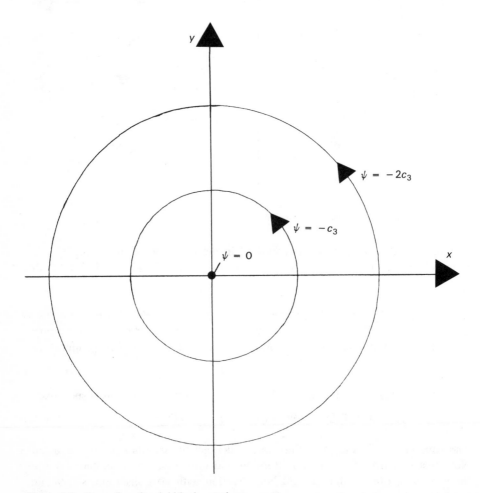

Figure 2.9. Streamlines for rigid body rotation.

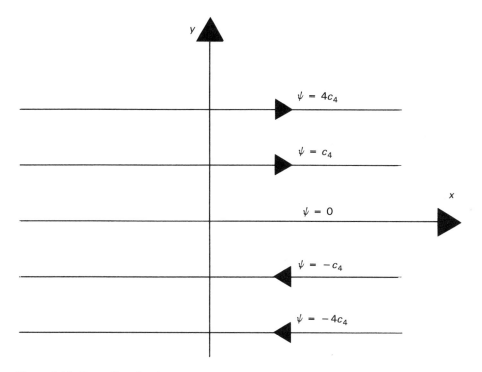

Figure 2.10. Streamlines for simple shear flow.

$$\mathbf{u} = \nabla_{\wedge}\mathbf{\Psi} \qquad \mathbf{\Psi} = \psi\mathbf{i}_z \qquad \psi = \tfrac{1}{2}\gamma y^2 \qquad (2.5\text{-}37)$$

The streamlines are shown in Fig. 2.10; c_4 denotes an arbitrary positive constant. Note that this flow involves straining and rotation: it is called a *simple shear flow*; γ is called the *shear rate* (see Section 5.4).

Note that, in these three examples, we have used the velocity field $\mathbf{u}$ to determine the kinematics of the flow, that is, to determine $\mathbf{e}$, $\boldsymbol{\omega}$, and $\nabla \cdot \mathbf{u}$, and hence to express $\mathbf{u}$ in terms of ϕ or ψ or both, as appropriate. In a typical flow problem, in contrast, we usually use the kinematics to express $\mathbf{u}$ (which is unknown) in an appropriate way and then solve the equations of motion to determine $\mathbf{u}$.

THREE
STRESS

3.1 BODY AND SURFACE FORCES

We can, as a rule, distinguish between two kinds of forces acting on a body:

1. Long-range forces, such as gravitational forces, which generally vary only slowly (if at all) over the body.
2. Short-range forces, such as viscous and elastic forces, which have a molecular origin and are, as a result, generally negligible unless there is physical contact between the interacting parts of the body.

Long-range forces are called body forces or volume forces. They are usually easy to incorporate into a mathematical description of a deforming body. Thus, for example, gravitational forces lead to an acceleration **g** in the linear momentum conservation equation. Short-range forces can be approximated by forces on the surface of each part of the body; it is with such surface forces or contact forces that we are concerned here.

3.2 CAUCHY'S FUNDAMENTAL THEOREM FOR STRESS

Surface forces lead to the concept of stress in a body. We show this by considering an elementary tetrahedron of the body (see Fig. 3.1). Let S_i denote the area of

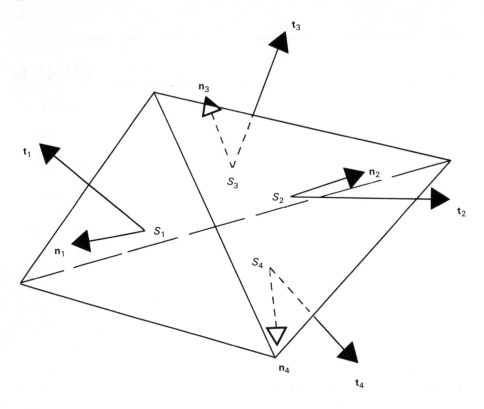

Figure 3.1. Elementary tetrahedron of a body.

face i ($i = 1, 2, 3,$ or 4) and $\mathbf{n}_i$ denote the unit outer normal to face i (i.e., the vector of unit magnitude which is normal to the face and oriented outward from the tetrahedron). Then the force $\mathbf{F}_i$ on face i is given by

$$\mathbf{F}_i = \int_{S_i} \mathbf{t}_i \, dS_i \approx \mathbf{t}_i S_i \qquad (3.2\text{-}1)$$

if S_i is small enough, where $\mathbf{t}_i$ denotes the *stress vector* on face i. We adopt the convention that $\mathbf{t}_i$ and $\mathbf{F}_i$ are exerted *by* material on the side of face i *to* which $\mathbf{n}_i$ points *on* material on the side of face i *from* which $\mathbf{n}_i$ points. Thus $\mathbf{t}_i$ and $\mathbf{F}_i$ are exerted *by* material outside the tetrahedron *on* material inside the tetrahedron. The reason for introducing $\mathbf{t}_i$ is that it acts at a point (see Section 1.1) whereas $\mathbf{F}_i$ does not. Because, however, $\mathbf{t}_i$ varies with the orientation of S_i (i.e., with $\mathbf{n}_i$), it is not a property at a point. We now remedy this deficiency.

A force balance on the material in the tetrahedron leads to

$$S_1\mathbf{t}_1 + S_2\mathbf{t}_2 + S_3\mathbf{t}_3 + S_4\mathbf{t}_4 \to 0 \qquad (3.2\text{-}2)$$

in the limit as all the $S_i \to 0$. Note that forces due to stresses (i.e., surface forces) acting on the tetrahedron vary as the surface area of the tetrahedron; inertial and body forces vary as its volume. If Δ denotes a typical linear dimension of the tetrahedron, then the surface area is of order Δ^2 and the volume is of order Δ^3. As $\Delta \to 0$, the forces that vary as the volume of the tetrahedron become negligible compared with those that vary as its surface area. This is why there are no inertial and body force terms in Eq. (3.2-2).

Because the surface of the tetrahedron is closed, that is, faces 1 to 4 completely bound the tetrahedron, it follows that

$$S_1 \mathbf{n}_1 + S_2 \mathbf{n}_2 + S_3 \mathbf{n}_3 + S_4 \mathbf{n}_4 = \mathbf{0} \qquad (3.2\text{-}3)$$

In general, if $\mathbf{n}$ denotes the unit outer normal to an element dS of a closed surface S, $\int_S \mathbf{n}\, dS = \mathbf{0}$. This may be shown by applying Gauss' divergence theorem [see Eq. (A1.3-32) of Appendix A] to the unit tensor field $\mathbf{I}$ [see Eq. (A1.2-18) of Appendix A]. We now fix our interest arbitrarily on face 4, and put $\mathbf{t} = \mathbf{t}_4$, $\mathbf{n} = \mathbf{n}_4$, and $S = S_4$. Then, in the limit as all the $S_i \to 0$,

$$\mathbf{t} = -\frac{1}{S}(S_1 \mathbf{t}_1 + S_2 \mathbf{t}_2 + S_3 \mathbf{t}_3) \qquad (3.2\text{-}4)$$

$$\mathbf{n} = -\frac{1}{S}(S_1 \mathbf{n}_1 + S_2 \mathbf{n}_2 + S_3 \mathbf{n}_3) \qquad (3.2\text{-}5)$$

One of several equivalent ways of defining a tensor $\mathbf{T}$ is that the following must be satisfied:

$$(s\mathbf{v}) \cdot \mathbf{T} = s(\mathbf{v} \cdot \mathbf{T}) \qquad (\mathbf{v}_1 + \mathbf{v}_2) \cdot \mathbf{T} = (\mathbf{v}_1 \cdot \mathbf{T}) + (\mathbf{v}_2 \cdot \mathbf{T}) \qquad (3.2\text{-}6)$$

where s is any scalar and $\mathbf{v}$, $\mathbf{v}_1$, and $\mathbf{v}_2$ are any vectors. It follows from Eqs. (3.2-4) and (3.2-5) that there exists a tensor field $\boldsymbol{\tau}$, the *stress tensor* field, such that

$$\mathbf{t} = \mathbf{n} \cdot \boldsymbol{\tau} = \boldsymbol{\tau}^T \cdot \mathbf{n} \qquad (3.2\text{-}7)$$

This is Cauchy's fundamental theorem for stress. It asserts the existence of the stress tensor field $\boldsymbol{\tau}$. Note that the stress tensor $\boldsymbol{\tau}$ at a point is independent of the orientation of any surface through the point. It is, therefore, a property at a point, unlike the stress vector $\mathbf{t}$.

In an orthogonal coordinate system (ξ_2, ξ_2, ξ_3) (see Section 1.2),

$$\begin{aligned}
\boldsymbol{\tau} =\ & \tau_{\xi_1\xi_1} \mathbf{i}_{\xi_1} \mathbf{i}_{\xi_1} + \tau_{\xi_1\xi_2} \mathbf{i}_{\xi_1} \mathbf{i}_{\xi_2} + \tau_{\xi_1\xi_3} \mathbf{i}_{\xi_1} \mathbf{i}_{\xi_3} \\
& + \tau_{\xi_2\xi_1} \mathbf{i}_{\xi_2} \mathbf{i}_{\xi_1} + \tau_{\xi_2\xi_2} \mathbf{i}_{\xi_2} \mathbf{i}_{\xi_2} + \tau_{\xi_2\xi_3} \mathbf{i}_{\xi_2} \mathbf{i}_{\xi_3} \\
& + \tau_{\xi_3\xi_1} \mathbf{i}_{\xi_3} \mathbf{i}_{\xi_1} + \tau_{\xi_3\xi_2} \mathbf{i}_{\xi_3} \mathbf{i}_{\xi_2} + \tau_{\xi_3\xi_3} \mathbf{i}_{\xi_3} \mathbf{i}_{\xi_3}
\end{aligned} \qquad (3.2\text{-}8)$$

The components of $\boldsymbol{\tau}$ may be interpreted as follows: $\tau_{\xi_1\xi_1}$ is the ξ_1 component of force per unit area exerted across a plane surface element normal to the ξ_1 direction; $\tau_{\xi_1 \xi_2}$ is the ξ_2 component of force per unit area exerted across a plane surface element normal to the ξ_1 direction; etc.

We refer to $\tau_{\xi_1 \xi_1}$, $\tau_{\xi_2 \xi_2}$, and $\tau_{\xi_3 \xi_3}$ as the normal stress components of $\boldsymbol{\tau}$ and to the other components as its shear stress components. The sign convention for stress is that $\mathbf{t} \cdot \mathbf{n} = (\mathbf{n} \cdot \boldsymbol{\tau}) \cdot \mathbf{n} > 0$ corresponds to tension and $\mathbf{t} \cdot \mathbf{n} = (\mathbf{n} \cdot \boldsymbol{\tau}) \cdot \mathbf{n} < 0$ to compression. For this reason, the tensor $-\boldsymbol{\tau}$ is sometimes called the pressure tensor.

FOUR

CONSERVATION EQUATIONS

4.1 CONSERVATION EQUATIONS AND CONTROL VOLUMES

In continuum mechanics we postulate that mass, linear momentum, angular momentum, and energy are conserved in any motion or flow of a body. In order to write down mathematical statements of conservation (i.e., conservation equations), we consider a control volume V (see Fig. 4.1) in the body which is fixed with respect to a reference frame. For the conservation of mass and energy, any reference frame can be chosen. For the conservation of linear and angular momentum, in contrast, it is necessary that the reference frame is inertial (see Section 1.2). The volume V must have an orientable surface S, so that S is two sided: thus there exist two regions in the body, the volume V and the rest of the body. Because S is orientable, the unit outer normal $\mathbf{n}$ to an element dS of S is defined unambiguously.

4.2 MASS CONSERVATION

Conservation of mass leads to the requirement that:

Rate of accumulation of mass in V + flux of mass out through $S = 0$

Now (1) the rate of accumulation of mass in V is given by

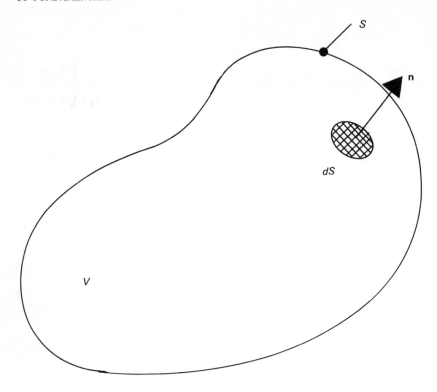

Figure 4.1. Fixed control volume with an orientable surface.

$$\frac{\partial}{\partial t} \int_V \rho \, dV$$

and hence by

$$\int_V \frac{\partial \rho}{\partial t} \, dV$$

since V is fixed and is thus independent of time t; and (2) the flux of mass out through S is given by $\int_S \rho \mathbf{u} \cdot \mathbf{n} \, dS$. Thus mass conservation yields

$$\int_V \frac{\partial \rho}{\partial t} \, dV + \int_S \rho \mathbf{u} \cdot \mathbf{n} \, dS = 0 \qquad (4.2\text{-}1)$$

Gauss' divergence theorem [see Eq. (A1.3-31) of Appendix A] may now be used to convert the surface integral (over S) in Eq. (4.2-1) into a volume integral (over V), whence

$$\int_V \frac{\partial \rho}{\partial t} \, dV + \int_V \boldsymbol{\nabla} \cdot (\rho \mathbf{u}) \, dV = 0 \qquad (4.2\text{-}2)$$

or

∂ Nree

$$\int_V \left[\frac{\partial \rho}{\partial t} + \nabla \cdot (\rho \mathbf{u}) \right] dV = 0 \qquad (4.2\text{-}3)$$

Although the volume V is fixed, it is otherwise entirely arbitrary. Thus the integrand in Eq. (4.2-3) must vanish, that is,

$$\frac{\partial \rho}{\partial t} + \nabla \cdot (\rho \mathbf{u}) = 0 \qquad (4.2\text{-}4)$$

Now $\nabla \cdot (\rho \mathbf{u}) = \mathbf{u} \cdot \nabla \rho + \rho \nabla \cdot \mathbf{u}$ [see Eq. (A1.3-28) of Appendix A], so,

$$\frac{\partial \rho}{\partial t} + \mathbf{u} \cdot \nabla \rho + \rho \nabla \cdot \mathbf{u} = 0 \qquad (4.2\text{-}5)$$

or

$$\frac{D\rho}{Dt} + \rho \nabla \cdot \mathbf{u} = 0 \qquad (4.2\text{-}6)$$

Equations (4.2-4), (4.2-5), and (4.2-6) are often referred to as forms of the *continuity equation*.

If a material is incompressible, the density ρ of each material point is constant with respect to time t. Hence $D\rho/Dt = 0$ and

$$\nabla \cdot \mathbf{u} = 0 \qquad (4.2\text{-}7)$$

Equation (4.2-7) is given in component form in rectangular, cylindrical polar, and spherical polar coordinates in Section A2.2 of Appendix B. It will be seen that Eq. (4.2-7) is precisely the same as Eq. (2.5-3). Thus, as was claimed in Section 2.5, any flow of an incompressible material is *solenoidal*. In fact, when we speak of a flow of an incompressible material, we should strictly speak of an isochoric (i.e., volume-preserving) flow. This is because a material may be incompressible, in the sense that a change of pressure does not affect its density, and yet not be in an isochoric flow. Temperature variations, for example, may affect its density.

Note that it is not necessary that the densities of all material points be the same for Eq. (4.2-7) to hold, merely that the density of each material point be independent of time. Nevertheless, by an incompressible material, we will henceforth always mean one for which the densities of all material points are the same. Thus, if we refer to a body composed of an incompressible material, we mean that there are no variations of density ρ with time t or position $\mathbf{x}$ within the body.

4.3 LINEAR MOMENTUM CONSERVATION

In an inertial reference frame, conservation of linear momentum, often referred to as momentum (where "linear" is understood) leads to the requirement that

Rate of accumulation of linear momentum in V
+ flux of linear momentum out through S
= rate of gain of linear momentum due to body forces
+ rate of gain of linear momentum due to surface stresses

Now,

1. The rate of accumulation of linear momentum in V is given by

$$\frac{\partial}{\partial t} \int_V \rho \mathbf{u} \, dV$$

and hence by

$$\int_V \frac{\partial}{\partial t} (\rho \mathbf{u}) \, dV$$

since V is independent of t.
2. The flux of linear momentum out through S is given by $\int_S \rho \mathbf{n} \cdot \mathbf{u}\mathbf{u} \, dS$.
3. The rate of gain of linear momentum due to body forces is given by $\int_V \rho \mathbf{g} \, dV$ assuming, as we always will henceforth, that the only body forces are gravitational in origin.
4. The rate of gain of linear momentum due to surface stresses is given by $\int_S \mathbf{t} \, dS$ and hence by $\int_S \mathbf{n} \cdot \boldsymbol{\tau} \, dS$ by Cauchy's fundamental theorem for stress (see Section 3.2).

Thus linear momentum conservation yields

$$\int_V \frac{\partial}{\partial t} (\rho \mathbf{u}) \, dV + \int_S \rho \mathbf{n} \cdot \mathbf{u}\mathbf{u} \, dS = \int_V \rho \mathbf{g} \, dV + \int_S \mathbf{n} \cdot \boldsymbol{\tau} \, dS \qquad (4.3\text{-}1)$$

Use of Gauss' divergence theorem leads to

$$\int_V \frac{\partial}{\partial t} (\rho \mathbf{u}) \, dV + \int_V \boldsymbol{\nabla} \cdot (\rho \mathbf{u}\mathbf{u}) \, dV = \int_V \rho \mathbf{g} \, dV + \int_V \boldsymbol{\nabla} \cdot \boldsymbol{\tau} \, dV \qquad (4.3\text{-}2)$$

Because the volume V is arbitrary,

$$\frac{\partial}{\partial t} (\rho \mathbf{u}) + \boldsymbol{\nabla} \cdot (\rho \mathbf{u}\mathbf{u}) = \rho \mathbf{g} + \boldsymbol{\nabla} \cdot \boldsymbol{\tau} \qquad (4.3\text{-}3)$$

or, using Eq. (A1.3-26) of Appendix A,

$$\mathbf{u} \frac{\partial \rho}{\partial t} + \rho \frac{\partial \mathbf{u}}{\partial t} + \rho \mathbf{u} \cdot \boldsymbol{\nabla} \mathbf{u} + \mathbf{u} \boldsymbol{\nabla} \cdot (\rho \mathbf{u}) = \rho \mathbf{g} + \boldsymbol{\nabla} \cdot \boldsymbol{\tau} \qquad (4.3\text{-}4)$$

Mass conservation means, however, that

$$\mathbf{u} \frac{\partial \rho}{\partial t} + \mathbf{u} \boldsymbol{\nabla} \cdot (\rho \mathbf{u}) = \mathbf{0} \qquad (4.3\text{-}5)$$

[see Eq. (4.2-4)], so

$$\rho \frac{\partial \mathbf{u}}{\partial t} + \rho \mathbf{u} \cdot \boldsymbol{\nabla} \mathbf{u} = \rho \mathbf{g} + \boldsymbol{\nabla} \cdot \boldsymbol{\tau} \qquad (4.3\text{-}6)$$

or
$$\rho \frac{D\mathbf{u}}{Dt} = \rho \mathbf{g} + \boldsymbol{\nabla} \cdot \boldsymbol{\tau} \qquad (4.3\text{-}7)$$

Equations (4.3-6) and 4.3-7) are often referred to as forms of the *equation of motion;* they are both expressions of *Cauchy's first law of motion.*

We now decompose the total stress $\boldsymbol{\tau}$ arbitrarily thus:

$$\boldsymbol{\tau} = -q\mathbf{I} + \boldsymbol{\tau}_E \qquad (4.3\text{-}8)$$

where q is a scalar field, $\mathbf{I}$ is the unit tensor [see Eqs. (A1.2-17) and (A1.2-18) of Appendix A] and $\boldsymbol{\tau}_E$ is the extra stress tensor field. The extra stress $\boldsymbol{\tau}_E$ might be thought of as arising from the deformation of a material (see Section 5.4) if $-q\mathbf{I}$ is thought of as being the isotropic (i.e., nondirectional) stress in the material when it is not deforming. Note that this decomposition is totally arbitrary: we have merely identified a component $-q\mathbf{I}$ of $\boldsymbol{\tau}$ which is isotropic, and associated the rest of $\boldsymbol{\tau}$ with $\boldsymbol{\tau}_E$. We remove the arbitrariness in this decomposition thus:

$$\boldsymbol{\tau} = -p\mathbf{I} + \boldsymbol{\tau}_D \qquad \text{trace}(\boldsymbol{\tau}_D) = 0 \qquad (4.3\text{-}9)$$

where p is the pressure field and $\boldsymbol{\tau}_D$ is the deviatoric stress tensor field, so called because its deviator or trace vanishes. Thus, in an orthogonal coordinate system (ξ_1, ξ_2, ξ_3) [see Eq. (A1.2-13) of Appendix A],

$$\text{trace}(\boldsymbol{\tau}_D) = \tau_{D\xi_1\xi_1} + \tau_{D\xi_2\xi_2} + \tau_{D\xi_3\xi_3} = 0 \qquad (4.3\text{-}10)$$

Note that the isotropic component $-p\mathbf{I}$ of $\boldsymbol{\tau}$ is now identified with pressure. The decomposition of total stress in Eq. (4.3-9) thus reflects the traditional manner in which stress is thought of as comprising two components, one of which is associated with pressure and is positive in compression; the other is positive in tension. This accounts for the signs in Eq. (4.3-9) and is in accordance with the sign convention for stress in Section 3.2. Since, by definition, trace $(\boldsymbol{\tau}_D) = 0$, Eq. (4.3-9) implies that

$$\text{trace}(\boldsymbol{\tau}) = 3p \qquad (4.3\text{-}11)$$

The factor of three appears because space is three dimensional. Thus pressure is defined as one-third of the trace of the total stress tensor. Note that for incompressible materials (i.e., materials of constant density ρ) p is not the thermodynamic pressure, which is in fact undefined for such materials (see also Section 5.2). Except in Section 4.5, we will never henceforth use the total stress but only the deviatoric stress. Accordingly, we let $\boldsymbol{\tau}$ denote deviatoric stress (except where otherwise stated in Section 4.5), whence

$$\rho \frac{\partial \mathbf{u}}{\partial t} + \rho \mathbf{u} \cdot \boldsymbol{\nabla}\mathbf{u} = \rho \mathbf{g} - \boldsymbol{\nabla}p + \boldsymbol{\nabla} \cdot \boldsymbol{\tau}$$

$$\begin{array}{ccccc} \uparrow & \uparrow & \uparrow & \uparrow & \uparrow \\ (a) & (b) & (c) & (d) & (e) \end{array} \qquad (4.3\text{-}12)$$

[since $\boldsymbol{\nabla} \cdot (s\mathbf{I}) = \boldsymbol{\nabla}s$ for any scalar field s: see Eq. (A1.3-29) of Appendix A]. The terms in this equation may be interpreted (to within a factor of ρ) as follows:

- Term (*a*) is called the *transient term* and represents that part of the acceleration due to unsteadiness in an Eulerian sense (see Section 2.3), that is, to variations in velocity with time at a fixed position in space.
- Term (*b*) is called the *convection term;* terms (*a*) and (*b*) are together called the *inertial terms* and represent the acceleration due to unsteadiness in a Lagrangian sense (i.e., to variations in velocity with time following the motion of a material point).
- Term (*c*) is called the *gravitational term* and represents the acceleration caused by gravity.
- Term (*d*) is called the *pressure term* and represents the acceleration caused by pressure differences.
- Term (*e*) is called the *stress term* and represents the acceleration caused by (deviatoric) stress differences.

If a material is incompressible, we can eliminate the gravitational acceleration **g** from Eq. (4.3-12) by defining a modified pressure $\tilde{p}$:

$$\tilde{p} = p - \rho \mathbf{g} \cdot \mathbf{x} \tag{4.3-13}$$

Then

$$\nabla \tilde{p} = \nabla p - \rho \mathbf{g} \tag{4.3-14}$$

and so

$$\rho \frac{\partial \mathbf{u}}{\partial t} + \rho \mathbf{u} \cdot \nabla \mathbf{u} = -\nabla \tilde{p} + \nabla \cdot \boldsymbol{\tau} \tag{4.3-15}$$

In certain problems, however, pressure appears explicitly in the boundary conditions (see Chapter 6) and gravity cannot, therefore, be eliminated. Problems involving free surfaces are common examples of these.

If a material is incompressible and we take the curl of each of the terms in Eq. (4.3-12), after some algebraic manipulation we obtain the *vorticity transport equation:*

$$\rho \frac{\partial \boldsymbol{\omega}}{\partial t} + \rho \mathbf{u} \cdot \nabla \boldsymbol{\omega} - \rho \boldsymbol{\omega} \cdot \nabla \mathbf{u} = \nabla_\wedge (\nabla \cdot \boldsymbol{\tau}) \tag{4.3-16}$$

$$\uparrow \qquad \uparrow \qquad \uparrow \qquad \uparrow$$
$$(a) \qquad (b) \qquad (c) \qquad (d)$$

where the vorticity $\boldsymbol{\omega} = \nabla_\wedge \mathbf{u}$. The terms in this equation may be identified as the transient (*a*), convection (*b*), straining (*c*), and stress (*d*) terms. The importance of the straining term in turbulent flows is discussed in Section 12.1.

4.4 ANGULAR MOMENTUM CONSERVATION

Let **c** denote the body couple per unit mass and $\boldsymbol{\mu}$ denote the couple stress tensor field, defined in terms of the couple stress vector field **m** by

$$\mathbf{m} = \mathbf{n} \cdot \boldsymbol{\mu} \tag{4.4-1}$$

which is the couple analogue of Cauchy's fundamental theorem for stress. In an inertial reference frame, conservation of angular (or rotational) momentum about some point leads to the requirement that

> Rate of accumulation of angular momentum in V
> + flux of angular momentum out through S
> = rate of gain of angular momentum due to
> body forces and body couples
> + rate of gain of angular momentum due to
> surface stresses and surface couple stresses

If we now proceed as we did in Sections 4.2 and 4.3, we eventually obtain the following equation:

$$\mathbf{t}^A = \rho \mathbf{c} - \nabla \cdot \boldsymbol{\mu} \tag{4.4-2}$$

where $\mathbf{t}^A$ denotes the axial stress which is the vector field defined by

$$\mathbf{t}^A . \mathbf{v} = \mathbf{v} \cdot [\boldsymbol{\tau} - (\boldsymbol{\tau}^T)] \tag{4.4-3}$$

Thus $\mathbf{t}^A$ is the axial vector or pseudovector formed from the antisymmetric tensor $[\boldsymbol{\tau} - (\boldsymbol{\tau}^T)]$ in the same way as the vorticity vector $\boldsymbol{\omega}$ is the axial vector or pseudovector formed from the antisymmetric vorticity tensor $\mathbf{w}$ (see Section 2.4). Equation (4.4-2) is an expression of Cauchy's second law of motion.

A *polar* material is one in which there are couple stresses and body couples, that is, one for which (1) there are torques that are not the moments of forces, and (2) the angular momentum per unit mass is not the moment of the linear momentum per unit mass. A *nonpolar* material is one in which there are no couple stresses (so that $\mathbf{m} = \mathbf{0}$ and hence $\boldsymbol{\mu} = \mathbf{0}$) and no body couples (so that $\mathbf{c} = \mathbf{0}$), that is, one for which

$$\mathbf{t}^A = \mathbf{0} \tag{4.4-4}$$

Hence it follows from Eq. (4.4-3) that, for a nonpolar material,

$$\boldsymbol{\tau} = \boldsymbol{\tau}^T \tag{4.4-5}$$

that is, the stress tensor is symmetric. Most materials, and certainly most fluids, are nonpolar. Accordingly, we henceforth assume that conservation of angular momentum implies Eq. (4.4-5) (i.e., symmetry of the stress tensor). Note that because the pressure component $-p\mathbf{I}$ of the total stress is symmetric, symmetry of the deviatoric stress is equivalent to symmetry of the total stress.

The reason why the stress tensor is symmetric for a nonpolar material is that if it were not, the angular velocity of the material would become unbounded. To see this, suppose that in a rectangular coordinate system (x, y, z), $\boldsymbol{\tau}$ is such that

$$\boldsymbol{\tau} = \tau_{xy}\mathbf{i}_x\mathbf{i}_y + \tau_{yx}\mathbf{i}_y\mathbf{i}_x \tag{4.4-6}$$

Consider an elementary rectangular parallelepiped of material centered at (x, y, z) of length Δx in the x direction, Δy in the y direction, and Δz in the z direction (see Fig. 4.2). The force in the y direction on the face of the parallelepiped normal to the x direction at $x + \frac{1}{2}\Delta x$ is given by $\Delta y\, \Delta z\, \tau_{xy}|_{x+(1/2)\Delta x,y,z}\, \mathbf{i}_y$ to leading order. This force produces a torque given by $\frac{1}{2}\,\Delta x\, \Delta y\, \Delta z\, \tau_{xy}|_{x+(1/2)\Delta x,y,z}\, \mathbf{i}_z$ about an axis through the center of the parallelepiped and aligned in the z direction. The torque on the parallelepiped is thus given by

$$\tfrac{1}{2}\,\Delta x\, \Delta y\, \Delta z\,(\tau_{xy}|_{x+(1/2)\Delta x,y,z} + \tau_{xy}|_{x-(1/2)\Delta x,y,z} - \tau_{yx}|_{x,y+(1/2)\Delta y,z} - \tau_{yx}|_{x,y-(1/2)\Delta y,z})\,\mathbf{i}_z$$

In the limit as Δx and Δy vanish, the torque is given by $\Delta x\, \Delta y\, \Delta z\,(\tau_{xy}|_{x,y,z} - \tau_{yx}|_{x,y,z})\,\mathbf{i}_z$. For a nonpolar material, the torque equals the rate of change with time of the angular momentum of the parallelepiped, which is given by $I\, \partial\Omega/\partial t\, \mathbf{i}_z$, where I denotes the moment of inertia of the parallelepiped and Ω denote its angular velocity about an axis through $(x,\, y,\, z)$ aligned in the z direction. It is easy to show that

$$I = \rho\, \Delta x\, \Delta y\, \Delta z\, \tfrac{1}{12}\,(\Delta x^2 + \Delta y^2) \tag{4.4-7}$$

Hence,

$$\rho\, \tfrac{1}{12}\,(\Delta x^2 + \Delta y^2)\, \frac{\partial\Omega}{\partial t} = \tau_{xy}|_{x,y,z} - \tau_{yx}|_{x,y,z} \tag{4.4-8}$$

In the limit as Δx and Δy vanish, it follows that, unless $\partial\Omega/\partial t$ becomes unbounded, which is physically unreasonable,

$$\tau_{xy}|_{x,y,z} = \tau_{yx}|_{x,y,z} \tag{4.4-9}$$

that is, τ_{xy} and τ_{yx} are equal at any position $(x,\, y,\, z)$. Thus boundedness of angular

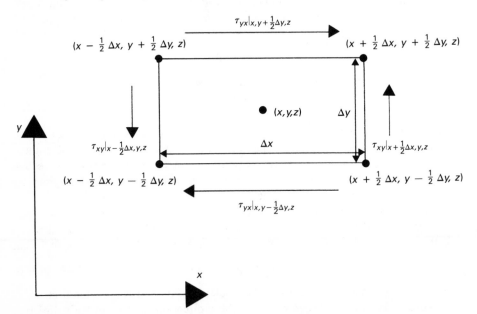

Figure 4.2. Section of an elementary rectangular parallelepiped.

acceleration and hence of angular velocity (i.e., the requirement that they are finite) means that the stress tensor $\boldsymbol{\tau}$ given by Eq. (4.4-6) is symmetric. The extension of this argument to a general stress tensor, with up to nine nonzero components, is obvious.

4.5 ENERGY CONSERVATION

If we form the dot product of $\mathbf{u}$ with each term in the linear momentum Eq. (4.3-12), we obtain

$$\mathbf{u} \cdot \left(\rho \frac{\partial \mathbf{u}}{\partial t} \right) + \mathbf{u} \cdot (\rho \, \mathbf{u} \cdot \nabla \mathbf{u}) = - \mathbf{u} \cdot \nabla p + \mathbf{u} \cdot (\rho \mathbf{g}) + \mathbf{u} \cdot (\nabla \cdot \boldsymbol{\tau}) \quad (4.5\text{-}1)$$

or

$$\tfrac{1}{2} \rho \frac{\partial}{\partial t} (\mathbf{u} \cdot \mathbf{u}) + \tfrac{1}{2} \rho \, \mathbf{u} \cdot \nabla(\mathbf{u} \cdot \mathbf{u}) = \rho \frac{D}{Dt} (\tfrac{1}{2} |\mathbf{u}|^2)$$
$$= - \mathbf{u} \cdot \nabla p + \rho \, \mathbf{u} \cdot \mathbf{g} + \mathbf{u} \cdot (\nabla \cdot \boldsymbol{\tau}) \quad (4.5\text{-}2)$$

where $\tfrac{1}{2} |\mathbf{u}|^2$ is the kinetic energy per unit mass. Equation (4.5-2) is often referred to as a conservation equation for mechanical energy. We note, however, that it is derived not by conservation of energy but rather by conservation of mass and linear momentum. It will, nonetheless, prove useful in what follows.

Let U denote the internal energy per unit mass and $-\mathbf{g} \cdot \mathbf{x}$ the potential energy per unit mass. The minus sign arises because displacement in the direction of the gravitational acceleration decreases the potential energy and vice versa. The energy of a material comprises internal, potential, and kinetic energy. Conservation of energy (in the absence of heat generation by chemical or nuclear reaction, etc.) means that

Rate of accumulation of energy + flux of energy out through S
   = flux of heat in through S + rate of gain of energy due to surface stresses

Note that the third term comprising the flux of heat in through S is a result perhaps of conduction and radiation (see Section 5.3) but not of convection, which is instead just the flux of energy out through S in the second term. Now,

1. The rate of accumulation of energy in V is given by

$$\frac{\partial}{\partial t} \int_V (\rho U + \tfrac{1}{2} \rho |\mathbf{u}|^2 - \rho \, \mathbf{g} \cdot \mathbf{x}) \, dV$$

and hence by

$$\int_V \frac{\partial}{\partial t} (\rho U + \tfrac{1}{2} \rho |\mathbf{u}|^2 - \rho \, \mathbf{g} \cdot \mathbf{x}) \, dV$$

since V is independent of t.

2. The flux of energy out through S is given by

$$\int_S (\rho U + \tfrac{1}{2} \rho |\mathbf{u}|^2 - \rho \, \mathbf{g} \cdot \mathbf{x}) \mathbf{u} \cdot \mathbf{n} \, dS$$

3. The flux of heat in through S is given by $- \int_S \mathbf{q} \cdot \mathbf{n} \, dS$, where $\mathbf{q}$ denotes the heat flux vector field.

4. The rate of gain of energy due to surface stresses is the rate at which the surface stresses do work on the material inside S and is given by $\int_S \mathbf{t} \cdot \mathbf{u} \, dS$, since work is the dot product of force and resultant displacement, and so the rate of doing work is the dot product of force and resultant velocity. Hence the rate of gain of energy due to surface stresses is given by $\int_S \mathbf{n} \cdot (\boldsymbol{\tau} \cdot \mathbf{u}) \, dS$ by Cauchy's fundamental theorem for stress, where $\boldsymbol{\tau}$ here denotes the total, and not the deviatoric, stress.

Thus energy conservation yields

$$\int_V \frac{\partial}{\partial t} (\rho U + \tfrac{1}{2} \rho |\mathbf{u}|^2 - \rho \, \mathbf{g} \cdot \mathbf{x}) \, dV + \int_S (\rho U + \tfrac{1}{2} \rho |\mathbf{u}|^2 - \rho \, \mathbf{g} \cdot \mathbf{x}) \mathbf{u} \cdot \mathbf{n} \, dS$$

$$= - \int_S \mathbf{q} \cdot \mathbf{n} \, dS + \int_S \mathbf{n} \cdot (\boldsymbol{\tau} \cdot \mathbf{u}) \, dS \tag{4.5-3}$$

Use of Gauss' divergence theorem leads to

$$\int_V \frac{\partial}{\partial t} (\rho U + \tfrac{1}{2} \rho |\mathbf{u}|^2 - \rho \, \mathbf{g} \cdot \mathbf{x}) \, dV + \int_V \boldsymbol{\nabla} \cdot [(\rho U + \tfrac{1}{2} \rho |\mathbf{u}|^2 - \rho \, \mathbf{g} \cdot \mathbf{x}) \mathbf{u}] \, dV$$

$$= - \int_V \boldsymbol{\nabla} \cdot \mathbf{q} \, dV + \int_V \boldsymbol{\nabla} \cdot (\boldsymbol{\tau} \cdot \mathbf{u}) \, dV \tag{4.5-4}$$

Because the volume V is arbitrary,

$$\frac{\partial}{\partial t} (\rho U + \tfrac{1}{2} \rho |\mathbf{u}|^2 - \rho \, \mathbf{g} \cdot \mathbf{x}) + \boldsymbol{\nabla} \cdot [(\rho U + \tfrac{1}{2} \rho |\mathbf{u}|^2 - \rho \, \mathbf{g} \cdot \mathbf{x}) \mathbf{u}]$$

$$= -\boldsymbol{\nabla} \cdot \mathbf{q} + \boldsymbol{\nabla} \cdot (\boldsymbol{\tau} \cdot \mathbf{u}) \tag{4.5-5}$$

Clearly,

$$\frac{\partial}{\partial t} (\rho U + \tfrac{1}{2} \rho |\mathbf{u}|^2) + \boldsymbol{\nabla} \cdot [(\rho U + \tfrac{1}{2} \rho |\mathbf{u}|^2) \mathbf{u}]$$

$$= \rho \frac{\partial}{\partial t} (U + \tfrac{1}{2} |\mathbf{u}|^2) + \rho \boldsymbol{\nabla} \cdot [(U + \tfrac{1}{2} |\mathbf{u}|^2) \mathbf{u}]$$

$$+ (U + \tfrac{1}{2} |\mathbf{u}|^2) \left(\frac{\partial \rho}{\partial t} + \boldsymbol{\nabla} \cdot (\rho \mathbf{u}) \right) \tag{4.5-6}$$

and

$$\frac{\partial}{\partial t} (\rho \, \mathbf{g} \cdot \mathbf{x}) + \nabla \cdot (\rho \, \mathbf{ug} \cdot \mathbf{x})$$

$$= \mathbf{g} \cdot \mathbf{x} \frac{\partial \rho}{\partial t} + \rho \frac{\partial}{\partial t} (\mathbf{g} \cdot \mathbf{x}) + \mathbf{g} \cdot \mathbf{x} \nabla \cdot (\rho \mathbf{u}) + \rho \, \mathbf{u} \cdot \nabla (\mathbf{g} \cdot \mathbf{x}) \qquad (4.5\text{-}7)$$

Hence, using the mass conservation equation (4.2-4),

$$\frac{\partial}{\partial t} (\rho U + \tfrac{1}{2} \rho \, |\mathbf{u}|^2) + \nabla \cdot [(\rho U + \tfrac{1}{2} \rho \, |\mathbf{u}|^2)\mathbf{u}] = \rho \frac{D}{Dt} (U + \tfrac{1}{2} |\mathbf{u}|^2) \qquad (4.5\text{-}8)$$

and

$$\frac{\partial}{\partial t} (\rho \, \mathbf{g} \cdot \mathbf{x}) + \nabla \cdot (\rho \, \mathbf{ug} \cdot \mathbf{x}) = \rho \, \mathbf{u} \cdot \mathbf{g} \qquad (4.5\text{-}9)$$

since the gravitational acceleration $\mathbf{g}$ is independent of time t and position $\mathbf{x}$. Thus Eq. (4.5-5) becomes

$$\rho \frac{D}{Dt} (U + \tfrac{1}{2} |\mathbf{u}|^2) = -\nabla \cdot \mathbf{q} + \nabla \cdot (\boldsymbol{\tau} \cdot \mathbf{u}) + \rho \, \mathbf{u} \cdot \mathbf{g} \qquad (4.5\text{-}10)$$

We now decompose total stress as in Eq. (4.3-9) and henceforth let $\boldsymbol{\tau}$ denote the deviatoric stress. Then

$$\rho \frac{D}{Dt} (U + \tfrac{1}{2} |\mathbf{u}|^2) = -\nabla \cdot \mathbf{q} + \nabla \cdot (\boldsymbol{\tau} \cdot \mathbf{u}) - \nabla \cdot (p\mathbf{u}) + \rho \, \mathbf{u} \cdot \mathbf{g} \qquad (4.5\text{-}11)$$

Substitution of Eq. (4.5-2) into Eq. (4.5-11) leads to

$$\rho \frac{DU}{Dt} = -\nabla \cdot \mathbf{q} - p\nabla \cdot \mathbf{u} + \boldsymbol{\tau} : \nabla \mathbf{u} \qquad (4.5\text{-}12)$$

where use has been made of Eq. (A1.3-30) of Appendix A. Equation (4.5-12) is an expression of the first law of thermodynamics.

If the internal energy U is independent of the kinematics of the flow, then classical thermodynamic (or, strictly, thermostatic) arguments can be applied, and U can be regarded as a function only of the volume per unit mass $v = 1/\rho$ and the entropy per unit mass S, so:

$$dU = -p \, dv + T \, dS \qquad (4.5\text{-}13)$$

or

$$dU = -pd \left(\frac{1}{\rho} \right) + T \, dS \qquad (4.5\text{-}14)$$

where T denotes the absolute temperature. If the material is incompressible, ρ is constant with respect to time and so it follows from Eq. (4.5-12) that

$$\frac{DS}{Dt} + \frac{1}{\rho T} \nabla \cdot \mathbf{q} - \frac{1}{\rho T} \boldsymbol{\tau} : \nabla \mathbf{u} = 0 \qquad (4.5\text{-}15)$$

where $(1/\rho T)\boldsymbol{\tau} : \nabla \mathbf{u}$ is called the *irreversible dissipation*. If the material is iso-thermal, that is, T is uniform in space and time, $\mathbf{q} = \mathbf{0}$ (as we will see in Section 5.3).

The second law of thermodynamics can be written in the form

$$\frac{DS}{Dt} \geq 0 \tag{4.5-16}$$

It follows that, for an incompressible, isothermal material,

$$\mathbf{\tau}:\nabla\mathbf{u} \geq 0 \tag{4.5-17}$$

(i.e., the irreversible dissipation is nonnegative). Note that this equation holds only if U is independent of the kinematics of the flow. Thus, for a material possessing some elasticity (see Section 5.4), U has an elastic component which, clearly, depends on the kinematics of the flow and Eq. (4.5-17) does not necessarily hold.

Because U can also be regarded as a function only of v and T,

$$dU = \frac{\partial U}{\partial v}\bigg|_T dv + \frac{\partial U}{\partial T}\bigg|_v dT \tag{4.5-18}$$

It then follows from Eq. (4.5-13) that

$$\frac{DU}{Dt} = \left(-p + T\frac{\partial S}{\partial v}\bigg|_T\right)\frac{Dv}{Dt} + \frac{\partial U}{\partial T}\bigg|_v \frac{DT}{Dt} \tag{4.5-19}$$

Since

$$\frac{\partial S}{\partial v}\bigg|_T = \frac{\partial p}{\partial T}\bigg|_v = \frac{\partial p}{\partial T}\bigg|_\rho \tag{4.5-20}$$

(which is a Maxwell equation) and since

$$\frac{Dv}{Dt} = \frac{D}{Dt}\left(\frac{1}{\rho}\right) = -\frac{1}{\rho^2}\frac{D\rho}{Dt} = \frac{1}{\rho}\nabla\cdot\mathbf{u} \tag{4.5-21}$$

[see Eq. (4.2-6)], it follows from Eq. (4.5-12) that

$$\rho c_v \frac{\partial}{\partial} + \rho c_v\, \mathbf{u}\cdot\nabla T = -\nabla\cdot\mathbf{q} + \mathbf{\tau}:\nabla\mathbf{u} - T\frac{\partial p}{\partial T}\bigg|_\rho \nabla\cdot\mathbf{u} \tag{4.5-22}$$
$$\uparrow \qquad\quad \uparrow \qquad\qquad \uparrow \qquad\quad \uparrow \qquad\qquad\quad \uparrow$$
$$(a) \qquad (b) \qquad\quad (c) \qquad (d) \qquad\qquad\quad (e)$$

where c_v denotes the specific heat evaluated at constant volume (and hence constant density) given by

$$c_v = \frac{\partial U}{\partial T}\bigg|_v = T\frac{\partial S}{\partial T}\bigg|_v \tag{4.5-23}$$

Let H denote the enthalpy per unit mass given by

$$H = U + pv = U + \frac{p}{\rho} \tag{4.5-24}$$

Then, using Eq. (4.5-21), Eq. (4.5-12) becomes

$$\rho \frac{DH}{Dt} = -\nabla \cdot \mathbf{q} + \frac{Dp}{Dt} + \tau : \nabla \mathbf{u} \tag{4.5-25}$$

Because H can be regarded as a function only of p and T, and since

$$\left. \frac{\partial S}{\partial p} \right|_T = - \left. \frac{\partial v}{\partial T} \right|_p = - \left. \frac{d(1/\rho)}{dT} \right|_p \tag{4.5-26}$$

(which is another Maxwell equation), Eq. (4.5-22) becomes

$$\rho c_p \frac{\partial T}{\partial t} + \rho c_p \, \mathbf{u} \cdot \nabla T = -\nabla \cdot \mathbf{q} + \tau : \nabla \mathbf{u} + \left. \frac{\partial \ln (1/\rho)}{\partial \ln T} \right|_p \frac{Dp}{Dt} \tag{4.5-27}$$

$$\quad\;\uparrow \qquad\qquad \uparrow \qquad\qquad \uparrow \qquad \uparrow \qquad\qquad\qquad \uparrow$$

$$\quad\,(a) \qquad\quad\;\; (b) \qquad\quad\;\; (c) \qquad (d) \qquad\qquad\qquad (e)$$

where c_p denotes the specific heat evaluated at constant pressure given by

$$c_p = \left. \frac{\partial H}{\partial T} \right|_p = T \left. \frac{\partial S}{\partial T} \right|_p \tag{4.5-28}$$

The terms in Eqs. (4.5-22) and (4.5-27) may be identified as the transient (a), convection (b), heat flux (c), dissipation (d), and compression (or expansion) work (e) terms. Mechanical energy is converted to thermal energy irreversibly by dissipation and reversibly by compression work.

The mass conservation equation for an incompressible material is [see Eq. (4.2-7)]

$$\nabla \cdot \mathbf{u} = 0 \tag{4.5-29}$$

In most motions or flows of *nearly* incompressible materials, and in particular of nearly incompressible fluids, density (or volume) variations, due for example to temperature variations, are small enough for the mass conservation Eq. (4.5-29) to hold with sufficient accuracy. The density variations are, however, often too large for the compression work term $-T(\partial \rho / \partial T)|_p \, \nabla \cdot \mathbf{u}$ in Eq. (4.5-22) to be neglected. On the other hand, pressure variations are usually small enough for the corresponding term

$$\left. \frac{\partial \ln (1/\rho)}{\partial \ln (T)} \right|_p \frac{Dp}{Dt}$$

to be neglected in Eq. (4.5-27). Accordingly, in most flows of nearly incompressible materials, the appropriate form of the energy conservtion equation is

$$\rho c_p \frac{\partial T}{\partial t} + \rho c_p \, \mathbf{u} \cdot \nabla T = -\nabla \cdot \mathbf{q} + \tau : \nabla \mathbf{u} \tag{4.5-30}$$

The appropriate specific heat is thus c_p and not c_v , paradoxical as that may seem.

FIVE

CONSTITUTIVE EQUATIONS (EQUATIONS OF STATE)

5.1 CONSTITUTIVE EQUATIONS

The conservation equations derived in Chapter 4 govern the motion of any material. Thus, for example, they govern the motion of fluids and of solids. The mass, linear momentum, and energy conservation equations describe how the density ρ, velocity $\mathbf{u}$, and internal energy U or temperature T vary during and throughout a body in motion. They do this, however, by reference to the otherwise unknown pressure p, heat flux $\mathbf{q}$, and stress $\boldsymbol{\tau}$ (though conservation of angular momentum does interrelate some components of $\boldsymbol{\tau}$).

We clearly need equations expressing p, $\mathbf{q}$, and $\boldsymbol{\tau}$ in terms of ρ, $\mathbf{u}$, and U or T (and, perhaps, each other) if we are to formulate well-posed problems. These equations cannot be deduced by arguments based on continuum mechanics alone. Instead, they must be obtained by phenomenological and molecular arguments corroborated by experiments. The equations will, in general, differ for different materials. Indeed, it is in these equations that the nature of a given material is manifested. The equations are called *constitutive equations* or, equivalently, *equations of state*, because they describe the constitution or state of a material.

5.2 PRESSURE

For compressible materials such as gases, we can identify pressure p with the thermodynamic pressure. Usually we then assume that p is a function only of density ρ and absolute temperature T. Thus, for example, for a perfect gas,

$$p = \frac{\rho RT}{M} \qquad (5.2\text{-}1)$$

where R denotes the universal gas constant and M denotes the molecular weight or relative molar mass of the gas.

Most liquids under normal conditions are nearly incompressible. For incompressible materials, the pressure p is *not* the thermodynamic pressure, which is in fact undefined (as we noted in Section 4.3). Instead, p is determined as part of the solution of the mass and linear momentum conservation equations [see Eqs. (4.2-7) and (4.3-12)]:

$$\nabla \cdot \mathbf{u} = 0 \qquad (5.2\text{-}2)$$

$$\rho \frac{\partial \mathbf{u}}{\partial t} + \rho \, \mathbf{u} \cdot \nabla \mathbf{u} = \rho \, \mathbf{g} - \nabla p + \nabla \cdot \boldsymbol{\tau} \qquad (5.2\text{-}3)$$

Thus Eq. (5.2-3) determines the velocity $\mathbf{u}$ only to within ∇p. The indeterminacy in $\mathbf{u}$ is eliminated by Eq. (5.2-2) which completely determines $\mathbf{u}$ but determines p only to within a scalar field p_0 which is independent of position $\mathbf{x}$ (though it may depend on time t). This is because it is ∇p which appears in Eq. (5.2-3) and $\nabla(p + p_0) = \nabla p$; also, while in certain problems pressure appears in the boundary conditions, it only usually in fact appears as a pressure difference, so that adding p_0 to p does not affect $\mathbf{u}$.

5.3 HEAT FLUX

Heat can be transmitted by three mechanisms: conduction, convection, and radiation. As noted in Section 4.5, the heat flux $\mathbf{q}$ does not incorporate convection and comprises, therefore, a component $\mathbf{q}_c$ due to conduction and a component $\mathbf{q}_r$ due to radiation:

$$\mathbf{q} = \mathbf{q}_c + \mathbf{q}_r \qquad (5.3\text{-}1)$$

Because $\mathbf{q}_r$ is found to vary as T^4, where T denotes absolute temperature, it follows that $\mathbf{q}_r$ may be neglected for most low temperature flows, and this is precisely what we will do henceforth. Most materials, in particular most fluids, are found under normal conditions to be *Fourier* materials, that is, to be such that $\mathbf{q}_c$ is proportional to the temperature gradient ∇T:

$$\mathbf{q}_c = -k \nabla T \qquad (5.3\text{-}2)$$

where k denotes the thermal conductivity of the fluid. The minus sign arises because heat is transmitted from hotter places to cooler ones. Henceforth, therefore, we assume that

$$\mathbf{q} = -k \nabla T \qquad (5.3\text{-}3)$$

5.4 STRESS

It is to the equation relating stress to the other flow variables that the term constitutive equation is most commonly applied. In fact, this equation relates *extra* stress τ_E to the other flow variables: as we noted in Section 4.3, τ_E may be thought of as arising from the deformation of a body. The *deviatoric* stress τ may or may not be the same as the extra stress, depending on the form of the constitutive equation. It is, however, easy to obtain τ from τ_E:

$$\tau = \tau_E - \tfrac{1}{3}\,\mathrm{trace}(\tau_E)\,\mathbf{I} \tag{5.4-1}$$

Materials may be classified schematically in terms of their stress behavior as in Fig. 5.1. Because the classification is schematic, it should not be interpreted too literally. The two extreme types of behavior are

1. The *rigid* solid, in which there is no strain and the stress is indeterminate. Because there is no strain, the only possible motion is a rigid body motion, that is, a combination of translation and rotation (see Section 2.4).
2. The *inviscid* fluid, in which there is no stress:

$$\tau_E = \mathbf{0} \tag{5.4-2}$$

and clearly $\tau = \tau_E$. Inviscid fluids are the idealized basis for most theories of gas dynamics, provided the gases are dense enough for the continuum hy-

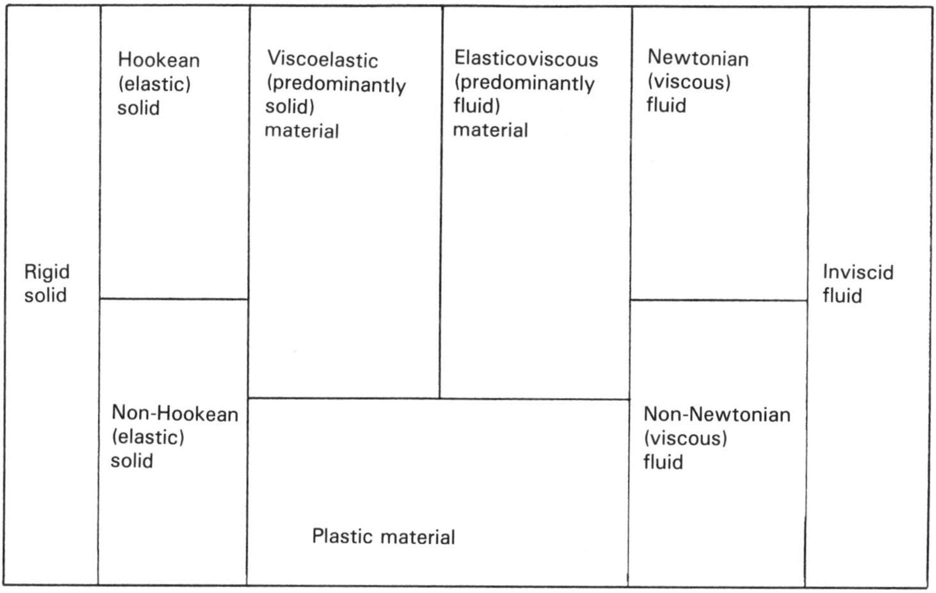

Rigid solid	Hookean (elastic) solid	Viscoelastic (predominantly solid) material	Elasticoviscous (predominantly fluid) material	Newtonian (viscous) fluid	Inviscid fluid
	Non-Hookean (elastic) solid	Plastic material		Non-Newtonian (viscous) fluid	

Figure 5.1. Schematic classification of the stress behavior of materials.

pothesis (see Section 1.1) to apply, and for theories of rapid liquid flows (see Chapter 9).

We have called the rigid material a solid and the inviscid material a fluid. Crudely, the distinction between a solid and a fluid is as follows. The stress in a solid depends on the relative displacements of (i.e., strains in) material points relative to some initial state or reference configuration. For a *Hookean* solid, the stress is directly proportional to the strain (suitably defined). For a *non-Hookean* solid, the stress is a nonlinear function of the strain. For both, the material is elastic. By contrast, there is no reference configuration for a fluid. Instead, the stress in a fluid depends on the relative velocities of material points. The distinction between a solid and a fluid is, however, not sharp: there are materials that combine solid-like and fluid-like behavior, as we will see.

We now concentrate on the constitutive equation for the extra stress in a fluid. For a viscous fluid, we expect that the extra stress τ_E at position x and time t may depend on the instantaneous velocity u, temperature T, etc., at all positions y (including x) in the fluid:

$$\tau_E(\mathbf{x}, t) = \tau_E(\mathbf{u}(\mathbf{y}, t), T(\mathbf{y}, t), \ldots) \qquad (5.4\text{-}3)$$

Of course, Eq. (5.4-3) is almost useless as it stands, since the functional form is unknown. To make it more specific and hence more useful, we argue as follows. We expect that

1. Stress depends not on velocity but on velocity difference, since it is the relative motion of neighboring material points that produces stress, not bulk motion.
2. Stress depends only on the flow field in the neighborhood of a material point. This is the *principle of local action*. Thus recalling that

$$\mathbf{u}(\mathbf{x} + \Delta\mathbf{x}) - \mathbf{u}(\mathbf{x}) \simeq \Delta\mathbf{x} \cdot \nabla\mathbf{u}(\mathbf{x}) \qquad (5.4\text{-}4)$$

   for small $|\Delta\mathbf{x}|$ [see Eq. (2.4-1)], instead of expecting that τ_E at position x depends on u and T at all positions y, we now expect that $\tau_E(\mathbf{x})$ depends on $\nabla\mathbf{u}(\mathbf{x})$ and $T(\mathbf{x})$.
3. Stress is independent of the coordinate system, which is guaranteed by our use of tensor notation.
4. Stress is independent of the reference frame, which is the *principle of frame indifference*.

We recall from Eq. (2.4-6) that

$$\nabla\mathbf{u} = \tfrac{1}{2}(\mathbf{e} + \mathbf{w}) \qquad (5.4\text{-}5)$$

where e denotes the rate of strain and w the vorticity (tensor, as opposed to ω, the vorticity vector). Though it is too lengthy to do so here, it can be shown that independence of reference frame means that τ_E depends on e and not on w, that is,

$$\boldsymbol{\tau}_E (\mathbf{x}, t) = \boldsymbol{\tau}_E (\mathbf{e} (\mathbf{x}, t), T (\mathbf{x}, t), \ldots) \tag{5.4-6}$$

That $\boldsymbol{\tau}_E$ depends on $\mathbf{e}$ and not on $\mathbf{w}$ accords with the physically obvious notions that stress arises only when material points move apart or together, and that for a fluid, it is the rate at which material points move (i.e., the rate of strain) that determines the stress. Stress does not depend on vorticity because rotation (or, more generally, any rigid body motion) cannot directly give rise to stress. It is primarily because of its relevance to stress in fluids that the decomposition of motion into translation, rotation, and straining was made in Section 2.4.

In order to make further progress, we now assume that $\boldsymbol{\tau}_E$ is a function of $\mathbf{e}$ only, so that variations of temperature T, etc., do not affect $\boldsymbol{\tau}_E$, that is, we develop a purely mechanical theory. Then

$$\boldsymbol{\tau}_E (\mathbf{x}, t) = \boldsymbol{\tau}_E (\mathbf{e}(\mathbf{x}, t)) \tag{5.4-7}$$

Subject to certain minor restrictions which need not concern us here, we may express the function $\boldsymbol{\tau}_E(\mathbf{e}(\mathbf{x}, t))$ as a series expansion in powers of $\mathbf{e}$:

$$\boldsymbol{\tau}_E(\mathbf{x}, t) = a_0 \mathbf{I} + a_1 \mathbf{e}(\mathbf{x}, t) + a_2 \mathbf{e}^2(\mathbf{x}, t) + a_3 \mathbf{e}^3(\mathbf{x}, t) + a_4 \mathbf{e}^4 (\mathbf{x}, t) + \ldots \tag{5.4-8}$$

where $\mathbf{e}^2 = \mathbf{e} \cdot \mathbf{e}$, etc., and the a_i ($i = 0, 1, \ldots$) are scalars. The *Cayley-Hamilton theorem* states that a tensor satisfies its own characteristic equation. Thus in three-dimensional space and for arbitrary n:

$$\mathbf{e}^n = \mathrm{I}_{\mathbf{e}} \mathbf{e}^{n-1} + \mathrm{II}_{\mathbf{e}} \mathbf{e}^{n-2} + \mathrm{III}_{\mathbf{e}} \mathbf{e}^{n-3} \tag{5.4-9}$$

where $\mathrm{I}_{\mathbf{e}}$, $\mathrm{II}_{\mathbf{e}}$, and $\mathrm{III}_{\mathbf{e}}$ denote the three invariants of $\mathbf{e}$ given by

$$\mathrm{I}_{\mathbf{e}} = \mathrm{trace}(\mathbf{e}) \qquad \mathrm{II}_{\mathbf{e}} = \tfrac{1}{2} \{\mathrm{trace}(\mathbf{e}^2) - [\mathrm{trace}(\mathbf{e})]^2\} \qquad \mathrm{III}_{\mathbf{e}} = \det(\mathbf{e}) \tag{5.4-10}$$

[The trace and determinant of a tensor are defined in Eqs. (A1.2-13) and (A1.2-2), respectively, of Appendix A.] The three invariants are thus scalars which are, in fact, related to the three eigenvalues of $\mathbf{e}$. It follows from Eq. (5.4-9) that

$$\mathbf{e}^3 = \mathrm{I}_{\mathbf{e}} \mathbf{e}^2 + \mathrm{II}_{\mathbf{e}} \mathbf{e} + \mathrm{III}_{\mathbf{e}} \mathbf{I} \tag{5.4-11}$$

and

$$\mathbf{e}^4 = \mathrm{I}_{\mathbf{e}} \mathbf{e}^3 + \mathrm{II}_{\mathbf{e}} \mathbf{e}^2 + \mathrm{III}_{\mathbf{e}} \mathbf{e} \tag{5.4-12}$$

and so

$$\mathbf{e}^4 = \mathrm{I}_{\mathbf{e}} (\mathrm{I}_{\mathbf{e}} \mathbf{e}^2 + \mathrm{II}_{\mathbf{e}} \mathbf{e} + \mathrm{III}_{\mathbf{e}} \mathbf{I}) + \mathrm{II}_{\mathbf{e}} \mathbf{e}^2 + \mathrm{III}_{\mathbf{e}} \mathbf{e} \tag{5.4-13}$$

that is,

$$\mathbf{e}^4 = (\mathrm{I}_{\mathbf{e}}^2 + \mathrm{II}_{\mathbf{e}}) \mathbf{e}^2 + (\mathrm{I}_{\mathbf{e}} \mathrm{II}_{\mathbf{e}} + \mathrm{III}_{\mathbf{e}}) \mathbf{e} + \mathrm{I}_{\mathbf{e}} \mathrm{III}_{\mathbf{e}} \mathbf{I} \tag{5.4-14}$$

We thus see that $\mathbf{e}^n$ can be expressed in terms of $\mathbf{e}^2$, $\mathbf{e}$, and $\mathbf{I}$ for all integers $n > 2$. Hence it follows from Eq. (5.4-8) that the most general form for $\boldsymbol{\tau}_E$ is

$$\boldsymbol{\tau}_E (\mathbf{x}, t) = c_0 \mathbf{I} + c_1 \mathbf{e}(\mathbf{x}, t) + c_2 \mathbf{e}^2(\mathbf{x}, t) \tag{5.4-15}$$

where c_0, c_1, and c_2 are functions of I_e, II_e, and III_e. Note that, in two-dimensional space, Eq. (5.4-9) becomes

$$\mathbf{e}^n = I_e \, \mathbf{e}^{n-1} + II_e \, \mathbf{e}^{n-2} \tag{5.4-16}$$

and so Eq. (5.4-15) becomes

$$\boldsymbol{\tau}_E \, (\mathbf{x}, \, t) = c_0 \, \mathbf{I} + c_1 \, \mathbf{e}(\mathbf{x}, \, t) \tag{5.4-17}$$

To make still further progress, we henceforth assume that the fluid is incompressible. Then mass conservation implies that

$$\nabla \cdot \mathbf{u} = 0 \tag{5.4-18}$$

[see Eq. (4.2-7)]. It is easy to show that

$$\text{trace}(\mathbf{e}) = 2 \, \nabla \cdot \mathbf{u} \tag{5.4-19}$$

Hence

$$\text{trace}(\mathbf{e}) = I_e = 0 \tag{5.4-20}$$

for an incompressible fluid. Furthermore, for an incompressible fluid, the isotropic term $c_0 \, \mathbf{I}$ in Eq. (5.4-15) can be absorbed into the isotropic pressure term $- p \, \mathbf{I}$ in the decomposition of total stress [see Eq. (4.3-9)]. Hence,

$$\boldsymbol{\tau}_E = c_1(\tfrac{1}{2}\text{trace}(\mathbf{e}^2), \, \det(\mathbf{e})) \, \mathbf{e} + c_2(\tfrac{1}{2}\text{trace}(\mathbf{e}^2), \, \det(\mathbf{e})) \, \mathbf{e}^2 \tag{5.4-21}$$

Note that, because $\mathbf{e}$ and $\mathbf{e}^2$ are symmetric, $\boldsymbol{\tau}_E$ and hence $\boldsymbol{\tau}$ are also symmetric. Thus, if the extra stress of a fluid is given by an equation of the form (5.4-21), that fluid is a *nonpolar* material (see Section 4.4).

The simplest fluids are those for which c_1 and c_2 vanish; such fluids are clearly inviscid [see Eq. (5.4-2)]. The next simplest fluids are those for which c_1 is a nonzero constant, μ say, and c_2 vanishes:

$$\boldsymbol{\tau}_E = \mu \mathbf{e} \tag{5.4-22}$$

This is the constitutive equation of a *Newtonian* fluid; μ denotes the viscosity of the fluid, sometimes called the *dynamic viscosity* and sometimes denoted by η. It should be noted that, if we deviate slightly from a purely mechanical theory, μ can be permitted to be a function of temperature T. It is often convenient to use the *kinematic viscosity* ν given by

$$\nu = \frac{\mu}{\rho} \tag{5.4-23}$$

instead of μ. A Newtonian fluid is one for which there is a linear relationship between $\boldsymbol{\tau}_E$ and $\mathbf{e}$. It is thus directly analogous to the Fourier material, for which there is a linear relationship between the heat flux $\mathbf{q}$ and the temperature gradient ∇T (see Section 5.3).

If Eq. (5.4-22) is substituted into the linear momentum conservation equation (4.3-12), we obtain

$$\rho \frac{\partial \mathbf{u}}{\partial t} + \rho \, \mathbf{u} \cdot \nabla \mathbf{u} = \rho \, \mathbf{g} - \nabla p + \mu \, \nabla \cdot \mathbf{e} \qquad (5.4\text{-}24)$$

It can be shown that, since $\nabla \cdot \mathbf{u} = 0$,

$$\nabla \cdot \mathbf{e} = \Delta \mathbf{u} \qquad (5.4\text{-}25)$$

whence

$$\rho \frac{\partial \mathbf{u}}{\partial t} + \rho \, \mathbf{u} \cdot \nabla \mathbf{u} = \rho \, \mathbf{g} - \nabla p + \mu \, \Delta \mathbf{u} \qquad (5.4\text{-}26)$$

These are the *Navier-Stokes equations*. They are given in component form in rectangular, cylindrical polar, and spherical polar coordinates in Section A2.3 of Appendix B. Apart from certain special cases, many of which are included in Chapters 8–12, the Navier-Stokes equations have not (yet) been solved, certainly not exactly. Numerical, computer-based methods have, however, been used extensively to obtain approximate solutions. There are two main reasons why it is difficult, if not impossible, to obtain exact solutions. The first is that the equations are nonlinear (in fact, they are quadratic) in $\mathbf{u}$. The second is that there is no separate equation for p.

We recall from Section 4.5 that the irreversible dissipation given by $(1/\rho T)$ $\boldsymbol{\tau}{:}\nabla \mathbf{u}$ must, for an incompressible, isothermal material, be nonnegative by the second law of thermodynamics. It follows by substitution of Eq. (5.4-22) into Eq. (5.4-20) that $\mathrm{trace}(\boldsymbol{\tau}_E)$ vanishes and hence that

$$\boldsymbol{\tau} = \mu \, \mathbf{e} \qquad (5.4\text{-}27)$$

It is easy to show that

$$\mathbf{e}{:}\nabla \mathbf{u} = \tfrac{1}{2}\, \mathbf{e}{:}\mathbf{e} = \tfrac{1}{2}\, \mathrm{trace}(\mathbf{e}^2) \qquad (5.4\text{-}28)$$

Thus the irreversible dissipation for an incompressible Newtonian fluid is given by

$$\frac{1}{2}\frac{\mu}{\rho T}\, \mathrm{trace}(\mathbf{e}^2).$$

Clearly, $\mathrm{trace}(\mathbf{e}^2) \geq 0$, $\rho > 0$, and $T \geq 0$. Thus it follows that for an incompressible, isothermal, Newtonian fluid,

$$\mu \geq 0 \qquad (5.4\text{-}29)$$

There is, of course, no particular reason why a fluid should be inviscid or Newtonian. Indeed, there are many fluids that are viscous but non-Newtonian, that is, they are viscous but their constitutive equation is different from, and presumably more complicated than, Eq. (5.4-22). Such fluids are almost without exception liquids, for example polymer melts. The simplest incompressible, non-Newtonian fluid has a variable viscosity and thus a constitutive equation of the form

$$\boldsymbol{\tau}_E = \mu(\tfrac{1}{2}\mathrm{trace}(\mathbf{e}^2), \det(\mathbf{e}))\,\mathbf{e} \qquad (5.4\text{-}30)$$

Note that μ is independent of $\mathrm{trace}(\mathbf{e})$ because $\mathrm{trace}(\mathbf{e})$ vanishes for an incompressible fluid [see Eq. (5.4-20)]. With a slight deviation from a purely mechanical theory, however, μ can be permitted to depend on T. The linear momentum conservation equation for such a fluid is given in component form in rectangular, cylindrical polar, and spherical polar coordinates in Section A2.4 of Appendix B. A *shear flow* may be defined as one for which

$$\det(\mathbf{e}) = \mathrm{III}_\mathbf{e} = 0 \qquad (5.4\text{-}31)$$

An example is the simple shear flow discussed in Section 2.5; another is the flow in a long, narrow pipe discussed in Section 10.2. Thus, for shear flow of an incompressible fluid, $\mathrm{I}_\mathbf{e}$ and $\mathrm{III}_\mathbf{e}$ both vanish but $\mathrm{II}_\mathbf{e}$ does not. Indeed, since $\mathrm{I}_\mathbf{e}$ vanishes,

$$\mathrm{II}_\mathbf{e} = \tfrac{1}{2}\,\mathrm{trace}(\mathbf{e}^2) = |\mathbf{e}|^2 \qquad (5.4\text{-}32)$$

(see Section A1.2 of Appendix A). The shear rate γ is defined by

$$\gamma = \sqrt{\mathrm{II}_\mathbf{e}} \qquad (5.4\text{-}33)$$

[This is consistent with the use of γ in Eq. (2.5-35).] It is observed in practice that, in a shear flow, many non-Newtonian fluids are shear thinning (i.e., their viscosity μ decreases as γ increases). Adequate agreement with experimental results is often achieved by assuming that

$$\mu = \mu_0\,\mathrm{II}_\mathbf{e}^{-n} = \mu_0\,\gamma^{-2n} \qquad (5.4\text{-}34)$$

Such a fluid is called a power-law fluid. For shear-thinning fluids, $n > 0$.

A more complex non-Newtonian fluid is the *Reiner-Rivlin* fluid, where c_1 and c_2 in Eq. (5.4-21) are both nonzero. Nothing like such a fluid has, however, ever yet been found in practice. Accordingly, we will not discuss it further.

The only reasonable way in a purely mechanical theory in which the constitutive equations of more complex non-Newtonian fluids can be obtained is to relax the assumption implicit in Eq. (5.4-3) that $\boldsymbol{\tau}_E(\mathbf{x}, t)$ depends only on the deformation at time t. We would not expect the future to determine the present, which is the *principle of determinism*, so we assume that $\boldsymbol{\tau}_E(\mathbf{x}, t)$ depends on the deformation at all times up to and including t. Because we expect that it is deformation in the neighborhood of a given material point X that determines the stress in X, which is again the *principle of local action*, it follows that

$$\hat{\boldsymbol{\tau}}_E(X, t) = \hat{\boldsymbol{\tau}}_E \text{ (deformation in the neighborhood of } X \text{ at all times } t^{\#} \le t) \quad (5.4\text{-}35)$$

Note that a Lagrangian specification is appropriate here (we use $\hat{\boldsymbol{\tau}}_E$, not $\boldsymbol{\tau}_E$) because we need to be able to determine the deformation history in the neighborhood of a material point X (i.e., we need to be able to follow X). A typical constitutive equation for such a material is

$$\hat{\boldsymbol{\tau}}_E (X, t) = \int_{-\infty}^{t} m(t^{\#}) \, \hat{\mathbf{S}}(X, t^{\#}) \, dt^{\#} \tag{5.4-36}$$

where $m(t^{\#})$ is a memory function and $\hat{\mathbf{S}}(X, t^{\#})$ is an appropriate measure of deformation. If the memory of the material is perfect, that is, if it remembers its initial state or reference configuration perfectly, then the material will always attempt to return to its initial state and the material is an elastic solid. If the memory of the material is quite imperfect, that is, if it remembers nothing of previous states (including its initial state), but only responds to the current rate at which it is being deformed, then it is a viscous liquid. In a material with an intermediate memory, the distant past is expected to have less effect than the recent past. A typical memory function for such a fading memory material is

$$m(t^{\#}) = e^{-(t-t^{\#})/t_r} \tag{5.4-37}$$

where t_r denotes the natural or relaxation time of the material. Such a material is thus described as being *elasticoviscous* or *viscoelastic*; the former term more properly describes materials that are predominantly fluid-like and the latter describes predominantly solid-like materials. The names imply that such materials combine certain fluid and solid features.

In fact, no material behaves purely as a fluid or as a solid: the time scale of the deformation process determines the behavior of any material. Thus glass is usually thought of as being a solid under normal conditions. Over a very long period of time glass is found to flow. The windows of very old buildings testify to this, being thicker at the bottom than at the top. Similarly, water, usually thought of as being a fluid under normal conditions, can be shown to behave like a solid. In experiments of very short duration, water is found to shatter like a brittle solid. The way to determine whether a given material is predominantly fluid-like or solid-like is to compare the relaxation time of the material t_r with the time scale t_d of the deformation:

- If t_r is much less than t_d, fluid-like behavior can be expected.
- If t_r is much greater than t_d, solid-like behavior can be expected.
- If t_r and t_d are comparable, intermediate behavior can be expected.

Physically, the relaxation time t_r can be identified with the time taken for molecules of the material to undertake some characteristic motion. Thus, for example, it might be the time taken for two molecules to move past each other and hence for flow to occur. We would expect that t_r gets larger as molecules get longer, since they then become more flexible and hence more likely to become entangled in one another. It is not surprising then that polymer melts, for example, have both fluid- and solid-like behavior, since polymer molecules are very long indeed. Note, incidentally, that this discussion of relaxation times is based on molecular (i.e., noncontinuum) arguments. This accords with the statement made in Section 5.1 that constitutive equations cannot be deduced by arguments based on continuum mechanics alone.

Materials even more complex than elasticoviscous and viscoelastic ones do exist. They are called plastic materials but are beyond the scope of our discussion here. The study of the stress behavior of such materials and, indeed, of the whole range of materials shown in Fig. 5.1, is called *rheology*. For this reason, the constitutive equation for stress is often referred to as the rheological equation of state.

BOUNDARY CONDITIONS

6.1 INITIAL AND BOUNDARY CONDITIONS

The conservation equations derived in Chapter 4 and the constitutive equations derived in Chapter 5 determine the behavior of a material within a flow domain. In order to determine the behavior of the material fully, however, additional conditions must be specified. The mass, linear momentum, and energy conservation equations describe the rate of change with time t of density ρ, velocity $\mathbf{u}$, and internal energy U [see Eqs. (4.2-6), (4.3-7), and (4.5-12]. It follows that initial conditions are required on ρ, $\mathbf{u}$, and U in order that the equations can be solved. Specification of initial conditions usually poses few problems in practice. Indeed, no *initial conditions* are needed if the flow is steady in an Eulerian sense, that is, if

$$\frac{\partial \rho}{\partial t} = 0 \qquad \frac{\partial \mathbf{u}}{\partial t} = \mathbf{0} \qquad \frac{\partial U}{\partial t} = 0 \qquad (6.1\text{-}1)$$

so that there are no variations in ρ, $\mathbf{u}$, and U with t at a fixed position in space. Because the conservation and constitutive equations apply within but not at the edge of the flow domain, *boundary conditions* are also needed to determine ρ, $\mathbf{u}$, and U completely. It is with specification of boundary conditions that we are concerned here.

The conservation equations describe the motion of a quite general material in quite general circumstances; the constitutive equations particularize the material but not the circumstances; it is the boundary and initial conditions that distinguish one circumstance from another, and hence one fluid mechanics problem from another.

6.2 RIGID IMPERMEABLE WALLS

At a rigid impermeable wall the normal component of the fluid velocity $\mathbf{u}$ at the wall equals the normal component of the wall velocity $\mathbf{u}_w$. If this were not the case, either fluid would penetrate the wall, which would not then be impermeable, or there would be an unbounded positive or negative accumulation of fluid by the wall. Thus,

$$\mathbf{u} \cdot \mathbf{n} = \mathbf{u}_w \cdot \mathbf{n} \qquad (6.2\text{-}1)$$

where $\mathbf{n}$ denotes the unit outer normal to the fluid at the wall (i.e., $\mathbf{n}$ is directed from the fluid toward the wall; see Fig. 6.1). If the wall is at rest in the reference frame being used, then $\mathbf{u} \cdot \mathbf{n}$ vanishes at the wall. Boundary condition (6.2-1) is referred to as the *no flow-through condition*. It is entirely kinematic in origin.

If the fluid is inviscid, then the fluid can slip over the wall and the tangential components of $\mathbf{u}$ cannot be specified. If, on the other hand, the fluid is viscous, then molecular arguments corroborated by experiments show that the tangential components of $\mathbf{u}$ at the wall equal the tangential components of $\mathbf{u}_w$, so that there is no slip at the wall, that is,

$$\mathbf{u}_{\wedge}\mathbf{n} = \mathbf{u}_{w\wedge}\mathbf{n} \qquad (6.2\text{-}2)$$

[Recall that the vector $\mathbf{v}_1{}_{\wedge}\mathbf{v}_2$ is normal to the vectors $\mathbf{v}_1$ and $\mathbf{v}_2$; see Eq. (A1.2-19) of Appendix A]. Combining Eq. (6.2-1) and (6.2-2) yields

$$\mathbf{u} = \mathbf{u}_w \qquad (6.2\text{-}3)$$

Boundary condition (6.2-2) is referred to as the *no slip condition*. It is dy-

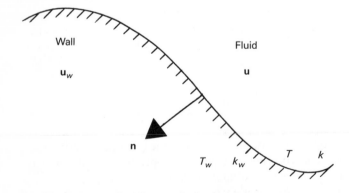

Figure 6.1. Rigid, impermeable wall.

namic in origin. Indeed, the following dynamic argument is often used to support it. If there were slip at a wall, there would be an unbounded velocity gradient, and hence an unbounded rate of strain, by the wall. For a viscous fluid, whether or not the viscosity is constant, this would lead to an unbounded stress in the fluid by the wall [see Eq. (5.4-30)] and hence to an unbounded force on the wall, which is physically unreasonable. This argument is flawed, however, because it assumes that the unbounded velocity gradient would occur in the fluid whereas it in fact occurs between the fluid and the wall. Stronger arguments for the validity of the no slip condition are molecular and empirical in nature. Molecular arguments are based on notions of thermodynamic equilibrium at a wall and hence equality of the velocities of the molecules of the fluid and those of the wall. Empirical arguments are based on the extensive experimental corroboration of the consequences of assuming that there is no slip at a wall.

Equation (6.2-1) provides just one boundary condition for an inviscid fluid. Equation (6.2-3) provides three for a viscous fluid. The reason for this difference is that the conservation equations, in particular the linear momentum conservation equation, are of lower order for an inviscid fluid than for a viscous one. Thus for the former,

inviscid

$$\rho \frac{\partial \mathbf{u}}{\partial t} + \rho\, \mathbf{u} \cdot \nabla \mathbf{u} = \rho \mathbf{g} - \nabla p \qquad (6.2\text{-}4)$$

which is first order in spatial derivatives of $\mathbf{u}$ (∇ is a first-order operator), while for the latter,

viscous

$$\rho \frac{\partial \mathbf{u}}{\partial t} + \rho\, \mathbf{u} \cdot \nabla \mathbf{u} = \rho \mathbf{g} - \nabla p + \mu\, \Delta \mathbf{u} \qquad (6.2\text{-}5)$$

which is second-order in spatial derivatives of $\mathbf{u}$ (Δ is a second-order operator).

Boundary conditions on temperature are based on two notions. The first is that the heat flux must be continuous normal to the wall, in order that energy is conserved. If it were not continuous, there would be a difference between the heat flux leaving the fluid and that entering the wall. A finite positive amount of heat causes an unbounded positive temperature, and a finite negative amount of heat an unbounded negative temperature, in an infinitesimal volume. Thus a difference in heat flux would cause the temperature in a layer of infinitesimal thickness at the wall to become unbounded, which is physically unreasonable. Hence, if the fluid is a Fourier material (see Section 5.3),

$$k\, \nabla T \cdot n = k_w\, \nabla T \cdot \mathbf{n} \qquad (6.2\text{-}6)$$

where k and k_w denote the thermal conductivities of the fluid and the wall, respectively. The second notion is that of thermodynamic equilibrium, which implies that the fluid temperature T at the wall is the same as the wall temperature T_w:

$$T = T_w \qquad (6.2\text{-}7)$$

Boundary conditions (6.2-2) and (6.2-7) hold for flows of most, but not all, fluids. Thus, for example, boundary condition (6.2-2) may not hold for flows of certain non-Newtonian, particularly elasticoviscous, fluids. Boundary conditions (6.2-1) and (6.2-6) hold for flows of all fluids, of course, since they are based on conservation, and not on constitutive, principles.

6.3 FLUID-FLUID INTERFACES

By analogous arguments to those of Section 6.2, boundary conditions at the interface between two mutually immiscible fluids A and B are

$$\mathbf{u}_A \cdot \mathbf{n} = \mathbf{u}_B \cdot \mathbf{n} \qquad (6.3\text{-}1)$$

for inviscid fluids and, for viscous fluids,

$$\mathbf{u}_A = \mathbf{u}_B \qquad (6.3\text{-}2)$$

where $\mathbf{u}_A$ and $\mathbf{u}_B$ denote the velocities at the interface of fluids A and B, respectively, and $\mathbf{n}$ denotes the unit normal to the interface directed from fluid A toward fluid B (see Fig. 6.2). Also,

$$k_A \, \nabla T_A \cdot \mathbf{n} = k_B \, \nabla T_B \cdot \mathbf{n} \qquad (6.3\text{-}3)$$

and

$$T_A = T_B \qquad (6.3\text{-}4)$$

where T_A and T_B denote the temperatures at the interface of fluids A and B, re-

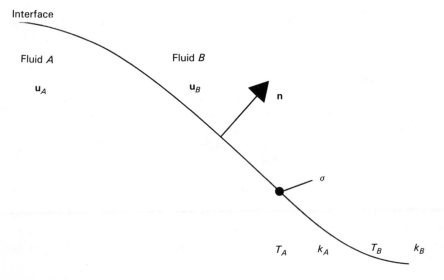

Figure 6.2. Fluid-fluid interface.

spectively, and k_A and k_B denote the thermal conductivities of fluids A and B, respectively.

The shape of the boundary is known in advance for a rigid wall. For a fluid-fluid interface, in contrast, the boundary is flexible. Its shape is, therefore, not known in advance and an additional boundary condition is needed. This is provided by requiring that linear momentum is conserved at the interface. Let σ denote the *interfacial tension* or *surface tension* of the two fluids and R_I and R_{II} denote the *principal radii of curvature* of the interface between fluids A and B, reckoned positive when the corresponding center of curvature lies in fluid A and negative when it is in fluid B. (Note that the introduction of interfacial tension implies a constitutive assumption about the interface.) In the absence of interfacial tension effects, conservation of linear momentum means that total stress must be continuous normal to the interface. Decomposing the total stress into a pressure component and a deviatoric stress component as in Eq. (4.3-9), conservation of linear momentum thus means that

$$\mathbf{n} \cdot (-p_A \mathbf{I} + \boldsymbol{\tau}_A) - \mathbf{n} \cdot (-p_B \mathbf{I} + \boldsymbol{\tau}_B) = \mathbf{0} \tag{6.3-5}$$

The reason for this is analogous to that for continuity of heat flux: if it were not so, a finite force would be applied to the interface which would cause it to have an unbounded acceleration, since it is of zero mass. When interfacial tension effects are included, Eq. (6.3-5) becomes

$$\mathbf{n} \cdot (-p_A \mathbf{I} + \boldsymbol{\tau}_A) - \mathbf{n} \cdot (-p_B \mathbf{I} + \boldsymbol{\tau}_B) = -\sigma \left(\frac{1}{R_I} + \frac{1}{R_{II}} \right) \mathbf{n} \tag{6.3-6}$$

To clarify boundary condition (6.3-6), we consider a specific example: a stationary sphere comprising fluid A of radius R surrounded by stationary fluid B (see Fig. 6.3). Because the fluids A and B are stationary, we expect that the only forces acting on the sphere are due to the interfacial tension σ of the two fluids and the uniform pressures p_A and p_B (modified, if necessary to incorporate gravitational effects; see Section 4.3) in fluids A and B, respectively. Consider a hypothetical plane passing through the center of the sphere. The force tending to separate the two halves of the sphere on either side of the plane is aligned normal to the plane; its magnitude is given by $(p_A - p_B) \pi R^2$. The force tending to keep both halves of the sphere together is also aligned normal to the plane; its magnitude is given by $2\pi R\sigma$. Clearly, these forces must balance, whence,

$$p_A - p_B = \frac{2\sigma}{R} \tag{6.3-7}$$

Thus the pressure inside the sphere exceeds that outside it by $2\sigma/R$. We now obtain this same result from Eq. (6.3-6). Because the center of curvature, which is clearly the center of the sphere, lies in fluid A, the principal radii of curvature of the interface between fluids A and B, R_I and R_{II}, are both positive; indeed,

$$R_I = R \qquad R_{II} = R \tag{6.3-8}$$

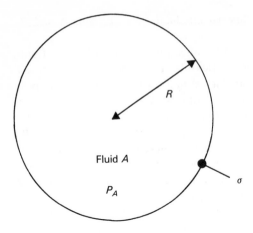

Fluid A

P_A

σ

R

Fluid B

P_B

Figure 6.3. Stationary sphere comprising one fluid surrounded by another fluid.

Because there is no motion in either fluid A or fluid B, we would expect that the deviatoric stress in fluids A and B vanishes, that is,

$$\tau_A = 0 \qquad \tau_B = 0 \tag{6.3-9}$$

As a result, it follows from Eq. (6.3-6) that

$$p_A - p_B = \frac{2\sigma}{R} \tag{6.3-10}$$

which is precisely the same as Eq. (6.3-7).

6.4 BOUNDARIES AT INFINITY

Consider flow past a body of typical dimension D which is at typical distances L_1 from other bodies in the flow and L_2 from the outer boundary of the flow (see Fig. 6.4). Suppose that L_1 and L_2 are both much larger than D. For a suspension of solids in a fluid, this corresponds to the *dilute* limit. Then we would expect to be able to neglect interaction from the other bodies and from the outer boundary when we determine the flow past the body. As a result, we can consider that the fluid round the body is unbounded, that is, it extends to infinity. The appropriate boundary conditions on velocity $\mathbf{u}$ and temperature T at infinity are then

$$\mathbf{u} \to \mathbf{u}_\infty \qquad T \to T_\infty \qquad \text{as } |\mathbf{x}| \to \infty \tag{6.4-1}$$

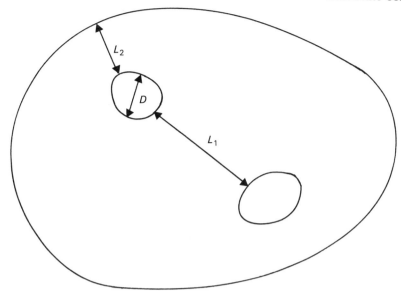

Figure 6.4. Bodies immersed in a fluid contained in an outer boundary.

where $\mathbf{u}_\infty$ and T_∞ denote the velocity and temperature fields at infinity, that is, at a great distance from the body; $\mathbf{x}$ denotes the position vector relative to an origin in or near the body.

6.5 SYMMETRY BOUNDARIES

In many flow problems, the boundaries imply a symmetry which we would expect to be manifested, purely kinematically, in the solution. Two cases are of particular importance: a symmetry line and a symmetry plane.

At a symmetry line, if $\mathbf{t}$ denotes a unit vector tangential to (i.e., along) the line (see Fig. 6.5) (and *not* the stress vector; see Section 3.2), then the boundary

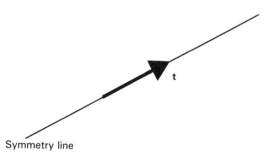

Symmetry line

Figure 6.5. Symmetry line.

conditions to be imposed at the line are that (1) the components of the velocity field **u** normal to the line vanish, otherwise mass would be created or destroyed at the line, that is,

$$\mathbf{u}_{\wedge}\mathbf{t} = \mathbf{0} \qquad\qquad (6.5\text{-}1)$$

(recall that the vector $\mathbf{v}_1 {\scriptstyle\wedge} \mathbf{v}_2$ is normal to the vectors $\mathbf{v}_1$ and $\mathbf{v}_2$); and (2) the normal gradients of the component of velocity tangential to the line vanish, otherwise the tangential velocity component would not vary smoothly and would have discontinuities in its normal gradients at the line, that is,

$$\boldsymbol{\omega}_{\wedge}\mathbf{t} = \mathbf{0} \qquad\qquad (6.5\text{-}2)$$

where $\boldsymbol{\omega} = \boldsymbol{\nabla}_{\wedge}\mathbf{u}$ denotes the vorticity field.

At a symmetry plane, if **n** denotes a unit vector normal to the plane (see Fig. 6.6), then the boundary conditions to be imposed are that (1) the component of velocity normal to the plane vanishes, that is,

$$\mathbf{u} \cdot \mathbf{n} = 0 \qquad\qquad (6.5\text{-}3)$$

and (2) the normal gradients of the components of velocity tangential to the plane vanish, that is,

$$\boldsymbol{\omega}_{\wedge}\mathbf{n} = \mathbf{0} \qquad\qquad (6.5\text{-}4)$$

In order to clarify these boundary conditions, we consider three examples, the third being a special, but common, instance of the second. The first example is flow along a pipe of circular cross section, in which we use an obvious cylindrical polar coordinate system (r, θ, z) with r denoting radial position with respect to the centerline of the pipe, θ denoting angular orientation about the centerline, and z denoting axial position along the centerline (see Fig. 6.7). We would expect the velocity field **u** given by

$$\mathbf{u} = u_r(r, \theta, z, t)\mathbf{i}_r + u_\theta(r, \theta, z, t)\mathbf{i}_\theta + u_z(r, \theta, z, t)\mathbf{i}_z \qquad (6.5\text{-}5)$$

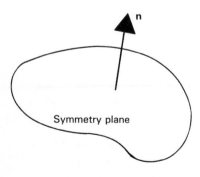

Symmetry plane

Figure 6.6. Symmetry plane.

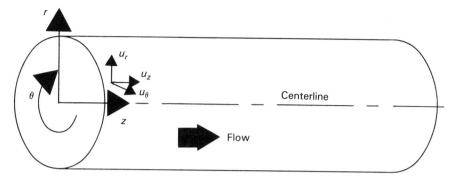

Figure 6.7. Geometry of flow in a pipe.

where t denotes time, to be *axisymmetric*, that is, independent of angular position about the centerline of the pipe, which comprises the axis of symmetry and hence a symmetry line. Thus we would expect that

$$\frac{\partial u_r}{\partial \theta} = 0, \qquad \frac{\partial u_\theta}{\partial \theta} = 0, \qquad \frac{\partial u_z}{\partial \theta} = 0 \qquad (6.5\text{-}6)$$

We would also expect the velocity field to be *swirl free* (i.e., to have no component about the axis of symmetry) because there is nothing to cause the flow to swirl about the centerline. Thus we would expect that

$$u_\theta = 0 \qquad (6.5\text{-}7)$$

and hence that

$$\mathbf{u} = u_r(r, z, t)\mathbf{i}_r + u_z(r, z, t)\mathbf{i}_z \qquad (6.5\text{-}8)$$

We would, therefore, expect to have to solve the equations of motion of the flow only in a single plane for some (any) fixed value of θ. The boundary conditions to be imposed at the centerline ($r = 0$), which follow directly from Eqs. (6.5-1) and (6.5-2), would be

$$u_r = 0 \qquad \frac{\partial u_z}{\partial r} = 0 \qquad \text{at } r = 0 \qquad (6.5\text{-}9)$$

The second example is flow along a channel of rectangular cross section, in which we use an obvious rectangular coordinate system (x, y, z) with x denoting axial position along the centerline of the channel, and y and z denoting transverse position with respect to the two centerplanes of the channel (see Fig. 6.8). We would expect the velocity field $\mathbf{u}$ given by

$$\mathbf{u} = u_x(x, y, z, t)\mathbf{i}_x + u_y(x, y, z, t)\mathbf{i}_y + u_z(x, y, z, t)\mathbf{i}_z \qquad (6.5\text{-}10)$$

where t again denotes time, to be symmetric about the two centerplanes which comprise symmetry boundaries and, therefore, we would have to solve the equations of motion of the flow only in one quadrant of the channel (say, $y \geq 0$, $z \geq$

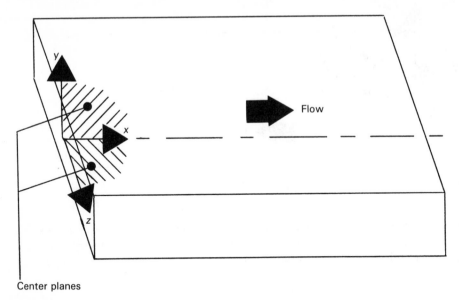

Center planes

Figure 6.8. Geometry of flow in a channel.

0). The boundary conditions to be imposed at one centerplane ($y = 0$), which follow directly from Eqs. (6.5-3) and (6.5-4), would be

$$u_y = 0 \qquad \frac{\partial u_x}{\partial y} = 0 \qquad \frac{\partial u_z}{\partial y} = 0 \qquad (6.5\text{-}11)$$

and at the other ($z = 0$) they would be

$$u_z = 0 \qquad \frac{\partial u_x}{\partial z} = 0 \qquad \frac{\partial u_y}{\partial z} = 0 \qquad (6.5\text{-}12)$$

The third example is flow along a very wide channel, with the same obvious rectangular coordinate system (x, y, z) as in the previous example, so that its dimension in the z direction, say, is very much greater than that in the y direction. As a result, the flow in the channel can nearly everywhere be very closely approximated by flow between two parallel flat plates and we would expect the velocity field **u** given by Eq. (6.5-10) to be *planar*. Thus we would expect it to be independent of transverse position in what we might call the wide direction, that is,

$$\frac{\partial u_x}{\partial z} = 0 \qquad \frac{\partial u_y}{\partial z} = 0 \qquad \frac{\partial u_z}{\partial z} = 0 \qquad (6.5\text{-}13)$$

We would also expect it to have no component in the wide direction, that is,

$$u_z = 0 \qquad (6.5\text{-}14)$$

and hence,

$$\mathbf{u} = u_x(x, y, t)\mathbf{i}_x + u_y(x, y, t)\mathbf{i}_y \qquad (6.5\text{-}15)$$

We would, therefore, expect to have to solve the equations of motion of the flow only in a single half-plane for some fixed value of z with $y \geq 0$. The boundary conditions to be imposed at the centerplane ($y = 0$) would be

$$u_y = 0 \qquad \frac{\partial u_x}{\partial y} = 0 \qquad (6.5\text{-}16)$$

We conclude by noting that, while it is common for an axisymmetric velocity field also to be swirl-free, this is not necessarily the case. Indeed, there are many instances, for example, in parts of axial turbines, pumps, and compressors, where the flow is inherently swirling and yet (at least very nearly) axisymmetric. Similarly while it is common for a velocity field that is independent of transverse position in what we have called the wide direction also to have no component in the wide direction, this is again not necessarily the case.

DIMENSIONLESS GROUPS

7.1 FLOW EQUATIONS

In Chapters 8 through 12 we consider problems involving the flow of a fluid that is of constant density ρ, viscosity μ (though in Section 10.2 we also consider flow of a fluid, the viscosity of which depends on shear rate and temperature), and thermal conductivity k. We thus consider flows of an incompressible, Newtonian, Fourier fluid. The conservation equations governing the flow of such a fluid are (see Chapters 4 and 5):

$$\nabla \cdot \mathbf{u} = 0 \tag{7.1-1}$$

$$\rho\,\frac{\partial \mathbf{u}}{\partial t} + \rho\,\mathbf{u} \cdot \nabla\mathbf{u} = \rho\mathbf{g} - \nabla p + \mu\Delta\mathbf{u} \tag{7.1-2}$$

$$\rho c\,\frac{\partial T}{\partial t} + \rho c\,\mathbf{u} \cdot \nabla T = k\nabla^2 T + \frac{1}{2}\mu\mathbf{e}{:}\mathbf{e} \tag{7.1-3}$$

where use has been made of Eq. (5.4-28). These equations are given in component form in rectangular, cylindrical polar, and spherical polar coordinates in Sections A2.2, A2.3, and A2.5 of Appendix B, respectively. For the reason given in Section 4.5, c in Eq. (7.1-3) can be taken to be c_p, the specific heat evaluated at constant pressure.

It appears to be impossible to develop general methods for obtaining exact solutions of Eqs. (7.1-1), (7.1-2), and (7.1-3), subject to appropriate boundary and initial conditions (see Chapter 6). There is, therefore, a need for a systematic method for solving them approximately. The development of such a method by the use of order-of-magnitude arguments based on dimensionless formulations is our concern here.

7.2 DIMENSIONAL ANALYSIS

The parameters occurring in Eqs. (7.1-1), (7.1-2), and (7.1-3) are density ρ, viscosity μ, specific heat c, thermal conductivity k, and the magnitude of the gravitational acceleration $g = |\mathbf{g}|$. The boundary conditions and initial conditions of a flow problem normally involve a characteristic length l_c (problems that do not involve a characteristic length generally admit *similarity solutions* while problems that involve more than one characteristic length require special consideration, as we will see in Chapters 8 through 12); a characteristic velocity, or more strictly a characteristic speed, u_c; and a characteristic temperature or temperature difference T_c.

Thus, for example, for flow in a long, narrow pipe, l_c might be the pipe radius, u_c might be the mean fluid velocity along the pipe, and T_c might be the temperature of the pipe wall. In all, therefore, there are eight parameters and four dimensions (mass, length, time, and temperature, say). We may now deduce from dimensional analysis that there are only four (eight minus four) truly independent parameters. To see this, we could make use of *Buckingham's pi theorem*. We prefer, however, an alternative approach and define dimensionless variables as follows:

$$\mathbf{u}^* = \frac{1}{u_c}\mathbf{u} \qquad p^* = \frac{1}{\rho u_c^2}p \qquad \mathbf{g}^* = \frac{1}{g}\mathbf{g} \qquad t^* = \frac{u_c}{l_c}t \qquad \nabla^* = l_c\nabla \tag{7.2-1}$$

$$\nabla^{2*} = l_c^2\nabla^2 \qquad \Delta^* = l_c^2\Delta \qquad T^* = \frac{1}{T_c}T \qquad \mathbf{e}^* = \frac{l_c}{u_c}\mathbf{e}$$

Substitution into Eq. (7.1-1) yields

$$\frac{u_c}{l_c}\nabla^* \cdot \mathbf{u}^* = 0 \tag{7.2-2}$$

that is,

$$\nabla^* \cdot \mathbf{u}^* = 0 \tag{7.2-3}$$

Equation (7.1-2) yields

$$\frac{\rho u_c^2}{l_c}\frac{\partial \mathbf{u}^*}{\partial t^*} + \frac{\rho u_c^2}{l_c}\mathbf{u}^* \cdot \nabla^*\mathbf{u}^* = \rho g\mathbf{g}^* - \frac{\rho u_c^2}{l_c}\nabla^* p^* + \frac{\mu u_c}{l_c^2}\Delta^*\mathbf{u}^* \tag{7.2-4}$$

that is,

$$\frac{\partial \mathbf{u}^*}{\partial t^*} + \mathbf{u}^* \cdot \nabla^* \mathbf{u}^* = \frac{1}{\text{Fr}} \mathbf{g}^* - \nabla^* p^* + \frac{1}{\text{Re}} \boldsymbol{\Delta}^* \mathbf{u}^* \qquad (7.2\text{-}5)$$

and Eq. (7.1-3) yields

$$\frac{\rho c u_c T_c}{l_c} \frac{\partial T^*}{\partial t^*} + \frac{\rho c u_c T_c}{l_c} \mathbf{u}^* \cdot \nabla^* T^* = \frac{kT_c}{l_c^2} \nabla^{2*} T^* + \frac{1}{2} \frac{\mu u_c^2}{l_c^2} \mathbf{e}^* : \mathbf{e}^* \qquad (7.2\text{-}6)$$

that is,

$$\text{Pe} \left(\frac{\partial T^*}{\partial t^*} + \mathbf{u}^* \cdot \nabla^* T^* \right) = \nabla^{2*} T^* + \text{Br } \mathbf{e}^* : \mathbf{e}^* \qquad (7.2\text{-}7)$$

where the Froude number Fr is given by

$$\text{Fr} = \frac{u_c^2}{g l_c} \qquad (7.2\text{-}8)$$

the Reynolds number Re is given by

$$\text{Re} = \frac{\rho u_c l_c}{\mu} \qquad (7.2\text{-}9)$$

the Péclet number Pe is given by

$$\text{Pe} = \frac{\rho c u_c l_c}{k} \qquad (7.2\text{-}10)$$

and the Brinkman number Br is given by

$$\text{Br} = \frac{\mu u_c^2}{2k T_c} \qquad (7.2\text{-}11)$$

There are, therefore, just four independent parameters or dimensionless groups: Fr, Re, Pe, and Br, as claimed earlier.

In order to determine the physical significance of these dimensionless groups, we argue as follows. Consider flow in a long, narrow pipe. If u_c denotes the mean fluid velocity along the pipe, we expect that $|\mathbf{u}|$ varies between zero at the pipe wall and something like u_c at its centerline. Hence $|\mathbf{u}^*|$ varies between zero at the pipe wall and something like one at its centerline. Thus $|\mathbf{u}^*|$ varies between zero and a value which is not small or large compared with one and we say that $\mathbf{u}^*$ is of unit order of magnitude. If two quantities α and β are of the same order of magnitude, we write

$$\alpha \sim \beta \qquad (7.2\text{-}12)$$

Provided the dimensionless quantities defined in Eqs. (7.2-1) are chosen sensibly, they are all of unit order of magnitude. The choice is often obvious and is usually straightforward. If the quantities are all of unit order of magnitude, it follows [see Eq. (7.2-5)] that the inertial term $\partial \mathbf{u}^*/\partial t^* + \mathbf{u}^* \cdot \nabla^* \mathbf{u}^* \sim 1$ and the gravitational term $(1/\text{Fr})\mathbf{g}^* \sim 1/\text{Fr}$ and hence that their ratio is Fr to an order of magnitude. Thus the Froude number may be interpreted as the ratio of inertial forces to gravitational forces. Similarly, the Reynolds number may be interpreted as the ratio of inertial forces to viscous forces; the Péclet number may be interpreted as the ratio of convection to conduction; and the Brinkman number may be interpreted as the ratio of heat generation by viscous dissipation to conduction [note that the factor of two in Eq. (7.2-11) is entirely arbitrary].

Of course, there may be more than eight parameters initially. If, for example, the flow problem involves a fluid-fluid interface, the interfacial tension σ increases the number of parameters from eight to nine. The number of dimensions remains unchanged at four. Accordingly, we expect that there are five truly independent parameters, the four already given and also, for example, the Weber number We given by

$$We = \frac{\rho u_c^2 l_c}{\sigma} \tag{7.2-13}$$

which may be interpreted as the ratio of inertial forces to interfacial tension forces. Note that there is no special reason, other than convention, for choosing the particular dimensionless groups that we have chosen here. Thus, for example, we could equally well choose not the Weber number but the ratio of the Weber number to the Reynolds number, which could be interpreted as the ratio of viscous forces to interfacial tension forces.

The question now arises: why make the flow equations dimensionless? There are three main reasons. The first is that, instead of eight parameters, the flow problem now contains just four. Thus there will be less work involved in solving the equations for given parameter ranges. Suppose that it is desired to solve the problem for n different values of each parameter. Then, instead of 8^n solutions only 4^n solutions are required, a reduction by a factor of 2^n. The second reason is as follows. Suppose that two flow problems have the same relative geometry (i.e., they are geometrically similar; an example is flow in two pipes of different size but of the same length to radius ratio), the same dimensionless boundary and initial conditions and the same values of the dimensionless groups Fr, Re, Pe, and Br. Then the dimensionless solutions of the two problems are identical (though the solutions may not, of course, be unique) and the flows are said to be dynamically similar. Clearly, we need to solve just one dimensionless problem to obtain solutions of the two (in fact, of infinitely many) dimensional problems. The third reason concerns making rational simplifications to the flow equations based on order-of-magnitude arguments, in order to obtain approximate solutions. This is our main concern here, and is discussed in what follows.

7.3 APPROXIMATE SOLUTIONS

Expressing the flow equations in dimensionless form provides a basis for making rational simplifications to them and hence for solving them approximately. Because we are not primarily interested here in heat transfer, we will not discuss simplifications to the energy conservation equation. When the mass conservation equation is expressed in dimensionless form (see Section 7.2), it does not involve any dimensionless groups. Accordingly, it cannot be simplified (except on the basis of symmetry; see Section 2.5). Thus we seek here simplifications to the linear momentum conservation equation, that is, the Navier-Stokes equations:

$$\rho \frac{\partial \mathbf{u}}{\partial t} + \rho \, \mathbf{u} \cdot \nabla \mathbf{u} = \rho \mathbf{g} - \nabla p + \mu \Delta \mathbf{u} \qquad (7.3\text{-}1)$$

$$\begin{array}{ccccc} \uparrow & \uparrow & \uparrow & \uparrow & \uparrow \\ (a) & (b) & (c) & (d) & (e) \end{array}$$

The terms in this equation may, as we have seen in Section 4.3, be identified as transient (a), convection (b), gravitational (c), pressure (d), and viscous (e) terms. Terms (a) and (b) together comprise the inertial terms; (e) is a particular form of the stress term.

We now consider under what circumstances one or more of these terms may be neglected and the equation correspondingly simplified. To be specific, we examine the relative importance of the inertial and the viscous terms and the role of the pressure term. In doing this, we show how Re plays a central role in making simplifications to the Navier-Stokes equations. We pay little attention to the gravitational term because whether or not it is important makes little difference to the ease of solution of the flow equations.

Re may, as we saw in Section 7.2, be interpreted as the ratio of inertial forces to viscous forces. We might expect, therefore, that the viscous term (e) in Eq. (7.3-1) could be neglected when Re is very high. Similarly, we might expect that the inertial terms (a) and (b) could be neglected when Re is very low. This is corroborated by an order-of-magnitude analysis of the terms in the Navier-Stokes equations (7.2-5) expressed in dimensionless form. Provided the dimensionless quantities defined in Eqs. (7.2-1) are of unit order of magnitude:

- The inertial terms $\partial \mathbf{u}^*/\partial t + \mathbf{u}^* \cdot \nabla^* \mathbf{u}^* \sim 1$.
- The gravitational term $(1/\mathrm{Fr})\mathbf{g}^* \sim 1/\mathrm{Fr}$.
- The pressure term $\nabla^* p^* \sim 1$.
- The viscous term $(1/\mathrm{Re})\Delta^* \mathbf{u}^* \sim 1/\mathrm{Re}$.

If Re is very high (i.e., Re $\gg$ 1), it follows that the viscous term should be negligible compared with the inertial terms. Hence, if Re $\gg$ 1, the Navier-Stokes equations should become approximately

$$\rho \frac{\partial \mathbf{u}}{\partial t} + \rho \, \mathbf{u} \cdot \nabla \mathbf{u} = \rho \mathbf{g} - \nabla p \qquad (7.3\text{-}2)$$

$$\begin{array}{cccc} \uparrow & \uparrow & \uparrow & \uparrow \\ (a) & (b) & (c) & (d) \end{array}$$

These are *Euler's equations*, the terms in which may be interpreted in just the same way as the corresponding ones in Eq. (7.3-1). Not surprisingly, Eq. (7.3-

2) is just the same as Eq. (6.2-4) for an inviscid fluid. Accordingly, if for a given flow the Reynolds number is high enough that viscous effects can be neglected, that flow is called an inviscid flow. Such flows are discussed in Chapter 9.

If, on the other hand, Re is very low (i.e., Re << 1), the inertial terms should be negligible compared with the viscous term in the Navier-Stokes equations which should, as a result, become approximately

$$0 = \rho \mathbf{g} + \mu \Delta \mathbf{u} \qquad (7.3\text{-}3)$$

There is no pressure term in this equation, which is clearly absurd because it implies that pressure is irrelevant in flows dominated by viscous forces. The reason for this absurdity is that pressure p is made dimensionless with respect to ρu_c^2 in Eqs. (7.2-1) and, as a result, the inertial and pressure terms in Eq. (7.2-5) are of the same order of magnitude. It is often difficult to choose a characteristic pressure p_c with which to make p dimensionless in flows of incompressible fluids (unless a pressure difference is specified, though this merely transfers the difficulty to choice of a characteristic velocity u_c). The difficulty stems from the fact that, for incompressible materials, pressure is determined only as part of the solution of the flow equations (see Section 5.2). Thus choice of p_c cannot be entirely arbitrary but must instead be consistent with the pressure determined by solving the flow equations. The choice

$$p_c = \rho u_c^2 \qquad (7.3\text{-}4)$$

is a good one for flows in which inertial forces are significant but a bad one when they are insignificant. A good choice when they are insignificant would lead to a different dimensionless pressure p^{**} such that $\nabla^* p^{**}$ is of the same order of magnitude as $\Delta^* \mathbf{u}^*$ and not as $\partial \mathbf{u}^*/\partial t^* + \mathbf{u}^* \cdot \nabla^* \mathbf{u}^*$. The choice

$$p_c = \frac{\mu u_c}{l_c} \qquad (7.3\text{-}5)$$

leads to the dimensionless pressure p^{**} given by

$$p^{**} = \frac{l_c}{\mu u_c} p = \mathrm{Re}\, p^* \qquad (7.3\text{-}6)$$

Suppose that pressure is made dimensionless as in Eq. (7.3-6) and all other variables are made dimensionless as in Eqs. (7.2-1). Then Eqs. (7.1-1) and (7.1-3) become Eqs. (7.2-3) and (7.2-7), respectively, as before but Eq. (7.1-2) becomes

$$\frac{\rho u_c^2}{l_c} \frac{\partial \mathbf{u}^*}{\partial t^*} + \frac{\rho u_c^2}{l_c} \mathbf{u}^* \cdot \nabla^* \mathbf{u}^* = \rho g \mathbf{g}^* - \frac{\mu u_c}{l_c^2} \nabla^* p^{**} + \frac{\mu u_c}{l_c^2} \Delta^* \mathbf{u}^* \qquad (7.3\text{-}7)$$

that is,

$$\mathrm{Re}\left(\frac{\partial \mathbf{u}^*}{\partial t^*} + \mathbf{u}^* \cdot \nabla^* \mathbf{u}^*\right) = \frac{\mathrm{Re}}{\mathrm{Fr}} \mathbf{g}^* - \nabla^* p^{**} + \Delta^* \mathbf{u}^* \qquad (7.3\text{-}8)$$

Provided the dimensionless quantities are of unit order of magnitude:

- The inertial terms Re $(\partial \mathbf{u}^*/\partial t^* + \mathbf{u}^* \cdot \nabla^* \mathbf{u}^*) \sim$ Re.
- The gravitational term $(\text{Re}/\text{Fr})\mathbf{g}^* \sim \text{Re}/\text{Fr}$.
- The pressure term $\nabla^* p^{**} \sim 1$.
- The viscous term $\Delta^* \mathbf{u}^* \sim 1$.

Thus the pressure and viscous terms are now both of the same order of magnitude, which shows that the choice of p_c in Eq. (7.3-5) is a good one for flows in which viscous forces are significant. We also see that, if Re $\ll 1$, the inertial terms should be negligible compared with the viscous term and hence that the Navier-Stokes equations should become approximately

$$0 = \rho \mathbf{g} - \nabla p + \mu \Delta \mathbf{u} \qquad (7.3\text{-}9)$$
$$\begin{array}{ccc} \uparrow & \uparrow & \uparrow \\ (c) & (d) & (e) \end{array}$$

These are *Stokes' equations,* the terms in which may be interpreted in just the same way as the corresponding ones in Eq. (7.3-1). If for a given flow the Reynolds number is low enough that inertial effects can be neglected, that flow is called a slow flow. Such flows are discussed in Chapter 8.

If Re is neither very high nor very low (i.e., Re ~ 1), viscous and inertial forces should both be comparable and hence both be significant. We would not, therefore, expect to be able to simplify the Navier-Stokes equations. The question does, however, arise of whether ρu_c^2 or $\mu u_c/l_c$ should be used as a characteristic pressure p_c. In fact, either can be used because

$$\rho u_c^2 = \text{Re} \, \frac{\mu u_c}{l_c} \qquad (7.3\text{-}10)$$

so that ρu_c^2 and $\mu u_c/l_c$ are of the same order of magnitude if Re ~ 1.

We have now developed a method for making simplifications to the Navier-Stokes equations in a rational way based on the magnitude of the Reynolds number. The method may be generalized so that all the flow equations, or indeed the equations governing any physical process, can be simplified and then solved approximately. The six steps in the method are

1. Make each variable dimensionless and of unit order of magnitude.
2. Make the exact flow equations dimensionless.
3. Identify the terms in the exact equations that are multiplied by small parameters, that is, by dimensionless groups that are much smaller than unity.
4. Obtain approximate equations by neglecting those terms multiplied by small parameters.
5. Solve the approximate equations exactly.
6. Check for consistency by using the exact solution of the approximate equations to estimate the orders of magnitude of those terms neglected from the exact equations.

All six steps in this method are crucial, though the last one is often omitted in practice. The method is based on two propositions, neither of which is flawless. The first is that

An exact solution of the approximate equations is an approximate solution of the exact equations.

While this is very often the case, there is no logical reason why it should be, and there are many instances when it is not, as we will see in Chapters 8 through 12. The second is that

It is possible to determine unambiguously whether a given dimensionless group is much smaller than unity.

Thus it is assumed that the statement $\alpha \ll 1$ has a precise meaning, where α denotes any dimensionless group. It is in fact impossible to assign such a precise meaning in general. While $\alpha < 1/10$ typically implies that $\alpha \ll 1$, this is by no means always the case. For example, in some instances $\alpha < 1/3$ implies that $\alpha \ll 1$ while in others $\alpha < 1/1000$ does not imply that $\alpha \ll 1$. Whether or not a given dimensionless group is much smaller than unity can usually only be determined in a specific case. We see then that, although we have developed a general method for obtaining approximate solutions of the flow equations, the method is by no means guaranteed to work. Fortunately, however, it works in the vast majority of cases and the cases when it does not are themselves of great interest, as we will see in Part 2.

PART
2

APPLICATIONS

In Part 2 we discuss applications of the fundamentals developed in Part 1 to specific fluid mechanics problems. Many but not all of these problems are of direct practical interest. All, however, provide insight into aspects of fluid mechanics and are, therefore, valuable even when they are (or seem to be) remote from what might be thought of as practical problems. We concentrate almost exclusively on analytical solutions of the flow equations. There are two reasons for this. The first is that a proper treatment of numerical computer-based solutions is beyond our scope here. The second is that numerical solutions often fail to reveal the essential physics of problems. Analytical solutions, on the other hand, do, although they are often more limited in their applicability. Moreover, analytical solutions (perhaps of related problems) can be very useful in determining the applicability of numerical solutions. Note, incidentally, that by a solution we mean not only expressions for the pressure and velocity fields but also expressions for such quantities as volumetric flow rate, drag force, and pressure drop. Thus we obtain both local and global features of a solution: the latter are usually of most immediate practical value, while the former are essential if a deep understanding is to be obtained.

We start in Chapter 8 by discussing slow flow, in which the Reynolds number is so small that inertial effects ought to be negligible. In Chapter 9, we go to the other extreme and discuss inviscid flow, in which the Reynolds number is so large that viscous effects ought to be negligible. In the following chapters, we discuss flows at intermediate Reynolds number. Thus in Chapter 10, we discuss what we call (nearly) nonaccelerating flow, in which inertial effects are nonexistent or negligible, even though the Reynolds number is not small. We go on in Chapter 11 to discuss boundary layer flow, in which the Reynolds number is sufficiently large that viscous effects are negligible everywhere except in the vicinity of boundaries such as walls. The flows discussed in Chapters 8 through 11 are laminar; in Chapter 12 we discuss turbulent flow, in which the Reynolds number is large and the flow variables fluctuate apparently at random. We conclude in Chapter 13 with a summary of the methodology used to solve many of the fluid mechanics problems here and some brief final remarks.

EIGHT

SLOW FLOW

8.1 STOKES' EQUATIONS

Suppose that a flow of an incompressible Newtonian fluid is such that the Reynolds number Re $\ll$ 1. It is then said to be a slow flow (or a creeping flow or Stokes flow). The equations governing such a flow are the mass conservation equation

$$\mathbf{\nabla} \cdot \mathbf{u} = 0 \qquad (8.1\text{-}1)$$

and the linear momentum conservation equation (i.e., the Navier-Stokes equations). As we saw in Section 7.3, these simplify to become Stokes' equations:

$$-\mathbf{\nabla}p + \mu\,\Delta\mathbf{u} = \mathbf{0} \qquad (8.1\text{-}2)$$

where gravitational effects, if any, are incorporated in the pressure term. Thus, instead of the pressure p, we should strictly use the modified pressure $\tilde{p}$ [see Eq. (4.3-13)].

It follows from Eq. (A1.3-22) of Appendix A that

$$\Delta\mathbf{u} = \mathbf{\nabla}(\mathbf{\nabla} \cdot \mathbf{u}) - \mathbf{\nabla}_\wedge(\mathbf{\nabla}_\wedge\mathbf{u}) \qquad (8.1\text{-}3)$$

Thus, using Eq. (8.1-1), Eq. (8.1-2) becomes

$$-\mathbf{\nabla}p - \mu\mathbf{\nabla}_\wedge(\mathbf{\nabla}_\wedge\mathbf{u}) = \mathbf{0} \qquad (8.1\text{-}4)$$

Hence, using Eq. (A1.3-24) of Appendix A,

$$\nabla^2 p = 0 \qquad (8.1\text{-}5)$$

which is a scalar Laplace equation for p; using Eq. (A1.3-23) of appendix A,

$$\nabla_\wedge \nabla_\wedge \nabla_\wedge \mathbf{u} = \mathbf{0} \qquad (8.1\text{-}6)$$

so, since $\nabla_\wedge \mathbf{u}$ is just the vorticity $\boldsymbol{\omega}$,

$$\nabla_\wedge \nabla_\wedge \boldsymbol{\omega} = \mathbf{0} \qquad (8.1\text{-}7)$$

and so, since $\nabla \cdot \boldsymbol{\omega} = \nabla \cdot (\nabla_\wedge \mathbf{u}) = 0$,

$$\Delta \boldsymbol{\omega} = \mathbf{0} \qquad (8.1\text{-}8)$$

which is a vector Laplace equation for $\boldsymbol{\omega}$.

We see, therefore, that we can decompose Stokes' equations (8.1-2) into two equations, one involving p alone and the other involving $\boldsymbol{\omega}$ alone. If boundary conditions are specified on pressure p, we could solve Eq. (8.1-5) for p and then substitute back into Eqs. (8.1-2) to obtain the velocity field $\mathbf{u}$. It is, however, rare for boundary conditions to be specified on p (one rather special example is discussed in Section 8.6). It is much more common for boundary conditions to be specified on $\mathbf{u}$ (see Chapter 6) and hence, in a sense, on $\boldsymbol{\omega}$. Thus we usually solve Eq. (8.1-8) for $\boldsymbol{\omega}$ and hence obtain $\mathbf{u}$ by integration, since $\boldsymbol{\omega}$ gives information on spatial derivatives of $\mathbf{u}$. We then substitute back into Eqs. (8.1-2) to obtain p.

8.2 FLOW AROUND A SOLID SPHERE

We consider a stationary, solid sphere of radius R in an unbounded, incompressible, Newtonian fluid of density ρ and viscosity μ. We suppose that the flow far away from the sphere (at infinity; see Section 6.4) is uniform with speed U_∞. Note that, because we take the sphere to be stationary, we are effectively fixing a reference frame (see Section 1.2) to the sphere. Because of the obvious geometry, we use spherical polar coordinates (r, θ, α) to analyze the problem, with origin at the center of the sphere and the r direction aligned with the direction of the flow at infinity when $\theta = 0$ (see Fig. 8.1). For convenience, we also use rectangular coordinates (x, y, z) with the same origin and the z direction aligned with the direction of the flow at infinity.

We now suppose that there is slow flow past the sphere, that is we assume that the Reynolds number given by

$$\mathrm{Re} = \frac{\rho U_\infty 2R}{\mu} \ll 1 \qquad (8.2\text{-}1)$$

Note that we have identified the characteristic velocity u_c with U_∞ and the characteristic length l_c with $2R$: U_∞ is an obvious choice; $2R$ is chosen instead of R because it is conventional to use the diameter of a sphere, cylinder, etc., and not its radius. Because $\mathrm{Re} \ll 1$, we expect that the inertial terms $\rho(\partial \mathbf{u}/\partial t) + \rho\, \mathbf{u} \cdot \nabla \mathbf{u}$ in the Navier-Stokes equations (5.4-26) are negligible (we would expect the tran-

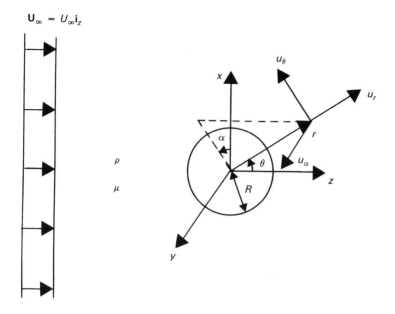

Figure 8.1. Geometry of flow past a solid sphere.

sient term $\rho(\partial\mathbf{u}/\partial t)$ to be negligible for the additional reason that we have no reason to expect the flow to be unsteady). We also expect the flow to be axisymmetric, because there is nothing that we can detect to make the flow asymmetric. Thus we expect that $\mathbf{u}$ and p are independent of α. Furthermore, we expect that the flow is swirl-free, (i.e., that $\mathbf{u}$ has no α component) because there is nothing that we can detect to impart swirl to the flow. As a result, there exists a Stokes stream function ψ_s [see Eqs. (2.5-20) and (2.5-21) and note the (arbitrary) change of sign] such that

$$u_r = -\frac{1}{r^2 \sin \theta} \frac{\partial \psi_s}{\partial \theta} \qquad u_\theta = \frac{1}{r \sin \theta} \frac{\partial \psi_s}{\partial r} \qquad (8.2\text{-}2)$$

Henceforth we omit the subscript s on the Stokes stream function since it is superfluous; no confusion should arise by doing this. In spherical polar coordinates, the r, θ and α components of vorticity $\boldsymbol{\omega}$ (ω_r, ω_θ, and ω_α, respectively) are given [see Eqs. (A2.1-12) of Appendix B] by

$$\omega_r = \frac{1}{r \sin \theta} \left[\frac{\partial}{\partial \theta} (\sin \theta \, u_\alpha) - \frac{\partial u_\theta}{\partial \alpha} \right] \qquad \omega_\theta = \frac{1}{r \sin \theta} \frac{\partial u_r}{\partial \alpha} - \frac{1}{r} \frac{\partial}{\partial r} (r u_\alpha)$$

$$\qquad (8.2\text{-}3)$$

$$\omega_\alpha = \frac{1}{r} \frac{\partial}{\partial r} (r u_\theta) - \frac{1}{r} \frac{\partial u_r}{\partial \theta}$$

Substitution of Eq. (8.2-2) into Eq. (8.2-3) yields $\omega_r = 0$, $\omega_\theta = 0$, and $\omega_\alpha \neq 0$ (i.e., ω has just one nonzero component); in fact,

$$\omega = \left(\frac{1}{r \sin \theta} \frac{\partial^2 \psi}{\partial r^2} + \frac{1}{r^3 \sin \theta} \frac{\partial^2 \psi}{\partial \theta^2} - \frac{\cos \theta}{r^3 \sin^2 \theta} \frac{\partial \psi}{\partial \theta} \right) i_\alpha \qquad (8.2\text{-}4)$$

We recall from Eq. (8.1-8) that

$$\Delta \omega = 0 \qquad (8.2\text{-}5)$$

Substitution of Eq. (8.2-4) into Eq. (8.2-5) and using Eq. (A1.3-21) of Appendix A yields

$$E^2(E^2\psi) = 0 \qquad E^2 = \frac{\partial^2}{\partial r^2} + \frac{\sin \theta}{r^2} \frac{\partial}{\partial \theta} \left(\frac{1}{\sin \theta} \frac{\partial}{\partial \theta} \right) \qquad (8.2\text{-}6)$$

Note that Eq. (8.2-6) is often written $E^4\psi = 0$. We prefer here to be more explicit: the form given in Eq. (8.2-6) implies that the operator E^2 operating on a quantity vanishes; that quantity is the operator E^2 operating on ψ.

Equation (8.2-6) comprises a fourth-order partial differential equation for ψ: because it is fourth order, we need four boundary conditions in order to solve it. Boundary conditions on velocity $\mathbf{u}$ are relatively straightforward to specify. No slip and no flow-through at the surface of the sphere (see Section 6.2) mean that $\mathbf{u} = \mathbf{0}$ at $r = R$, that is,

$$u_r = 0 \qquad u_\theta = 0 \qquad \text{at } r = R \qquad (8.2\text{-}7)$$

Similarly, uniform flow at infinity (see Section 6.4) means that $\mathbf{u} \to \mathbf{U}_\infty = U_\infty \mathbf{i}_z$ as $r \to \infty$, that is,

$$u_r \to U_\infty \cos \theta \qquad u_\theta \to -U_\infty \sin \theta \qquad \text{as } r \to \infty \qquad (8.2\text{-}8)$$

It then follows that the boundary conditions on ψ are

$$-\frac{1}{r^2 \sin \theta} \frac{\partial \psi}{\partial \theta} = 0 \qquad \frac{1}{r \sin \theta} \frac{\partial \psi}{\partial r} = 0 \qquad \text{at } r = R \qquad (8.2\text{-}9)$$

that is,

$$\frac{\partial \psi}{\partial \theta} = 0 \qquad \frac{\partial \psi}{\partial r} = 0 \qquad \text{at } r = R \qquad (8.2\text{-}10)$$

and

$$-\frac{1}{r^2 \sin \theta} \frac{\partial \psi}{\partial \theta} \to U_\infty \cos \theta \qquad \frac{1}{r \sin \theta} \frac{\partial \psi}{\partial r} \to -U_\infty \sin \theta \qquad \text{as } r \to \infty \qquad (8.2\text{-}11)$$

from which it follows by integration that

$$\psi \to -\tfrac{1}{2} U_\infty r^2 \sin^2 \theta \qquad \text{as } r \to \infty \qquad (8.2\text{-}12)$$

Note that we have set to zero the constant obtained on integration of Eqs. (8.2-11); this is permissible since it is only gradients of ψ that have physical significance; ψ is determinate only to within a constant. Note also that the two boundary conditions (8.2-11) have become the one boundary condition (8.2-12). No loss of information has, however, resulted from this, since Eqs. (8.2-11) can be recovered immediately from Eq. (8.2-12) by differentiation. Thus, while there appear to be three boundary conditions (8.2-10) and (8.2-12), there are in fact four, as required for the solution of Eq. (8.2-6).

We now proceed to solve Eq. (8.2-6) subject to boundary conditions (8.2-10) and (8.2-12). We do this by noting that the Navier-Stokes equations, and hence Stokes' equations and hence Eq. (8.2-6), apply quite generally, irrespective of the geometry of a given flow (see Section 6.1):

The geometry and hence any symmetry of a flow are determined by the boundary conditions (and, in an unsteady flow, by the initial conditions).

We now note that the boundary conditions apply at particular values of r (at $r = R$ and as $r \rightarrow \infty$) but not at particular values of θ: θ is arbitrary. The only information on the θ variation of the flow variables is given by the θ variation of ψ in the boundary conditions. Thus, motivated by the fact that the only specified θ variation of ψ is in boundary condition (8.2-12) (i.e., that ψ varies as $\sin^2 \theta$), we assume that

$$\psi(r, \theta) = f(r) \sin^2 \theta \qquad (8.2-13)$$

where f is a function of r which is yet to be determined.

We use the form of the boundary conditions to suggest the form of the solution of the differential equation.

If the assumed form for ψ is justified, we will find no inconsistency when we substitute Eq. (8.2-13) into Eqs. (8.2-6), (8.2-10), and (8.2-12) [though we would, of course, expect no inconsistency with Eq. (8.2-12) since it was that equation which motivated Eq. (8.2-13)]. If our assumption is unjustified, we will find an inconsistency. Substituting Eq. (8.2-13) into Eqs. (8.2-6), (8.2-10), and (8.2-12), we obtain:

$$\left(\frac{d^4}{dr^4} - \frac{4}{r^2} \frac{d^2}{dr^2} + \frac{8}{r^3} \frac{d}{dr} - \frac{8}{r^4} \right) f = 0 \qquad (8.2-14)$$

$$f = 0 \qquad \frac{df}{dr} = 0 \qquad \text{at } r = R \qquad (8.2-15)$$

$$f \rightarrow -\frac{1}{2} U_\infty r^2 \qquad \text{as } r \rightarrow \infty \qquad (8.2\text{-}16)$$

We note immediately that there is no inconsistency: Eqs. (8.2-14), (8.2-15), and (8.2-16) involve r and f (which is a function just of r) only: they do not involve θ in any way. So our assumption of the validity of Eq. (8.2-13) was justified. We now note that Eq. (8.2-14) is a linear, homogeneous, fourth-order ordinary differential equation for f. A trial solution,

$$f = c_n r^n \tag{8.2-17}$$

where c_n is a constant, when substituted into Eq. (8.2-14), yields

$$[n(n-1)(n-2)(n-3) - 4n(n-1) + 8n - 8]c_n r^{n-4} = 0 \tag{8.2-18}$$

or

$$(n+1)(n-1)(n-2)(n-4)c_n r^{n-4} = 0 \tag{8.2-19}$$

and so

$$n = -1, \ 1, \ 2, \ \text{or} \ 4 \tag{8.2-20}$$

Hence,

$$\psi = \left(\frac{c_1}{r} + c_2 r + c_3 r^2 + c_4 r^4 \right) \sin^2 \theta \tag{8.2-21}$$

It now follows from boundary condition (8.2-16) that

$$c_3 = -\tfrac{1}{2} U_\infty \qquad c_4 = 0 \tag{8.2-22}$$

[If $c_4 \neq 0$, ψ would vary as r^4 as $r \to \infty$ and thus become unbounded faster than, and hence dominate, the variation as r^2 implied by Eq. (8.2-16).] Boundary conditions (8.2-15) then yield

$$c_1 = -\tfrac{1}{4} U_\infty R^3 \qquad c_2 = \tfrac{3}{4} U_\infty R \tag{8.2-23}$$

Note that boundary condition (8.2-16) determines the two constants c_3 and c_4, just as boundary conditions (8.2-15) determine the two constants c_1 and c_2; this confirms our earlier argument about the total number of boundary conditions. It now follows that:

$$\psi = \left(-\frac{U_\infty R^3}{4r} + \frac{3U_\infty rR}{4} - \frac{U_\infty r^2}{2} \right) \sin^2 \theta \tag{8.2-24}$$

Note that Eq. (8.2-6) and boundary conditions (8.2-10) and (8.2-12) are linear in ψ. Thus, because Eq. (8.2-24) is a solution for ψ, it can be shown that it is the unique solution. Substitution of Eq. (8.2-24) into Eqs. (8.2-2) yields

$$\mathbf{u} = U_\infty \cos \theta \left(1 - \frac{3}{2} \frac{R}{r} + \frac{1}{2} \frac{R^3}{r^3} \right) \mathbf{i}_r - U_\infty \sin \theta \left(1 - \frac{3}{4} \frac{R}{r} - \frac{1}{4} \frac{R^3}{r^3} \right) \mathbf{i}_\theta \tag{8.2-25}$$

If we now substitute Eq. (8.2-25) back into Stokes' equations (8.1-2) and integrate, we obtain

$$p = p_\infty - \frac{3\mu U_\infty R \cos \theta}{2r^2} \tag{8.2-26}$$

where p_∞ is the (constant) pressure at infinity, that is, the value of p as $r \to \infty$; hence the subscript ∞.

We now calculate the force $\mathbf{F}$ exerted on the sphere by the fluid, which is given by:

$$\mathbf{F} = \int_{S} -\mathbf{n} \cdot (-p\mathbf{I} + \boldsymbol{\tau})\, dS \tag{8.2-27}$$

where $\mathbf{n}$ is the unit outer normal to an element dS of the surface S of the sphere. By outer normal we mean that $\mathbf{n}$ is directed from the fluid into the sphere, that is, it is the outer normal with respect to the fluid. A symmetry argument immediately enables us to put

$$\mathbf{F} = F\mathbf{i}_{z} \tag{8.2-28}$$

that is, $\mathbf{F}$ is aligned in the z direction, which is the direction of the flow at infinity: $\mathbf{F}$ is thus a *drag force* (if $\mathbf{F}$ were aligned normal to the z direction, it would be a *lift force*). The area dS of an elementary annular strip about the z axis on the surface of the sphere (see Fig. 8.2) is given by the product of the perimeter ($2\pi R \sin \theta$) and width ($R\, d\theta$) of the strip. Thus,

$$F = \int_{0}^{\pi} [(-p + \tau_{rr})\cos \theta - \tau_{r\theta} \sin \theta] 2\pi R \sin \theta\, R\, d\theta \Big|_{r=R} \tag{8.2-29}$$

For Newtonian fluid, $\boldsymbol{\tau} = \mu\, \mathbf{e}$. Using Eqs. (A2.1-10) of Appendix B:

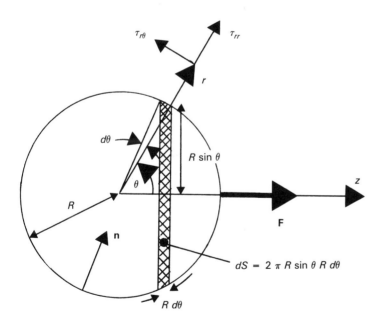

Figure 8.2. Elementary annular strip on the surface of a sphere. The directions shown associated with the stress components are the directions of the forces; the unit areas on which these forces act are both normal to the radial direction.

$$\tau_{rr} = \mu e_{rr} = 2\mu \frac{\partial u_r}{\partial r} \tag{8.2-30}$$

It follows from Eq. (8.2-25) that $\partial u_r/\partial r = 0$ at $r = R$, so

$$\tau_{rr} = 0 \qquad \text{at } r = R \tag{8.2-31}$$

In fact, it also follows from Eq. (8.2-25) and Eqs. (A2.1-10) of Appendix B that

$$\tau_{\theta\theta} = 0 \qquad \tau_{\alpha\alpha} = 0 \qquad \text{at } r = R \tag{8.2-32}$$

(i.e., all the three normal stress components vanish at the surface of the sphere). This result may be deduced in an alternative way, which does not involve prior determination of the velocity field, as follows:

- At the surface of the sphere, no flow-through means that u_r vanishes, and no slip means that both u_θ and u_α vanish (note that we do not assume that u_α vanishes everywhere: our argument does not rely on symmetry).
- Since the surface of the sphere is at $r = R$, with θ and α arbitrary, it follows that $\partial u_\theta/\partial\theta$ and $\partial u_\alpha/\partial\alpha$ both vanish at the surface of the sphere.
- For a Newtonian fluid, Eqs. (8.2-32) then follow immediately from Eqs. (A2.1-10) of Appendix B.
- For an incompressible fluid, it follows that the mass conservation equation (A2.2-4) of Appendix B becomes

$$\frac{1}{r^2}\frac{\partial}{\partial r}(r^2 u_r) = 0 \qquad \text{at } r = R \tag{8.2-33}$$

that is, since u_r vanishes at $r = R$,

$$\frac{\partial u_r}{\partial r} = 0 \qquad \text{at } r = R \tag{8.2-34}$$

and, for a Newtonian fluid, Eq. (8.2-31) follows immediately.

More generally, it can be shown that all three normal stress components vanish at any rigid impermeable wall in an incompressible Newtonian fluid. Returning to the determination of F and again using Eqs. (A2.1-10) of Appendix B.

$$\tau_{r\theta} = \mu e_{r\theta} = \mu \left(\frac{1}{r}\frac{\partial u_r}{\partial\theta} - \frac{1}{r}u_\theta + \frac{\partial u_\theta}{\partial r} \right) \tag{8.2-35}$$

Since there is no slip and no flow-through at $r = R$, it follows that

$$\tau_{r\theta} = \mu \frac{\partial u_\theta}{\partial r} \qquad \text{at } r = R \tag{8.2-36}$$

Thus Eq. (8.2-29) yields

$$F = 2\pi R^2 \int_0^\pi \left(-p_\infty \cos\theta + \frac{3\mu U_\infty \cos^2\theta}{2R} + \frac{3\mu U_\infty \sin^2\theta}{2R} \right) \sin\theta \, d\theta \qquad (8.2\text{-}37)$$

whence

$$F = 6\pi\mu U_\infty R \qquad (8.2\text{-}38)$$

Note that there is no contribution to F from p_∞, which is to have been expected: a *uniform* pressure applied around a body will not give rise to a force on that body. Note also that one-third of F is contributed by the pressure p in Eq. (8.2-29): this is called the *form drag* component. The remaining two-thirds is contributed by the shear stress $\tau_{r\theta}$: this is called the *friction drag* component. The expression for F in Eq. (8.2-38) is called *Stokes' law*; experimentally, it is found to be valid for Re < 0.1 or so. In dimensionless form, Stokes' law is

$$C_D = \frac{24}{\text{Re}} \qquad (8.2\text{-}39)$$

where the drag coefficient C_D, which is a dimensionless drag force, is given by

$$C_D = \frac{F}{\rho U_\infty^2 \, \pi R^2 / 2} \qquad (8.2\text{-}40)$$

Here, $\rho U_\infty^2/2$ is a characteristic stress or pressure (the factor of one-half is arbitrary) and πR^2 is a characteristic area, so that $\rho U_\infty^2 \, \pi R^2/2$ is a characteristic force. In fact, as we saw in Section 7.3 $\rho U_\infty^2/2$ is a good choice for a characteristic pressure in a flow in which inertial effects are significant, but a bad one when, as here, they are insignificant. A good choice when they are insignificant would be $\mu U_\infty/R$; this would lead to a different drag coefficient which would be independent of Re (indeed, it would be constant). Thus a good choice would remove the spurious dependence of the drag coefficient on the Reynolds number.

Instead of flow at infinity past a stationary sphere (with $\mathbf{u} \to U_\infty \mathbf{i}_z$ as $r \to \infty$ and $\mathbf{u} = \mathbf{0}$ at $r = R$), we now consider a sphere moving in a fluid stationary at infinity [with $\mathbf{u} \to \mathbf{0}$ as $r \to \infty$ and $\mathbf{u} = -U_\infty \mathbf{i}_z$ at $r = R$ (note the minus sign)]. Because we are making a *Galilean transformation* of reference frame, that is, are transforming from one reference frame to another which is *not* accelerating with respect to the first, if we subtract the velocity field $U_\infty \mathbf{i}_z = U_\infty \cos\theta \, \mathbf{i}_r - U_\infty \sin\theta \, \mathbf{i}_\theta$ from $\mathbf{u}$ given by Eq. (8.2-25) we obtain the resultant velocity field $\mathbf{u}$ given by

$$\mathbf{u} = U_\infty \cos\theta \left(-\frac{3}{2}\frac{R}{r} + \frac{1}{2}\frac{R^3}{r^3} \right) \mathbf{i}_r - U_\infty \sin\theta \left(-\frac{3}{4}\frac{R}{r} - \frac{1}{4}\frac{R^3}{r^3} \right) \mathbf{i}_\theta \qquad (8.2\text{-}41)$$

The drag force F is still given by Eqs. (8.2-28) and (8.2-38).

We now determine the terminal velocity (strictly, speed) U_t of a solid sphere that has a density ρ^+, in slow flow under the action of gravity. When a body is

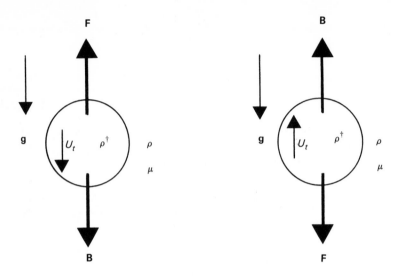

Figure 8.3. Forces acting on a solid sphere moving at its terminal velocity in a fluid when the density of the solid is (*a*) greater and (*b*) less than that of the fluid.

moving at its terminal velocity, the sum of the drag force on the body and the gravitational force (i.e., the buoyancy force), on the body vanishes; because there is no net force, the body does not accelerate. The drag force **F** is aligned in the opposite direction to that in which the sphere moves which, in turn, is aligned with the gravitational acceleration **g** if $\rho^\dagger > \rho$ and in the opposite direction if $\rho^\dagger < \rho$ (see Fig. 8.3). The magnitude F of **F** is given by Stokes' law, that is,

$$F = 6\pi\mu U_t R \qquad (8.2\text{-}42)$$

The buoyancy force **B** is aligned in the same direction as that in which the sphere moves; its magnitude B is given by

$$B = \tfrac{4}{3}\pi R^3 \, |\rho^\dagger - \rho| g \qquad (8.2\text{-}43)$$

where $g = |\mathbf{g}|$. Equating F and B, so that the sphere does not accelerate, yields

$$U_t = \frac{2}{9} \frac{R^2 g \, |\rho^\dagger - \rho|}{\mu} \qquad (8.2\text{-}44)$$

If there is a net force on the sphere, so that it is accelerating and hence not at its terminal velocity, it can be shown that the equation of motion of the sphere is

$$\frac{4}{3}\pi R^3 \rho^\dagger \frac{dU}{dt} = \frac{4}{3}\pi R^3 (\rho^\dagger - \rho)g - 6\pi\mu\,UR - \frac{2}{3}\pi R^3 \rho \frac{dU}{dt}$$

$$\begin{array}{cccc} \uparrow & \uparrow & \uparrow & \uparrow \\ (a) & (b) & (c) & (d) \end{array}$$

$$-6R^2\sqrt{\pi\rho\mu}\ \int_{-\infty}^{t}\frac{dU^{\#}}{dt^{\#}}\ \frac{1}{\sqrt{t-t^{\#}}}\ dt^{\#}\qquad(8.2\text{-}45)$$

$$\uparrow$$
$$(e)$$

where U denotes the magnitude of the velocity of the sphere in the direction of the gravitational acceleration $\mathbf{g}$ at time t. [If the sphere has a component of velocity perpendicular to $\mathbf{g}$ then Eq. (8.2-45) applies to the component of velocity in the direction of $\mathbf{g}$ and an equation similar to Eq. (8.2-45)—but with the term involving $g = |\mathbf{g}|$ omitted—applies to the component of velocity perpendicular to $\mathbf{g}$.]

Term (a) in Eq. (8.2-45) may be identified as the rate of change of momentum of the sphere (i.e., the product of the mass and the acceleration of the sphere). Terms $(b,)$ $(c,)$ $(d,)$ and (e) may be identified as the buoyancy force, viscous drag force, *inertial added mass* force and *viscous added mass* force (or *Basset history integral* term), respectively. The inertial added mass term arises because fluid has to undergo a net acceleration (which requires a force) when the sphere accelerates, in order to move from the region in front of the sphere to the region behind it (i.e., to move from the region to which the sphere is moving to the region from which the sphere is moving); it will be discussed more fully in Section 9.2. The viscous added mass term arises because Stokes' law gives the viscous drag only for a steady nonaccelerating flow. When the sphere accelerates, extra viscous drag arises.

Note that the integral in this term is from infinitely long ago to time t: if there is no motion for $t < 0$ (which defines the time origin), the lower limit on the integral becomes 0, not $-\infty$. We note, incidentally, that the presence of the viscous added mass term in Eq. (8.2-45) makes it an integrodifferential equation and consequently rather difficult to solve except in special asymptotic cases.

We conclude by considering not a solid sphere but a fluid sphere moving relative to an immiscible fluid that is stationary at infinity. It follows that we must solve the flow equations not only outside but also inside the sphere. We assume that there is no tendency for the flow to deform the sphere: as we will see, this assumption is consistent with our other assumptions. If the flow is slow, Stokes' equations (8.1-2) still apply. Thus, if $\rho^{\dagger}$ and $\mu^{\dagger}$ denote the density and viscosity of the fluid sphere, respectively, the flow is slow outside the sphere if Eq. (8.2-1) holds and the flow is slow inside the sphere if

$$\mathrm{Re}^{\dagger}=\frac{\rho^{\dagger}\,U_{\infty}2R}{\mu^{\dagger}}\ll 1\qquad(8.2\text{-}46)$$

where U_{∞} denotes the speed of the sphere (its velocity is $-U_{\infty}\,\mathbf{i}_z$; see Fig. 8.4). If the flow is slow everywhere (outside and inside the sphere), it follows that Eq. (8.2-6) holds outside the sphere and that

$$E^2(E^2\psi^{\dagger})=0\qquad(8.2\text{-}47)$$

inside the sphere, where $\psi^{\dagger}$ denotes the Stokes stream function inside the sphere. The boundary conditions on ψ and $\psi^{\dagger}$ are (1) zero velocity at infinity:

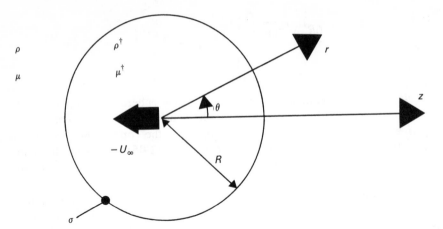

Figure 8.4. Geometry of flow past a fluid sphere.

$$u_r = -\frac{1}{r^2 \sin \theta} \frac{\partial \psi}{\partial \theta} \to 0 \qquad u_\theta = \frac{1}{r \sin \theta} \frac{\partial \psi}{\partial r} \to 0 \qquad \text{as } r \to \infty \quad (8.2\text{-}48)$$

(2) finite velocity near the origin:

$$|u_r^\dagger| = \left| -\frac{1}{r^2 \sin \theta} \frac{\partial \psi^\dagger}{\partial \theta} \right| < \infty \qquad |u_\theta^\dagger| = \left| \frac{1}{r \sin \theta} \frac{\partial \psi^\dagger}{\partial r} \right| < \infty \qquad \text{as } r \to 0 \quad (8.2\text{-}49)$$

(3) no flow-through at the interface (i.e., no flow across the interface):

$$u_r = -\frac{1}{r^2 \sin \theta} \frac{\partial \psi}{\partial \theta} = -U_\infty \cos \theta$$

$$(8.2\text{-}50)$$

$$u_r^\dagger = -\frac{1}{r^2 \sin \theta} \frac{\partial \psi^\dagger}{\partial \theta} = -U_\infty \cos \theta \qquad \text{at } r = R$$

(4) no slip at the interface:

$$u_\theta = \frac{1}{r \sin \theta} \frac{\partial \psi}{\partial r} = u_\theta^\dagger = \frac{1}{r \sin \theta} \frac{\partial \psi^\dagger}{\partial r} \qquad \text{at } r = R \quad (8.2\text{-}51)$$

(note that u_θ and $u_\theta^\dagger$ are not necessarily equal to $U_\infty \sin \theta$); and (5) continuity of total stress normal to the interface [see Section 6.3 and, in particular, Eq. (6.3-6)]:

$$\tau_{r\theta} = \mu \left(\frac{1}{r} \frac{\partial u_r}{\partial \theta} - \frac{1}{r} u_\theta + \frac{\partial u_\theta}{\partial r} \right)$$

$$= \mu \left[-\frac{1}{r} \frac{\partial}{\partial \theta} \left(\frac{1}{r^2 \sin \theta} \frac{\partial \psi}{\partial \theta} \right) + r \frac{\partial}{\partial r} \left(\frac{1}{r^2 \sin \theta} \frac{\partial \psi}{\partial r} \right) \right]$$

$$= \tau_{r\theta}^\dagger = \mu^\dagger \left(\frac{1}{r} \frac{\partial u_r^\dagger}{\partial \theta} - \frac{1}{r} u_\theta^\dagger + \frac{\partial u_\theta^\dagger}{\partial r} \right)$$

$$= \mu^\dagger \left[-\frac{1}{r} \frac{\partial}{\partial \theta} \left(\frac{1}{r^2 \sin \theta} \frac{\partial \psi^\dagger}{\partial \theta} \right) + r \frac{\partial}{\partial r} \left(\frac{1}{r^2 \sin \theta} \frac{\partial \psi^\dagger}{\partial r} \right) \right] \quad \text{at } r = R \quad (8.2\text{-}52)$$

and

$$p - \tau_{rr} + \frac{2\sigma}{R} = p^\dagger - \tau_{rr}^\dagger \quad \text{at } r = R \quad (8.2\text{-}53)$$

where $p^\dagger$, $\tau_{rr}^\dagger$, and $\tau_{r\theta}^\dagger$ denote the pressure, radial normal stress, and radial shear stress in the fluid in the sphere, and σ denotes the interfacial tension. By invoking arguments similar to those used to obtain Eq. (8.2-24), we can now show that

$$\psi = -\frac{U_\infty r^2 \sin^2 \theta}{4(\mu_r + 1)} \left[-\frac{R}{r} (2\mu_r + 3) + \frac{R^3}{r^3} \right] \quad (8.2\text{-}54)$$

and

$$\psi^\dagger = -\frac{U_\infty r^2 \sin^2 \theta}{4(\mu_r + 1)} \left[-(2 + 3\mu_r) + \mu_r \frac{r^2}{R^2} \right] \quad (8.2\text{-}55)$$

where the viscosity ratio μ_r is given by

$$\mu_r = \frac{\mu}{\mu^\dagger} \quad (8.2\text{-}56)$$

Note that ψ and $\psi^\dagger$ are independent of the interfacial tension σ. In fact, the interfacial tension merely relates the pressure p outside the sphere to the pressure $p^\dagger$ inside it, that is, boundary condition (8.2-53) is actually a condition on the pressure difference $(p - p^\dagger)$. It might be supposed that for the sphere not to be deformed by the flow, interfacial tension forces should be much larger than viscous forces (inertial forces are, of course, negligible since the flow is slow). In fact, however, provided the sphere moves steadily (i.e., does not accelerate) and slowly (i.e., Re $\ll$ 1 and Re† $\ll$ 1), it can be shown that, whatever the magnitude of σ, there is no tendency for the sphere to deform, as we assumed earlier. Thus boundary condition (8.2-53) is compatible with Eqs. (8.2-54) and (8.2-55): the problem is *not* overspecified.

The streamlines for the flows outside and inside the sphere are shown schematically in Fig. 8.5. Note the presence of the closed streamlines, implying internal circulation, within the sphere; note also that this internal circulation is obvious only when the streamlines are shown relative to a stationary sphere. Relative to fluid stationary at infinity, the internal circulation is not at all apparent, which illustrates the importance of the choice of reference frame in a flow problem. The force **F** exerted on the sphere by the fluid is still given by Eq. (8.2-28) but *F* is given not by Eq. (8.2-38) but instead, as is easy to show, by

$$F = 6\pi\mu U_\infty R \, \frac{1 + (2/3)\,\mu_r}{1 + \mu_r} \quad (8.2\text{-}57)$$

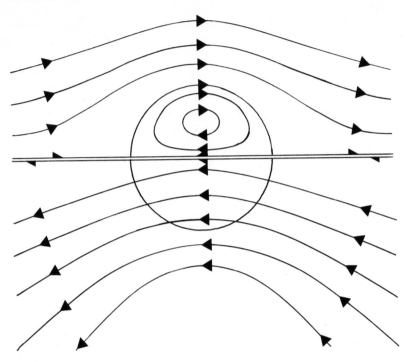

Figure 8.5. Streamlines for flow past a fluid sphere. Those above the plane of symmetry are relative to a stationary sphere, those below it are relative to fluid stationary at infinity.

We note immediately that, for a solid sphere, $\mu^\dagger$ is unbounded and so $\mu_r = 0$; thus Eq. (8.2-57) yields Eq. (8.2-38), as we would have expected.

For a gas sphere (gas bubble) in a viscous liquid, $\mu^\dagger$ is negligible compared with μ and so $\mu_r \to \infty$. Thus Eq. (8.2-57) yields

$$F = 4\pi\mu\, U_\infty R \qquad (8.2\text{-}58)$$

It then follows, by analogy with Eq. (8.2-44), that the terminal velocity of a gas bubble in a viscous liquid is given by

$$U_t = \frac{R^2 g \rho}{3\mu} \qquad (8.2\text{-}59)$$

Experimentally, it is observed that Eq. (8.2-59) holds for large bubbles, that is for bubbles for which the *Eotvos number* (Eo) is large. Eo, which may be interpreted as the ratio of gravitational or buoyancy forces to interfacial tension forces, is given by

$$\text{Eo} = \frac{|\rho^\dagger - \rho|\, g R^2}{\sigma} \qquad (8.2\text{-}60)$$

Thus Eq. (8.2-59) is found to hold for Eo $\gg$ 1. For Eo $\ll$ 1, on the other hand, that is for small bubbles, it is found that Eq. (8.2-44) with $\rho^\dagger \ll \rho$ holds instead; that is,

$$U_t = \frac{2R^2 g\rho}{9\mu} \qquad (8.2\text{-}61)$$

For intermediate values of Eo, the value of U_t is between those given by Eqs. (8.2-59) and (8.2-61). The reason why Eq. (8.2-59) does not hold unless Eo $\gg$ 1 is probably a result of the presence of surface-active agents which are present in all but the purest liquids. Any surface-active agent in the liquid around the bubble diffuses to the bubble surface and forms a shear-resisting layer there, so that the bubble surface has a certain rigidity. Indeed, for Eo $\ll$ 1, the rigidity of the bubble appears to be so great that the bubble acts as if it is completely rigid (i.e., like a solid sphere).

8.3 FLOW AROUND A SOLID CYLINDER

We consider an infinitely long, stationary, solid cylinder of radius R in an un-bounded, incompressible, Newtonian fluid of density ρ and viscosity μ. (By a cylinder, we always mean a right circular cylinder, so that it has a circular cross section normal to its axis or generator.) We suppose that the flow far away from the cylinder (at infinity) is uniform with speed U_∞; the direction of the flow at infinity is perpendicular to the axis of the cylinder. Because of the obvious ge-ometry, we use cylindrical polar coordinates (r, θ, z) to analyze the problem, with origin on the axis of the cylinder, the z direction aligned with the axis and the r direction aligned with the direction of the flow at infinity when $\theta = 0$ (see Fig. 8.6). We also use rectangular coordinates (x, y, z) with the same origin, the same z direction and the x direction aligned with the direction of the flow at infinity.

We now suppose that there is slow flow past the cylinder, that is we assume that the Reynolds number Re given by

$$Re = \frac{\rho U_\infty 2R}{\mu} \ll 1 \qquad (8.3\text{-}1)$$

Thus we expect that the inertial terms in the Navier-Stokes equations (5.4-26) are negligible. Because the cylinder is infinitely long, we also expect the flow to be *planar*: we expect **u** to have no z component and we expect that the r and θ components of **u** (u_r and u_θ, respectively) are independent of z; we also expect p to be independent of z. Thus there exists a stream function ψ [see Eqs. (2.5-5) and (2.5-6)] such that

$$u_r = \frac{1}{r}\frac{\partial\psi}{\partial\theta} \qquad u_\theta = -\frac{\partial\psi}{\partial r} \qquad (8.3\text{-}2)$$

In cylindrical polar coordinates, the r, θ, and z components of vorticity $\boldsymbol{\omega}$ (ω_r,

ω_θ, and ω_z, respectively) are given [see Eqs. (A2.1-8) of Appendix B] by

$$\omega_r = \frac{1}{r}\frac{\partial u_z}{\partial \theta} - \frac{\partial u_\theta}{\partial z} \qquad \omega_\theta = \frac{\partial u_r}{\partial z} - \frac{\partial u_z}{\partial r} \qquad \omega_z = \frac{1}{r}\frac{\partial}{\partial r}(r u_\theta) - \frac{1}{r}\frac{\partial u_r}{\partial \theta} \qquad (8.3\text{-}3)$$

Substitution of Eq. (8.3-2) into Eq. (8.3-3) yields

$$\boldsymbol{\omega} = \left(-\frac{\partial^2 \psi}{\partial r^2} - \frac{1}{r}\frac{\partial \psi}{\partial r} - \frac{1}{r^2}\frac{\partial^2 \psi}{\partial \theta^2} \right) \mathbf{i}_z \qquad (8.3\text{-}4)$$

Substitution of Eq. (8.3-4) into Eq. (8.1-8) using Eq. (A1.3-20) of Appendix A yields

$$F^2\left(F^2 \psi\right) = 0 \qquad F^2 = \frac{\partial^2}{\partial r^2} + \frac{1}{r}\frac{\partial}{\partial r} + \frac{1}{r^2}\frac{\partial^2}{\partial \theta^2} \qquad (8.3\text{-}5)$$

Note that F^2 is just the two-dimensional form of the three-dimensional scalar Laplacian operator ∇^2 in cylindrical polar coordinates [see Eq. (A1.3-17) of Appendix A].

Equation (8.3-5) comprises a fourth-order partial differential equation for ψ. We would, therefore, expect to need four boundary conditions in order to solve it. In fact, we need five, as we will see. No slip and no flow-through at the surface of the cylinder mean that $\mathbf{u} = \mathbf{0}$ at $r = R$ and so two boundary conditions are

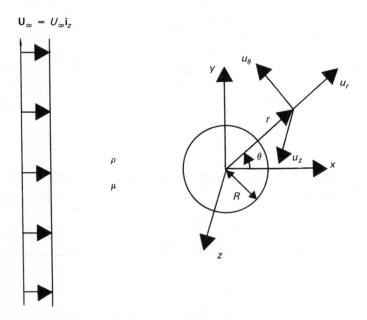

Figure 8.6. Geometry of flow past a solid cylinder.

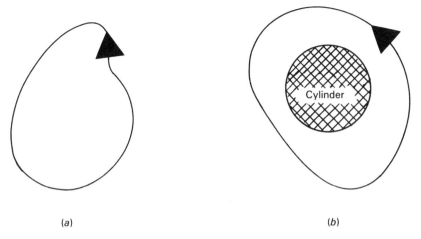

(a)   (b)

Figure 8.7. Closed curves in a fluid which in case (*a*) can and in case (*b*) cannot be continuously shrunk to a point without leaving the fluid.

$$\frac{\partial \psi}{\partial \theta} = 0 \qquad \frac{\partial \psi}{\partial r} = 0 \qquad \text{at } r = R \qquad (8.3\text{-}6)$$

Similarly, uniform flow at infinity means that $\mathbf{u} \to \mathbf{U}_\infty = U_\infty \mathbf{i}_x$ as $r \to \infty$, and so two more boundary conditions are

$$\frac{1}{r}\frac{\partial \psi}{\partial \theta} \to U_\infty \cos \theta \qquad -\frac{\partial \psi}{\partial r} \to - U_\infty \sin \theta \text{ as } r \to \infty \qquad (8.3\text{-}7)$$

or, equivalently,

$$\psi \to U_\infty r \sin \theta \qquad \text{as } r \to \infty \qquad (8.3\text{-}8)$$

An additional boundary condition is needed because the cylinder is infinitely long. Thus the region occupied by fluid is multiply connected, that is, closed curves can be drawn in the fluid which cannot be continuously shrunk to a point without leaving the fluid (see Fig. 8.7). To eliminate solutions that involve circulation (i.e., closed streamlines) around the cylinder, we invoke a symmetry condition and require that the solution is symmetric about the plane $y = 0$.

We now proceed to solve Eq. (8.3-5) subject to boundary conditions (8.3-6) and (8.3-8), and the requirement of symmetry about the plane $y = 0$. Motivated by the form of boundary condition (8.3-8), which contains the only information on the θ variation of ψ (i.e., that ψ varies as $\sin \theta$), we follow the argument given in Section 8.2 and assume that

$$\psi(r, \theta) = f(r) \sin \theta \qquad (8.3\text{-}9)$$

which satisfies the requirement of symmetry about the plane $y = 0$. Substituting Eq. (8.3-9) into Eqs. (8.3-5), (8.3-6), and (8.3-8), we obtain

$$\left(\frac{d^4}{dr^4} + \frac{2}{r}\frac{d^3}{dr^3} - \frac{3}{r^2}\frac{d^2}{dr^2} + \frac{3}{r^3}\frac{d}{dr} - \frac{3}{r^4}\right)f = 0 \qquad (8.3\text{-}10)$$

$$f = 0 \qquad \frac{df}{dr} = 0 \qquad \text{at } r = R \qquad (8.3\text{-}11)$$

$$f \to U_\infty r \qquad \text{as } r \to \infty \qquad (8.3\text{-}12)$$

We note immediately that Eqs. (8.3-10), (8.3-11), and (8.3-12) involve r and f only; they do not involve θ in any way. So our assumption of the validity of Eq. (8.3-9) was justified. A trial solution

$$f = c_n r^n \qquad (8.3\text{-}13)$$

where c_n is a constant reveals that

$$n = -1, 1, \text{ or } 3 \qquad (8.3\text{-}14)$$

hence determining three of the four solutions of Eq. (8.3-10); the fourth is obtained by using the trial solution

$$f = c_n r^n \ln r \qquad (8.3\text{-}15)$$

since a solution of the form $r^n \ln r$ is effectively the next simplest to one of the form r^n. The presence of a natural logarithm is a recurring feature of solutions of fluid mechanics problems.

When a solution in terms of algebraic functions is found to be incomplete, the presence of natural logarithms must be suspected.

Equation (8.3-15) then reveals that

$$n = 1 \qquad (8.3\text{-}16)$$

and so

$$\psi = \left(\frac{c_1}{r} + c_2 r + c_3 r \ln r + c_4 r^3\right) \sin \theta \qquad (8.3\text{-}17)$$

It now follows from boundary condition (8.3-12) that

$$c_2 = U_\infty \qquad c_3 = 0 \qquad c_4 = 0 \qquad (8.3\text{-}18)$$

Boundary conditions (8.3-11) cannot now both be satisfied, however, by *any* choice of c_1 [which we might have suspected because boundary condition (8.3-12) determines the three constants c_2, c_3, and c_4, and not two, as we might reasonably have anticipated]. We note, however, that the solution that satisfies the no slip and no flow-through conditions at $r = R$ exactly and that diverges from the uniform flow condition at infinity most slowly is

$$\psi = c_3 R \sin \theta \left[\frac{r}{R}\ln\left(\frac{r}{R}\right) - \frac{r}{2R} + \frac{R}{2r}\right] \qquad (8.3\text{-}19)$$

We thus see that there is no flow of the type we are seeking; in particular, there can be no steady, slow flow past a cylinder. By slow flow, we mean one in which inertial effects are completely neglected. This is normally referred to as Stokes' paradox. Of course, it is not a real paradox, since it is clearly possible for fluid to flow past a cylinder at an arbitrarily small Reynolds number; it merely means that we have overconstrained the problem by making at least one assumption too many. If we insist that the flow is steady, then the overconstraining assumption is that inertial effects are completely negligible. Thus, however small the Reynolds number, the inertial terms in the Navier-Stokes equations are never small compared with the viscous terms, at least not everywhere within the flow. An order-of-magnitude analysis of the Navier-Stokes equations based on the (admittedly imperfect) solution (8.3-19) leads to the improved estimate that the ratio of inertial forces to viscous forces is given by $\text{Re}\, r/R \ln (r/R)$, which should be compared with the initial estimate Re (see Section 7.2). Based on this improved estimate, we note that we may neglect inertial forces when $r/R \sim 1$ (i.e., r and R are of the same order of magnitude), and more generally when $r/R \ln (r/R)$ $\ll 1/\text{Re}$, because $\text{Re} \ll 1$. But we may not neglect inertial forces when r/R $\ln (r/R) \lll 1/\text{Re}$, which can happen *whatever* the value of Re (provided only that it does not vanish) because r/R becomes unbounded as $r \rightarrow \infty$. So, while the flow is slow near the cylinder, it is not slow far from the cylinder. We do, however, know the velocity field at infinity, albeit approximately. It is given by

$$\mathbf{u} \rightarrow U_\infty \mathbf{i}_x = \mathbf{U}_\infty \qquad \text{as } r \rightarrow \infty \qquad (8.3\text{-}20)$$

This suggests that we might approximate the inertial term in the Navier-Stokes equations by $\rho\, \mathbf{U}_\infty \cdot \nabla \mathbf{u}$ and hence replace Stokes' equations (8.1-2) by

$$-\nabla p + \mu\, \Delta \mathbf{u} = \rho\, \mathbf{U}_\infty \cdot \nabla \mathbf{u} \qquad (8.3\text{-}21)$$

These are *Oseen's equations* and can be solved for flow around a solid cylinder, though to do so is beyond our scope here. The solution apparently cannot be given exactly in closed form, though its very existence means that there is now no Stokes' paradox. The solution can, however, be given approximately in closed form; near the cylinder, that is for $r \sim R$,

$$\psi \simeq U_\infty R \sin \theta \left(\frac{1}{2C} \left\{ \frac{r}{R} \left[2 \ln \left(\frac{r}{R} \right) - 1 \right] + \frac{R}{r} \right\} \right.$$

$$+ \text{Re} \left[\frac{1}{8C} \left(\frac{r}{R} \right)^2 \ln \left(\frac{r}{R} \right) - \frac{1}{8} \left(\frac{r}{R} \right)^2 - \left(\frac{1}{16C} - \frac{1}{4} \right) \right.$$

$$\left. \left. + \left(\frac{1}{16C} - \frac{1}{8} \right) \left(\frac{R}{r} \right)^2 \right] \cos \theta \right) \qquad (8.3\text{-}22)$$

while far from the cylinder, that is for $r \gg R$,

$$\psi \simeq U_\infty R \left[\frac{r}{R} \sin \theta + \frac{4}{\text{Re}\, C} \sin \theta \right.$$

$$-\frac{1}{C}\sqrt{\frac{2\pi}{\mathrm{Re}}\frac{r}{R}}\int_0^\theta (1 + \cos\theta^\#)\, e^{-(1/4)\mathrm{Re}(r/R)(1-\cos\theta^\#)}d\theta^\# \Bigg] \qquad (8.3\text{-}23)$$

where C is a function of Re given by

$$C = \frac{1}{2} - \gamma + \ln\frac{8}{\mathrm{Re}} \qquad (8.3\text{-}24)$$

and γ denotes Euler's constant

$$\gamma \approx 0.5772157 \qquad (8.3\text{-}25)$$

A symmetry argument means that the force per unit length $\mathbf{F}'$ exerted on the cylinder by the fluid arises from drag and is given by

$$\mathbf{F}' = F'\mathbf{i}_x \qquad (8.3\text{-}26)$$

(For a cylinder of finite length $L >> R$, end effects that cause the flow near the ends of the cylinder not to be planar are negligible and so the drag force $\mathbf{F}$ on the cylinder is given by $L\mathbf{F}'$.) It can then be shown that

$$F' \simeq \frac{4\pi\mu\, U_\infty}{C} \qquad (8.3\text{-}27)$$

where C is given by Eq. (8.3-24). We note, incidentally, that ψ is not an even function of $(\theta - \pi/2)$ or, equivalently, of $(\theta - 3\pi/2)$ [see Eqs. (8.3-22) and (8.3-23)]. As a result, the solution of Oseen's equations displays a fore and aft asymmetry about the plane $x = 0$ (see Fig. 8.6), which presumably indicates the onset of wake formation downstream of the cylinder. This is in contrast to the symmetry that we would expect to be displayed by the solution of Stokes' equations, were it in fact to exist, and which is implicit in Eq. (8.3-9).

We conclude by noting that the reason why there is no Stokes' paradox for flow past a sphere is that, while inertial forces cannot be neglected for large r compared with viscous forces even for a sphere, inertial forces become important *sufficiently slowly* with increasing r that the Stokes stream function ψ, obtained by neglecting inertial effects, can satisfy boundary conditions both on the sphere (at $r = R$) and at infinity (as $r \to \infty$). In fact, we might have anticipated Stokes' paradox by the following dimensional argument. For a sphere,

- We expect the magnitude of the drag force F to be a function of radius R, velocity (strictly, speed) U_∞, and viscosity μ (but not of density ρ: inertial effects are unimportant).
- Dimensional analysis then yields the single dimensionless group $F/\mu U_\infty R$.
- Because there is a single dimensionless group, that group must be a constant [in fact, it follows from Eq. (8.2-38) that the constant is 6π].

For a cylinder,

- We expect the magnitude of the drag force per unit length F' to be a function again of R, U_∞, and μ.
- Dimensional analysis then yields the single dimensionless group $F'/\mu U_\infty$, which must, therefore, be a constant.
- This is implausible since it implies that F' is independent of R and hence independent of the size of the cylinder.

Thus we might expect that the complete neglect of inertial effects in flow past a cylinder could give rise to difficulties, as indeed it does. If, in fact, inertia is not neglected, so that F' is a function not just of R, μ, and U_∞ but also of ρ, dimensional analysis yields two dimensionless groups, say $F'/\mu U_\infty$ and Re $= \rho U_\infty 2R/\mu$. Thus $F'/\mu U_\infty$ must now not be a constant but instead be a function of Re (alone), which is in precise accordance with Eq. (8.3-27).

8.4 FLOW ABOUT AN OSCILLATING CYLINDER

We consider an infinitely long, solid cylinder of radius R which undergoes periodic, transverse oscillations of amplitude ϵR and frequency ω in an unbounded, incompressible, Newtonian fluid of density ρ and viscosity μ. Because of the obvious geometry, we use cylindrical polar coordinates (r, θ, z) to analyze the problem. We take the origin to be on, and the z direction to be aligned with, the axis of the cylinder in its average position and the r direction to be aligned with the direction of oscillation when $\theta = 0$ (see Fig. 8.8). We also use rectangular coordinates (x, y, z) with the same origin, the same z direction and the x direction aligned with the direction of oscillation.

We now suppose that the flow about the cylinder is slow. In order to be able to specify precisely what we mean here by slow flow, we note that, if the flow is slow, we expect that the convection term (i.e., the nonlinear inertial term) ρ $\mathbf{u} \cdot \nabla \mathbf{u}$ in the Navier-Stokes equations (5.4-26) is negligible and that the transient term (i.e., the linear inertial term) $\rho \, \partial \mathbf{u}/\partial t$ is not negligible, because a flow only exists in the fluid as a result of the transient (oscillatory) motion of the cylinder.

We now note that the characteristic length l_c associated with the flow in the fluid is the obvious geometric length scale R, the characteristic time t_c is the period of the oscillations of the cylinder $2\pi/\omega$, and the characteristic velocity u_c is the typical velocity of the cylinder $\omega \epsilon R$. Thus, to an order of magnitude $\partial/\partial t \sim \omega$, $|\nabla| \sim 1/R$, $|\Delta| \sim 1/R^2$ and $|\mathbf{u}| \sim \omega\epsilon R$. Hence,

$$\left|\rho\frac{\partial\mathbf{u}}{\partial t}\right| \sim \rho\omega^2\epsilon R, \qquad |\rho\mathbf{u}\cdot\nabla\mathbf{u}| \sim \rho\omega^2\epsilon^2 R \qquad |\mu\Delta\mathbf{u}| \sim \frac{\mu\omega\epsilon}{R} \qquad (8.4\text{-}1)$$

It thus follows that, if nonlinear inertial effects are to be negligible compared with viscous effects but linear inertial effects are not, the Reynolds number given by

$$\mathrm{Re} = \frac{\rho\omega\epsilon R^2}{\mu} \qquad (8.4\text{-}2)$$

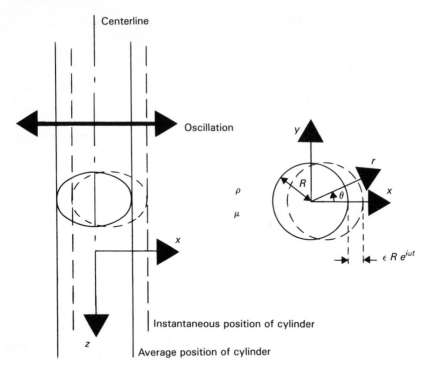

Figure 8.8. Geometry of flow about an oscillating cylinder.

must be very small (i.e., Re $\ll$ 1) while the quantity Ω given by

$$\Omega = \frac{\rho \omega R^2}{\mu} = \frac{Re}{\epsilon} \qquad (8.4\text{-}3)$$

must not be very small (i.e., $\Omega \ll 1$).

Because the cylinder is infinitely long, we expect that the velocity field $\mathbf{u}$ is planar, that is we expect $\mathbf{u}$ to have no z component and we expect that the r and θ components of $\mathbf{u}$ (u_r and u_θ, respectively) are independent of z; we also expect p to be independent of z. Thus there exists a stream function ψ [see Eqs. (2.5-5) and (2.5-6)] such that

$$u_r = \frac{1}{r} \frac{\partial \psi}{\partial \theta} \qquad u_\theta = -\frac{\partial \psi}{\partial r} \qquad (8.4\text{-}4)$$

Simplifying Eqs. (A2.3-3) of Appendix B by omitting the convection terms, the time-dependent Stokes' equations

$$\rho \frac{\partial \mathbf{u}}{\partial t} = -\nabla p + \mu \Delta \mathbf{u} \qquad (8.4\text{-}5)$$

(where gravitational effects, if any, are incorporated in the pressure term) become

$$\rho \frac{1}{r} \frac{\partial^2 \psi}{\partial t \partial \theta} = -\frac{\partial p}{\partial r} + \mu \left(\frac{1}{r} \frac{\partial^3 \psi}{\partial r^2 \partial \theta} - \frac{1}{r^2} \frac{\partial^2 \psi}{\partial r \partial \theta} + \frac{1}{r^3} \frac{\partial^3 \psi}{\partial \theta^3} + \frac{2}{r^2} \frac{\partial^2 \psi}{\partial r \partial \theta} \right) \quad (8.4\text{-}6)$$

$$-\rho \frac{\partial^2 \psi}{\partial t \partial r} = -\frac{1}{r} \frac{\partial p}{\partial \theta} + \mu \left(-\frac{\partial^3 \psi}{\partial r^3} - \frac{1}{r} \frac{\partial^2 \psi}{\partial r^2} + \frac{1}{r^2} \frac{\partial \psi}{\partial r} - \frac{1}{r^2} \frac{\partial^3 \psi}{\partial \theta^2 \partial r} + \frac{2}{r^3} \frac{\partial^2 \psi}{\partial \theta^2} \right) \quad (8.4\text{-}7)$$

If we now cross differentiate and subtract the terms in equations (8.4-6) and (8.4-7), that is we differentiate all the terms in Eq. (8.4-7) (multiplied by r) with respect to r and then subtract all the terms in Eq. (8.4-6) differentiated with respect to θ, we eliminate p from the equations of motion and obtain the following single equation for ψ:

$$\rho \left(\frac{1}{r} \frac{\partial^3 \psi}{\partial \theta^2 \partial t} + \frac{\partial^2 \psi}{\partial r \partial t} + r \frac{\partial^3 \psi}{\partial r^2 \partial t} \right) = \mu \left(\frac{2}{r} \frac{\partial^4 \psi}{\partial r^2 \partial \theta^2} - \frac{2}{r^2} \frac{\partial^3 \psi}{\partial r \partial \theta^2} \right.$$
$$\left. + \frac{1}{r^3} \frac{\partial^4 \psi}{\partial \theta^4} + 2 \frac{\partial^3 \psi}{\partial r^3} + r \frac{\partial^4 \psi}{\partial r^4} - \frac{1}{r} \frac{\partial^2 \psi}{\partial r^2} + \frac{1}{r^2} \frac{\partial \psi}{\partial r} + \frac{4}{r^3} \frac{\partial^2 \psi}{\partial \theta^2} \right) \quad (8.4\text{-}8)$$

We now consider the conditions that we must specify to solve Eq. (8.4-8). We will not seek to determine the motion in the fluid from some arbitrary initial state; instead, we will seek a solution that represents a periodic motion in the fluid that we presume to be valid when transients associated with the development from the initial state have decayed. Accordingly, we will not impose any initial conditions. We must, however, impose boundary conditions. Because the region occupied by the fluid is multiply connected, we require the solution to be symmetric about the plane $y = 0$ (see Section 8.3). Far from the cylinder, at infinity, the fluid should be motionless, that is,

$$u_r \to 0 \qquad u_\theta \to 0 \qquad \text{as } r \to \infty \qquad (8.4\text{-}9)$$

Thus, at infinity, gradients of ψ should vanish and ψ should tend to a constant value, which we may choose arbitrarily. Accordingly, we require that

$$\psi \to 0 \qquad \text{as } r \to \infty \qquad (8.4\text{-}10)$$

At the moving surface of the cylinder, no slip and no flow-through mean that the velocity of the fluid is the same as that of the solid (see Section 6.2). The velocity of the solid cylinder is just the time derivative of its displacement, which we require to be a periodic function of time. It is convenient in this, as in many problems involving periodic motions, to let the displacement of the cylinder in the x direction be $\epsilon R e^{i\omega t}$ where $i^2 = -1$. Of course, only the real part of this complex displacement is of physical significance. We now assume that the amplitude of the oscillations is much less than the radius of the cylinder, that is,

$$\epsilon \ll 1 \qquad (8.4\text{-}11)$$

(Note that this condition on ϵ follows from the requirement that $Re \ll 1$ and that $\Omega = Re/\epsilon \lll 1$. Thus we strictly require that $\epsilon \ggg Re \ll 1$.) As a result, the boundary conditions to be imposed at the instantaneous position of the surface of the cylinder may be transferred (by means of a Taylor series expansion) and imposed instead at its average position. It can then be shown (though it is too lengthy to do so here) that, to leading order, the boundary conditions on ψ at the surface of the cylinder are

$$\frac{\partial \psi}{\partial \theta} = ri\omega\epsilon R e^{i\omega t} \cos \theta \qquad \frac{\partial \psi}{\partial r} = i\omega\epsilon R e^{i\omega t} \sin \theta \qquad \text{at } r = R \qquad (8.4\text{-}12)$$

In order to solve Eq. (8.4-8) subject to boundary conditions (8.4-10) and (8.4-12), it is convenient to define the following dimensionless variables:

$$\tau = \frac{t\mu}{\rho R^2} \qquad \sigma = \frac{r}{R}, \qquad \Psi = \frac{\psi}{\epsilon\omega R^2} \qquad (8.4\text{-}13)$$

whence

$$\frac{1}{\sigma}\frac{\partial^3 \Psi}{\partial \theta^2 \partial \tau} + \frac{\partial^2 \Psi}{\partial \sigma \partial \tau} + \sigma\frac{\partial^3 \Psi}{\partial \sigma^2 \partial \tau} = \frac{2}{\sigma}\frac{\partial^4 \Psi}{\partial \sigma^2 \partial \theta^2} - \frac{2}{\sigma^2}\frac{\partial^3 \Psi}{\partial \sigma \partial \theta^2}$$

$$+ \frac{1}{\sigma^3}\frac{\partial^4 \Psi}{\partial \theta^4} + 2\frac{\partial^3 \Psi}{\partial \sigma^3} + \sigma\frac{\partial^4 \Psi}{\partial \sigma^4} - \frac{1}{\sigma}\frac{\partial^2 \Psi}{\partial \sigma^2} + \frac{1}{\sigma^2}\frac{\partial \Psi}{\partial \sigma} + \frac{4}{\sigma^3}\frac{\partial^2 \Psi}{\partial \theta^2} \qquad (8.4\text{-}14)$$

and

$$\left.\begin{array}{l} \dfrac{\partial \Psi}{\partial \theta} = \sigma i e^{i\Omega\tau} \cos \theta \\[4mm] \dfrac{\partial \Psi}{\partial \sigma} = i e^{i\Omega\tau} \sin \theta \end{array}\right\} \quad \text{at } \sigma = 1$$

$$\Psi \to 0 \qquad \qquad \text{as } \sigma \to \infty \qquad (8.4\text{-}15)$$

where Ω is given by Eq. (8.4-3).

Motivated by the form of boundary conditions (8.4-15), which contain the only information on the variation of Ψ with θ and τ, we assume that

$$\Psi = f(\sigma) \sin \theta \, e^{i\Omega\tau} \qquad (8.4\text{-}16)$$

which satisfies the requirement of symmetry about the plane $y = 0$. Substituting Eq. (8.4-16) into Eq. (8.4-14) and (8.4-15), we obtain

$$\left[\sigma\frac{d^4}{d\sigma^4} + 2\frac{d^3}{d\sigma^3} - \frac{3}{\sigma}\frac{d^2}{d\sigma^2} + \frac{3}{\sigma^2}\frac{d}{d\sigma} - \frac{3}{\sigma^3} \right.$$

$$\left. - i\Omega\left(\sigma\frac{d^2}{d\sigma^2} + \frac{d}{d\sigma} - \frac{1}{\sigma} \right) \right] f = 0 \qquad (8.4\text{-}17)$$

$$f = i \qquad \frac{df}{d\sigma} = i \qquad \text{at } \sigma = 1 \qquad f \to 0 \qquad \text{as } \sigma \to \infty \qquad (8.4\text{-}18)$$

which assures us of the validity of Eq. (8.4-16). We now note that Eq. (8.4-17) can be recast in the form

$$\left(\frac{d^2}{d\sigma^2} + \frac{1}{\sigma}\frac{d}{d\sigma} - \frac{1}{\sigma^2} - i\Omega\right)\left(\frac{d^2f}{d\sigma^2} + \frac{1}{\sigma}\frac{df}{d\sigma} - \frac{1}{\sigma^2}f\right) = 0 \qquad (8.4\text{-}19)$$

Thus we put

$$f = f_1 + f_2 \qquad (8.4\text{-}20)$$

where

$$\frac{d^2f_1}{d\sigma^2} + \frac{1}{\sigma}\frac{df_1}{d\sigma} - \frac{1}{\sigma^2}f_1 = 0 \qquad (8.4\text{-}21)$$

$$\frac{d^2f_2}{d\sigma^2} + \frac{1}{\sigma}\frac{df_2}{d\sigma} - \frac{1}{\sigma^2}f_2 - i\Omega f_2 = 0 \qquad (8.4\text{-}22)$$

The solution of Eq. (8.4-21) is readily seen to be

$$f_1 = \frac{c_1}{\sigma} + c_2\sigma \qquad (8.4\text{-}23)$$

where c_1 and c_2 are constants. The solution of Eq. (8.4-22) is

$$f_2 = c_3[\text{ber}_1(\sigma\sqrt{\Omega}) + i\,\text{bei}_1(\sigma\sqrt{\Omega})] + c_4[\text{ker}_1(\sigma\sqrt{\Omega}) + i\,\text{kei}_1(\sigma\sqrt{\Omega})] \qquad (8.4\text{-}24)$$

where c_3 and c_4 are constants; $\text{ber}_1(\ )$, $\text{bei}_1(\ )$, $\text{ker}_1(\ )$, and $\text{kei}_1(\ )$ denote first-order Kelvin functions. Because $f \to 0$ as $\sigma \to \infty$, it follows that

$$c_2 = 0 \qquad c_3 = 0 \qquad (8.4\text{-}25)$$

[since σ, $\text{ber}_1(\sigma\sqrt{\Omega})$, and $\text{bei}_1(\sigma\sqrt{\Omega})$ become unbounded as $\sigma \to \infty$]. Hence

$$f = \frac{c_1}{\sigma} + c_4[\text{ker}_1(\sigma\sqrt{\Omega}) + i\,\text{kei}_1(\sigma\sqrt{\Omega})] \qquad (8.4\text{-}26)$$

Because $f = i$ and $df/d\sigma = i$ at $\sigma = 1$, it follows that

$$\mathscr{Re}(c_1) = -\sqrt{\frac{2}{\Omega}\frac{\text{ker}_0(\sqrt{\Omega}) + \text{kei}_0(\sqrt{\Omega})}{\text{ker}_0^2(\sqrt{\Omega}) + \text{kei}_0^2(\sqrt{\Omega})}}\,\text{kei}_1(\sqrt{\Omega})$$

$$\quad -\sqrt{\frac{2}{\Omega}\frac{\text{ker}_0(\sqrt{\Omega}) - \text{kei}_0(\sqrt{\Omega})}{\text{ker}_0^2(\sqrt{\Omega}) + \text{kei}_0^2(\sqrt{\Omega})}}\,\text{ker}_1(\sqrt{\Omega})$$

$$\mathscr{Im}(c_1) = 1 - \sqrt{\frac{2}{\Omega}\frac{\text{ker}_0(\sqrt{\Omega}) - \text{kei}_0(\sqrt{\Omega})}{\text{ker}_0^2(\sqrt{\Omega}) + \text{kei}_0^2(\sqrt{\Omega})}}\,\text{kei}_1(\sqrt{\Omega}) \qquad (8.4\text{-}27)$$

$$\quad + \sqrt{\frac{2}{\Omega}\frac{\text{ker}_0(\sqrt{\Omega}) + \text{kei}_0(\sqrt{\Omega})}{\text{ker}_0^2(\sqrt{\Omega}) + \text{kei}_0^2(\sqrt{\Omega})}}\,\text{ker}_1(\sqrt{\Omega})$$

$$\mathscr{Re}(c_4) = \sqrt{\frac{2}{\Omega}\frac{\text{ker}_0(\sqrt{\Omega}) - \text{kei}_0(\sqrt{\Omega})}{\text{ker}_0^2(\sqrt{\Omega}) + \text{kei}_0^2(\sqrt{\Omega})}}$$

$$\mathscr{I}m(c_4) = -\sqrt{\frac{2}{\Omega} \frac{ker_0(\sqrt{\Omega}) + kei_0(\sqrt{\Omega})}{ker_0^2(\sqrt{\Omega}) + kei_0^2(\sqrt{\Omega})}}$$

where $\mathscr{R}e$ and $\mathscr{I}m$ denote real and imaginary parts, respectively.

Given the solution for Ψ in Eqs. (8.4-16), (8.4-26), and (8.4-27), we now determine the force per unit length of cylinder $\mathbf{F}'$ exerted by the cylinder on the fluid. A symmetry argument means that $\mathbf{F}'$ is aligned in the x direction (see Fig. 8.8); its magnitude F' is given by

$$F' = -\int_0^{2\pi} [(-p + \tau_{rr}) \cos \theta - \tau_{r\theta} \sin \theta] R \, d\theta \big|_{r=R} \tag{8.4-28}$$

(see Fig. 8.9), where, using Eqs. (A2.1-6) of Appendix B;

$$\tau_{rr} = \mu e_{rr} = 2\mu \frac{\partial u_r}{\partial r} = \mu \left(\frac{2}{r} \frac{\partial^2 \psi}{\partial r \partial \theta} - \frac{2}{r^2} \frac{\partial \psi}{\partial \theta^2} \right) \tag{8.4-29}$$

$$\tau_{r\theta} = \mu e_{r\theta} = \mu \left(\frac{1}{r} \frac{\partial u_r}{\partial \theta} - \frac{1}{r} u_\theta + \frac{\partial u_\theta}{\partial r} \right) = \mu \left(-\frac{\partial^2 \psi}{\partial r^2} + \frac{1}{r} \frac{\partial \psi}{\partial r} + \frac{1}{r^2} \frac{\partial^2 \psi}{\partial \theta^2} \right) \tag{8.4-30}$$

and p is given by integration with respect to θ of the terms in Eq. (8.4-7). It follows that

$$\tau_{rr} = \mu \omega \epsilon \frac{2}{\sigma} e^{i\Omega\tau} \cos \theta \left(\frac{df}{d\sigma} - \frac{1}{\sigma} f \right) \tag{8.4-31}$$

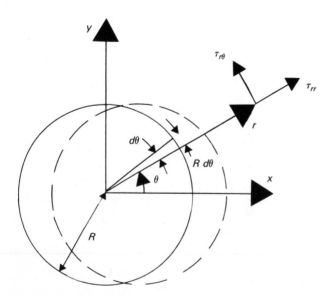

Figure 8.9. Elementary strip on the surface of a cylinder. The directions shown associated with the stress components are the directions of the forces; the unit areas on which these forces act are both normal to the radial direction.

$$\tau_{r\theta} = \mu\omega\epsilon \frac{1}{\sigma} e^{i\Omega\tau} \sin\theta \left(-\sigma \frac{d^2 f}{d\sigma^2} + \frac{df}{d\sigma} - \frac{1}{\sigma} f \right) \tag{8.4-32}$$

$$p = p_0 + \mu\omega\epsilon \frac{1}{\sigma} e^{i\Omega\tau} \cos\theta \left(\sigma^2 \frac{d^3 f}{d\sigma^3} + \sigma \frac{d^2 f}{d\sigma^2} - 2\frac{df}{d\sigma} + \frac{2}{\sigma} f - i\Omega\sigma^2 \frac{df}{d\sigma} \right) \tag{8.4-33}$$

where p_0 is an integration constant, the value of which is irrelevant to the determination of F' since an arbitrary constant change in the pressure in the fluid about a body does not alter the force exerted on that body. It then follows that

$$\tau_{rr} = 0 \qquad \text{at } r = R \tag{8.4-34}$$

$$\tau_{r\theta} = -\mu\omega\epsilon e^{i\Omega\tau} \sin\theta \left. \frac{d^2 f}{d\sigma^2} \right|_{\sigma=1} \qquad \text{at } r = R \tag{8.4-35}$$

$$p = p_0 + \mu\omega\epsilon e^{i\Omega\tau} \cos\theta \left. \left(\frac{d^3 f}{d\sigma^3} + \frac{d^2 f}{d\sigma^2} + \Omega \right) \right|_{\sigma=1} \qquad \text{at } r = R \tag{8.4-36}$$

whence

$$F' = -\mu\omega\epsilon R \int_0^{2\pi} \left[-e^{i\Omega\tau} \left(\frac{d^3 f}{d\sigma^3} + \frac{d^2 f}{d\sigma^2} + \Omega \right) \cos^2\theta \right.$$
$$\left. + e^{i\Omega\tau} \frac{d^2 f}{d\sigma^2} \sin^2\theta \right] d\theta \bigg|_{\sigma=1} \tag{8.4-37}$$

so

$$F' = \mu\omega\epsilon R\pi e^{i\Omega\tau} \left. \left(\frac{d^3 f}{d\sigma^3} + \Omega \right) \right|_{\sigma=1} \tag{8.4-38}$$

and so

$$F' = \mu\omega\epsilon R\pi\Omega e^{i\Omega\tau}(1 + 2ic_1) \tag{8.4-39}$$

where c_1 is given by Eqs. (8.4-27). We now express F' in the form

$$F' = \rho\pi R^2 (-\omega^2 \epsilon Re^{i\omega t} k_1 + i\omega\epsilon Re^{i\omega t} \omega k_2) \tag{8.4-40}$$

which defines the quantities k_1 and k_2. Because the velocity and acceleration of the cylinder in the direction of oscillation are given by $i\omega\epsilon Re^{i\omega t}$ and $-\omega^2\epsilon Re^{i\omega t}$, respectively, it follows that the component of F' involving k_1 may be thought of as the product of the acceleration of the cylinder and an added mass (strictly, an inertial added mass) per unit length m':

$$m' = \rho\pi R^2 k_1 \tag{8.4-41}$$

[see Eqs. (8.2-45) and (9.2-32) and the discussions following them]; and that the component of F' involving k_2 may be thought of as the product of the velocity of the cylinder and a viscous drag per unit length d':

$$d' = \rho\pi R^2 \omega k_2 \tag{8.4-42}$$

It follows from Eq. (8.4-40) that

$$F' = \mu\omega\epsilon R\pi\Omega e^{i\Omega\tau}(-k_1 + ik_2) \qquad (8.4\text{-}43)$$

and hence from Eq. (8.4-39) that

$$k_1 = -1 + 2\,\mathcal{I}m(c_1) \qquad k_2 = 2\,\mathcal{R}e(c_1) \qquad (8.4\text{-}44)$$

where $\mathcal{R}e(c_1)$ and $\mathcal{I}m(c_1)$ are given by Eqs. (8.4-27); k_1 and k_2 are tabulated as functions of Ω in Table 8.1. For $\Omega \gg 1$, it is possible to obtain asymptotic expressions for k_1 and k_2:

$$k_1 \simeq 1 + \frac{4}{\sqrt{2\Omega}} + \frac{1}{2\Omega\sqrt{2\Omega}} - \frac{1}{2\Omega^2} + \frac{25}{32\Omega^2\sqrt{2\Omega}} \qquad (8.4\text{-}45)$$

$$k_2 \simeq \frac{4}{\sqrt{2\Omega}} + \frac{2}{\Omega} - \frac{1}{2\Omega\sqrt{2\Omega}} + \frac{25}{32\,\Omega^2\sqrt{2\Omega}} \qquad (8.4\text{-}46)$$

Note that

$$k_1 \to 1 \qquad k_2 \to 0 \qquad \text{as } \Omega \to \infty \qquad (8.4\text{-}47)$$

which means that, for very large Ω, the only force exerted by the cylinder on the fluid is a result of the inertial added mass of the fluid: there is no viscous drag force. This is not surprising because, although Re $\ll$ 1 means that nonlinear inertial effects are negligible compared with viscous effects, $\Omega \gg 1$ means that viscous effects are themselves negligible compared with linear inertial effects. Note, incidentally, that when viscous effects are negligible the inertial added mass (per unit length) is precisely the mass (per unit length) of the fluid displaced by the cylinder. In contrast, for a sphere, the inertial added mass is precisely half the mass of the fluid displaced by the sphere when viscous effects are negligible, as is implicit in Eq. (8.2-45) and as we will show in Section 9.2.

   We conclude by noting that time-dependent flow about an oscillating cylinder is not subject to Stokes' paradox, whereas steady (and hence time-dependent) flow about a stationary cylinder (which we analyzed in Section 8.3) is. There are two related reasons for this. The first is that inertial effects are not completely ne-glected in the time-dependent flow; although nonlinear effects [arising from the term $\rho\,\mathbf{u}\cdot\nabla\mathbf{u}$ in the Navier-Stokes equations (5.4-26)] are neglected, linear effects (arising from the term $\rho\,\partial\mathbf{u}/\partial t$) are not. The second reason is that the neglected nonlinear inertial effects become important sufficiently slowly near the cylinder (where they are largest) that the solution, obtained by neglecting such effects, can satisfy the boundary conditions both on the cylinder and at infinity. We note, incidentally, that, if nonlinear inertial effects are small but not negligibly so, the flow comprises an additional small component called *streaming*. Streaming is dominated by a flow that is symmetrical fore and aft about the plane $x = 0$ (see Fig. 8.8), and hence makes no net contribution to the force per unit length F' exerted by the cylinder on the fluid.

Table 8.1 Values of k_1 and k_2 as functions of Ω

Ω	k_1	k_2
0.04	19.700	48.630
0.09	12.431	25.839
0.16	9.166	16.726
0.25	7.333	12.040
0.36	6.166	9.258
0.49	5.360	7.446
0.64	4.771	6.185
0.81	4.322	5.264
1.00	3.968	4.567
1.21	3.683	4.022
1.44	3.447	3.586
1.69	3.250	3.231
1.96	3.082	2.936
2.25	2.938	2.688
2.56	2.812	2.477
2.89	2.702	2.295
3.24	2.604	2.137
3.61	2.517	1.998
4.00	2.439	1.876
4.41	2.369	1.767
4.84	2.305	1.669
5.29	2.247	1.582
5.76	2.194	1.503
6.25	2.146	1.431
6.76	2.101	1.365
7.29	2.059	1.306
7.84	2.021	1.251
8.41	1.985	1.200
9.00	1.952	1.153
9.61	1.920	1.110
10.24	1.891	1.069
10.89	1.864	1.032
11.56	1.838	0.996
12.25	1.814	0.964
12.96	1.791	0.933
13.69	1.769	0.904
14.44	1.749	0.877
15.21	1.730	0.851
16.00	1.711	0.827
16.81	1.694	0.804
17.64	1.677	0.782
18.49	1.661	0.762
19.36	1.646	0.742
20.25	1.631	0.724
21.16	1.618	0.706
22.09	1.604	0.689
23.04	1.592	0.673
24.01	1.580	0.658
25.00	1.568	0.643

8.5 FLOW THROUGH A POROUS MEDIUM

When a fluid flows as a result of a pressure difference through a porous medium, that is, a consolidated assemblage of solid particles, each material point of the fluid traces out a devious path as it passes through the irregularly arranged interstices between the particles. Let l_c denote a characteristic dimension of the interstices (i.e., a typical distance between neighboring particles) and u_c a characteristic velocity of fluid through the interstices (see Fig. 8.10). Then, for an incompressible Newtonian fluid of density ρ and viscosity μ, the Reynolds number is usually such that

$$\text{Re} = \frac{\rho u_c l_c}{\mu} \ll 1 \qquad (8.5\text{-}1)$$

(i.e., the flow is usually slow). This is because u_c and particularly l_c are usually very small in practice.

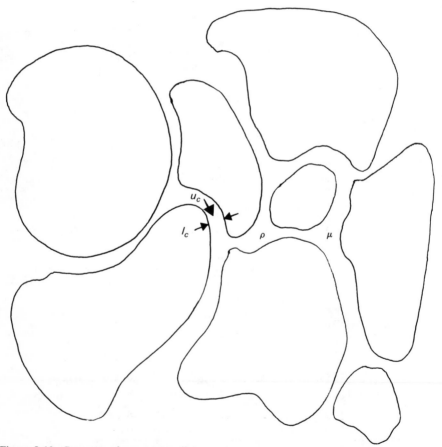

Figure 8.10. Geometry of a porous medium.

To solve the eqations for flow in a porous medium exactly, we would need to know the precise geometry of the pores or interstices. This is certainly impractical, if not impossible. So, instead, we use a statistical description of the flow and define average flow quantities. Thus, if f denotes some flow variable such as $\mathbf{u}$ or p, we define the pore volume average $\bar{f}$:

$$\bar{f} = \frac{1}{v_f} \int_{v_f} f \, dv \qquad (8.5\text{-}2)$$

where the averaging volume v_f is that part of a volume v of the porous medium that is occupied by pores that are accessible to fluid (i.e., excluding closed pores). The volume v_f and v are large compared with a typical pore volume l_c^3 and small compared with the volume V of the whole porous medium. Thus, if ϵ denotes the porosity of the medium (i.e., the volume fraction of the medium that is accessible to fluid),

$$l_c^3 \ll v_f = \epsilon v \leq v \ll V \qquad (8.5\text{-}3)$$

We note that the inequalities (8.5-3) are precisely analogous to inequalities (1.1-1) used to define a continuum and a material point. We also note that we might define the superficial volume average $\bar{\bar{f}}$:

$$\bar{\bar{f}} = \frac{1}{v} \int_{v_f} f \, dv \qquad (8.5\text{-}4)$$

instead of $\bar{f}$. The choice is arbitrary. Clearly,

$$\bar{\bar{f}} = \epsilon \bar{f} \qquad (8.5\text{-}5)$$

Experimentally, it is found that the pore volume average velocity $\bar{\mathbf{u}}$ is related to the pore volume average pressure $\bar{p}$ (modified if necessary to incorporate any gravitational effects) and to the fluid viscosity μ by

$$\bar{\mathbf{u}} = -\frac{K}{\mu} \nabla \bar{p} \qquad (8.5\text{-}6)$$

which is a combination of the linear momentum conservation equation and a constitutive equation; it is known as *Darcy's law* and can (in a sense) be derived from Stokes' equations (8.1-2). The constant K is called the *permeability*. Provided all the accessible pore volume is occupied by a single fluid, K is a function just of the structure of the porous medium; an order-of-magnitude argument then yields

$$K \sim l_c^2 \qquad (8.5\text{-}7)$$

Henceforth, we assume that the porous medium is spatially uniform so that l_c and hence K do not vary. We note, incidentally, that it follows from Eq. (8.5-6) that

$$\bar{\omega} = \nabla \cdot \bar{\mathbf{u}} = -\frac{K}{\mu} \nabla \cdot \nabla \bar{p} = 0 \qquad (8.5\text{-}8)$$

[see Eq. (A1.3-23) of Appendix A], so that the average flow is irrotational and there exists a scalar potential $\bar{\phi}$ such that

$$\bar{\mathbf{u}} = \nabla\bar{\phi} \qquad (8.5\text{-}9)$$

[see Eq. (2.5-2)] where, clearly,

$$\bar{\phi} = -\frac{K}{\mu}\bar{p} \qquad (8.5\text{-}10)$$

It also follows that for nonisotropic porous media (whose properties differ in different directions) the scalar permeability K must be replaced by a tensor permeability $\mathbf{K}$, and Darcy's law becomes

$$\bar{\mathbf{u}} = -\frac{1}{\mu}\,\mathbf{K}\cdot\nabla\bar{p} \qquad (8.5\text{-}11)$$

We now consider a specific example. We suppose that an incompressible Newtonian fluid of viscosity μ flows in a spatially uniform, isotropic porous medium of permeability K containing a spherical hole of radius R which is much larger than l_c. The pressure p_h in the hole is constant; it will emerge from the analysis that follows that p_h is the pore volume average pressure that there would be in the volume occupied by the hole if there were no hole. We assume that the flow far from the hole (at infinity) is uniform with pore volume average speed U_∞. Because of the obvious geometry, we use spherical polar coordinates (r, θ, α) with origin at the center of the hole and the r direction aligned with the direction of the flow at infinity when $\theta = 0$ (see Fig. 8.11). We also use rectangular coordinates (x, y, z) with the same origin and the z direction aligned with the direction of the flow at infinity.

Because the fluid is incompressible, it follows that

$$\nabla\cdot\bar{\mathbf{u}} = 0 \qquad (8.5\text{-}12)$$

and so Darcy's law yields

$$\nabla^2\bar{p} = 0 \qquad (8.5\text{-}13)$$

that is, $\bar{p}$ is determined by a scalar Laplace equation [see Eq. (8.1-5)]. Using Eq. (A1.3-18) of Appendix A, and assuming that the flow is axisymmetric (about the z axis, so that $\bar{p}$ is independent of α), Eq. (8.5-13) yields

$$\frac{1}{r^2}\frac{\partial}{\partial r}\left(r^2\frac{\partial\bar{p}}{\partial r}\right) + \frac{1}{r^2\sin\theta}\frac{\partial}{\partial\theta}\left(\sin\theta\frac{\partial\bar{p}}{\partial\theta}\right) = 0 \qquad (8.5\text{-}14)$$

The boundary conditions are

$$\bar{p} = p_h \qquad \text{at } r = R \qquad (8.5\text{-}15)$$

and

$$\bar{\mathbf{u}} \to U_\infty\cos\theta\,\mathbf{i}_r - U_\infty\sin\theta\,\mathbf{i}_\theta \qquad \text{as } r \to \infty \qquad (8.5\text{-}16)$$

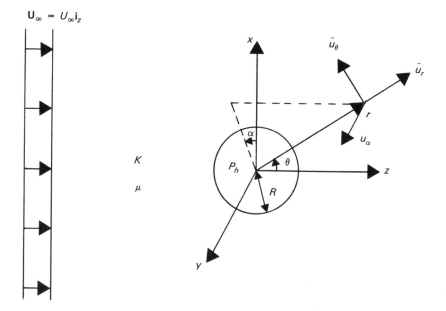

Figure 8.11. Geometry of flow through a porous medium containing a spherical hole.

Using Darcy's law:

$$\frac{\partial \bar{p}}{\partial r} \to -\frac{U_\infty \mu}{K} \cos \theta \qquad \frac{1}{r}\frac{\partial \bar{p}}{\partial \theta} \to \frac{U_\infty \mu}{K} \sin \theta \qquad \text{as } r \to \infty \qquad (8.5\text{-}17)$$

Motivated by the form of boundary conditions (8.5-15) and (8.5-17), we assume that

$$\bar{p}(r, \theta) = p_h + f(r) \cos \theta \qquad (8.5\text{-}18)$$

Substituting Eq. (8.5-18) into Eqs. (8.5-14), (8.5-15), and (8.5-17), we obtain

$$\left(\frac{d^2}{dr^2} + \frac{2}{r}\frac{d}{dr} - \frac{2}{r^2}\right)f = 0 \qquad (8.5\text{-}19)$$

$$f = 0 \qquad \text{at } r = R \qquad (8.5\text{-}20)$$

$$f \to -\frac{U_\infty \mu}{K} r \qquad \text{as } r \to \infty \qquad (8.5\text{-}21)$$

A trial solution:

$$f = c_n r^n \qquad (8.5\text{-}22)$$

reveals that

$$n = -2 \text{ or } 1 \qquad (8.5\text{-}23)$$

Hence,

$$f = \frac{c_1}{r^2} + c_2 r \tag{8.5-24}$$

It follows from boundary conditions (8.5-20) and (8.5-21) that

$$c_1 = \frac{U_\infty \mu R^3}{K} \qquad c_2 = -\frac{U_\infty \mu}{K} \tag{8.5-25}$$

Hence,

$$\bar{p} = p_h - \frac{U_\infty \mu}{K} r \cos\theta \left(1 - \frac{R^3}{r^3}\right) \tag{8.5-26}$$

and so, using Eq. (A1.3-5) of Appendix A and again using Darcy's law:

$$\bar{\mathbf{u}} = U_\infty \cos\theta \left(1 + \frac{2R^3}{r^3}\right) \mathbf{i}_r - U_\infty \sin\theta \left(1 - \frac{R^3}{r^3}\right) \mathbf{i}_\theta \tag{8.5-27}$$

The streamlines and isobars for the average flow are shown schematically in Fig. 8.12. The volumetric flow rate Q through the hole is given by

$$Q = \int_0^{\pi/2} \epsilon \bar{\mathbf{u}} \cdot \mathbf{i}_r 2\pi r \sin\theta r d\theta \big|_{r=R} \tag{8.5-28}$$

Substitution of $\bar{\mathbf{u}}$ from Eq. (8.5-27) yields

$$Q = 3\pi R^2 \epsilon U_\infty \tag{8.5-29}$$

Clearly, if there were no hole present, Q would be just $\pi R^2 \epsilon U_\infty$ (i.e., only one-third as large). Thus the hole effectively short circuits the porous medium by providing a flow path of lower resistance.

8.6 EXTENSIONAL FLOW OF A CYLINDRICAL FILAMENT

We consider the extensional flow of a cylindrical filament of radius R and length L in an immiscible, inviscid fluid at ambient pressure p_a. We suppose that the filament comprises an incompressible, Newtonian fluid of density ρ and viscosity μ and that it is fixed at one end and pulled at the other with an axial tensile force of magnitude T. We use cylindrical polar coordinates (r, θ, z) with origin at the fixed end of the axis of the filament and z direction aligned with the axis of the filament (see Fig. 8.13). We neglect end effects near $z = 0$ and $z = L$, and also gravitational and interfacial tension effects. We also suppose that the flow is slow, that is, that

$$\mathrm{Re} = \rho L \frac{dL/dt}{\mu} \ll 1 \tag{8.6-1}$$

Note that we have identified the characteristic length l_c with L and the characteristic velocity u_c with dL/dt.

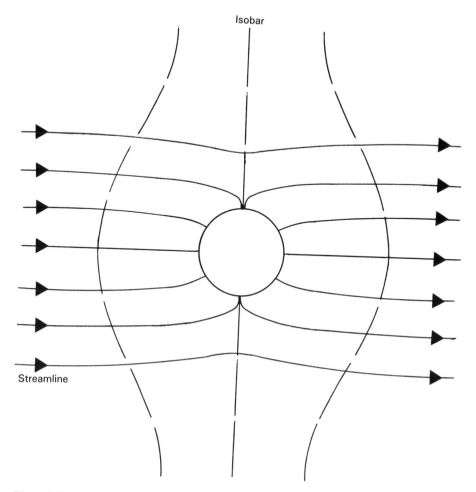

Isobar

Streamline

Figure 8.12. Pore volume average streamlines and isobars in a porous medium containing a spherical hole.

Symmetry means that the filament remains cylindrical on extension, that is, radius R is independent of axial position z, provided the flow is slow and gravitational effects are negligible. Thus symmetry implies that

$$\mathbf{u} = u_r(r, t)\, \mathbf{i}_r + u_z(z, t)\, \mathbf{i}_z \qquad (8.6\text{-}2)$$

that is, the flow is axisymmetric and swirl-free. Mass conservation implies [using Eq. (A2.2-3) of Appendix B] that

$$\frac{1}{r}\frac{\partial}{\partial r}(ru_r) + \frac{\partial u_z}{\partial z} = 0 \qquad (8.6\text{-}3)$$

Because u_r is independent of z, it follows that

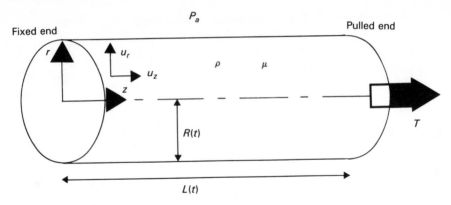

Figure 8.13. Geometry of a cylindrical filament.

$$\frac{1}{r}\frac{\partial}{\partial r}(ru_r)$$

is also independent of z. Similarly, it follows that $\partial u_z/\partial z$ is independent of r. It then follows from equation (8.6-3) that

$$\frac{1}{r}\frac{\partial}{\partial r}(ru_r) \qquad \text{and} \qquad \frac{\partial u_z}{\partial z}$$

are both at most functions of time t. Thus we put

$$\frac{\partial u_z}{\partial z} = \epsilon \tag{8.6-4}$$

and
$$\frac{1}{r}\frac{\partial}{\partial r}(ru_r) = -\epsilon \tag{8.6-5}$$

where ϵ is a function of time t. Because the end of the filament at $z = 0$ is fixed, it follows that $u_z = 0$ at $z = 0$. Thus Eq. (8.6-4) can be integrated to yield

$$u_z = \epsilon z \tag{8.6-6}$$

Symmetry means that $u_r = 0$ at $r = 0$; thus Eq. (8.6-5) can be integrated to yield

$$u_r = -\tfrac{1}{2}\epsilon r \tag{8.6-7}$$

Since dL/dt can be identified with u_z evaluated at $z = L$, and hence with ϵL [see Eq. (8.6-6)], we note that the condition that the flow is slow is that

$$\text{Re} = \frac{\rho \epsilon L^2}{\mu} << 1 \tag{8.6-8}$$

We also note that we have used symmetry and mass conservation to deduce the velocity field **u**; as yet, we have used *no* dynamic information, that is, we have not used linear momentum conservation [which is expressed in Stokes' equations (8.1-2)]. Thus we are solving this problem in a very different, and (in a sense) in a reverse, way from the way we solved the problems in Sections 8.2 through 8.5.

Using Eqs. (A2.1-6) of Appendix B, it follows from Eq. (8.6-2) that

$$\mathbf{e} = 2\frac{\partial u_r}{\partial r}\mathbf{i}_r\mathbf{i}_r + \frac{2}{r}u_r\mathbf{i}_\theta\mathbf{i}_\theta + 2\frac{\partial u_z}{\partial z}\mathbf{i}_z\mathbf{i}_z \tag{8.6-9}$$

and hence from Eqs. (8.6-6) and (8.6-7) that

$$\mathbf{e} = -\epsilon\mathbf{i}_r\mathbf{i}_r - \epsilon\mathbf{i}_\theta\mathbf{i}_\theta + 2\epsilon\mathbf{i}_z\mathbf{i}_z \tag{8.6-10}$$

Thus we see that rate-of-strain **e** is independent of position. Using Eqs. (A2.1-8) of Appendix B, it follows that the vorticity **ω** vanishes. Thus we see that the flow in the filament is linear (see Section 2.5). Indeed, comparison of Eqs. (8.6-6) and (8.6-7) with Eq. (2.5-29) shows that it is a simple uniaxial extensional flow with extension rate ϵ. Furthermore, because $\boldsymbol{\tau} = \mu\mathbf{e}$ for a Newtonian fluid, it follows that the stress $\boldsymbol{\tau}$ is, like **e**, independent of position. Hence,

$$\boldsymbol{\nabla}\cdot\boldsymbol{\tau} = \mathbf{0} \tag{8.6-11}$$

and so Stokes' equations, which may be written in the form:

$$-\boldsymbol{\nabla}p + \boldsymbol{\nabla}\cdot\boldsymbol{\tau} = \mathbf{0} \qquad \boldsymbol{\tau} = \mu\mathbf{e} \tag{8.6-12}$$

mean that

$$\boldsymbol{\nabla}p = \mathbf{0} \tag{8.6-13}$$

that is, p is also independent of position.

At the surface of the filament, at $r = R$, it follows from Eq. (6.3-6) that

$$-p\mathbf{n}\cdot\mathbf{I} + \mathbf{n}\cdot\boldsymbol{\tau} = -p_a\mathbf{n}\cdot\mathbf{I} \qquad \text{at } r = R \tag{8.6-14}$$

where the unit outer normal **n** is just $\mathbf{i}_r$, the unit vector in the r direction. Thus,

$$-p + \tau_{rr} = -p_a \qquad \text{at } r = R \tag{8.6-15}$$

Because, however, p and $\boldsymbol{\tau}$ are independent of position, as of course is p_a, we deduce that Eq. (8.6-15) holds not just at $r = R$ but for $0 \le r \le R$, that is,

$$p = \tau_{rr} + p_a \qquad \text{for } 0 \le r \le R \tag{8.6-16}$$

The magnitude of the axial tensile force T in the filament is given by

$$T = \pi R^2(-p + \tau_{zz}) - \pi R^2(-p_a) \tag{8.6-17}$$

whence, using Eq. (8.6-16),

$$T = \pi R^2(\tau_{zz} - \tau_{rr}) \tag{8.6-18}$$

But, for a Newtonian fluid,

$$\tau_{zz} = \mu e_{zz} \qquad \tau_{rr} = \mu e_{rr} \tag{8.6-19}$$

so, using Eq. (8.6-10),

$$T = 3\mu\epsilon\pi R^2 \qquad (8.6\text{-}20)$$

The extensional viscosity η, also referred to as the elongational viscosity or Trouton viscosity, is defined as the ratio of tensile stress to extension rate (in an analogous way, μ, often referred to as the shear viscosity, can be defined as the ratio of shear stress to shear rate). It follows from Eq. (8.6-20) that

$$\eta = \frac{T/\pi R^2}{\epsilon} = 3\mu \qquad (8.6\text{-}21)$$

Thus for simple uniaxial extensional flow, the extensional viscosity η is three times μ. In contrast, for simple biaxial extensional flow (see Section 2.5), η is six times μ. These results for an incompressible Newtonian fluid are directly analogous to those for an incompressible Hookean solid (see Section 5.4) where, in simple uniaxial extension for example, the extensional modulus E is three times the shear modulus G.

We now suppose that T is constant with respect to time (it is, of course, constant with respect to position in the filament). Then Eq. (8.6-20) implies that ϵR^2 is constant, and so

$$R^2 \frac{d\epsilon}{dt} + \epsilon 2R \frac{dR}{dt} = 0 \qquad (8.6\text{-}22)$$

so

$$\frac{d\epsilon}{dt} = -\frac{2\epsilon}{R}\frac{dR}{dt} \qquad (8.6\text{-}23)$$

We now note that u_r evaluated ar $r = R$ can be identified with dR/dt; thus Eq. (8.6-7) implies that

$$\frac{dR}{dt} = -\frac{1}{2}\epsilon R \qquad (8.6\text{-}24)$$

so that Eq. (8.6-23) yields

$$\frac{d\epsilon}{dt} = \epsilon^2 \qquad (8.6\text{-}25)$$

which can be integrated to yield

$$-\frac{1}{\epsilon} = t + c_1 \qquad (8.6\text{-}26)$$

where c_1 is a constant yet to be determined. At time $t = 0$, we apply the initial condition (see Section 6.1):

$$\epsilon = \epsilon_0 = \frac{T}{3\pi\mu R_0^2} \qquad (8.6\text{-}27)$$

where $R = R_0$ at $t = 0$, that is, R_0 denotes the initial radius of the filament. Hence $c_1 = -1/\epsilon_0$, and so

$$\frac{1}{\epsilon} = \frac{1}{\epsilon_0} - t \qquad (8.6\text{-}28)$$

We note immediately that $\epsilon \to \infty$ as $t \to 1/\epsilon_0$; ϵ is undefined for $t > 1/\epsilon_0$ for the following reason. Equations (8.6-24) and (8.6-28) together yield

$$\frac{dR}{dt} = -\frac{R}{2}\frac{1}{(1/\epsilon_0) - t} \qquad (8.6\text{-}29)$$

which can be integrated to yield

$$\ln R = \ln (c_2) + \frac{1}{2}\ln \left(\frac{1}{\epsilon_0} - t\right) \qquad (8.6\text{-}30)$$

where c_2 is another constant yet to be determined. Since $R = R_0$ at $t = 0$, it follows that $c_2 = R_0\sqrt{\epsilon_0}$ and so

$$R = R_0\sqrt{1 - \epsilon_0 t} \qquad (8.6\text{-}31)$$

Because the fluid is incompressible, the volume of the filament is constant and so

$$\pi R^2 L = \pi R_0^2 L_0 \qquad (8.6\text{-}32)$$

where L_0 denotes the length of the filament at time $t = 0$. Hence,

$$L = \frac{L_0}{1 - \epsilon_0 t} \qquad (8.6\text{-}33)$$

We see from Eqs. (8.6-31) and (8-6-33) that $R \to 0$ and $L \to \infty$ as $t \to 1/\epsilon_0$. We interpret this to mean that the filament breaks at time t_b given by

$$t_b = \frac{1}{\epsilon_0} \qquad (8.6\text{-}34)$$

where ϵ_0 is given by Eq. (8.6-27). Clearly, ϵ is undefined for $t > t_b$ (i.e., for $t > 1/\epsilon_0$).

We conclude by noting that Eq. (8.6-8) requires that

$$Re = \frac{Re_0}{(1 - \epsilon_0 t)^3} \ll 1 \qquad (8.6\text{-}35)$$

where Re_0 denotes the Reynolds number at time $t = 0$ given by

$$Re_0 = \frac{\rho \epsilon_0 L_0^2}{\mu} \qquad (8.6\text{-}36)$$

For $t \sim 1/\epsilon_0$, Re $\lll 1$ even if $\mathrm{Re}_0 \ll 1$. So the flow in the filament cannot be slow immediately before it breaks. This accords with the physically obvious fact that, when a constant tensile force is applied to a viscous filament, part (at least) of the filament will accelerate so much that inertial effects become important. In fact, the acceleration $\mathbf{a} = D\mathbf{u}/Dt$ is given by

$$\mathbf{a} = -\tfrac{1}{4}\epsilon^2 r \mathbf{i}_r + 2\epsilon^2 z \mathbf{i}_z \qquad (8.6\text{-}37)$$

and we see that $|\mathbf{a}| \to \infty$ as $t \to t_b$. If, however, Re_0 is small enough, Eq. (8.6-35) may be violated only for times t very close to t_b, in which case we anticipate that the flow will be slow over most of the interval $0 \leq t \leq t_b$, and hence that t_b will not differ significantly from $1/\epsilon_0$.

NINE

INVISCID FLOW

9.1 EULER'S EQUATIONS AND BERNOULLI'S EQUATION

Suppose that a flow of an incompressible Newtonian fluid is such that the Reynolds number is so high that viscous effects can be neglected. It is then said to be an inviscid flow. The equations governing such a flow are the mass conservation equation

$$\nabla \cdot \mathbf{u} = 0 \tag{9.1-1}$$

and the linear momentum conservation equation (i.e., the Navier-Stokes equations). As we saw in Section 7.3, these simplify to become Euler's equations:

$$\rho \frac{\partial \mathbf{u}}{\partial t} + \rho \, \mathbf{u} \cdot \nabla \mathbf{u} = \rho \mathbf{g} - \nabla p \tag{9.1-2}$$

It follows from Equation (A1.3-25) of Appendix A that

$$\nabla(\mathbf{u} \cdot \mathbf{u}) = 2 \, \mathbf{u} \cdot \nabla \mathbf{u} + 2 \, \mathbf{u}_\wedge \boldsymbol{\omega} \tag{9.1-3}$$

where the vorticity $\boldsymbol{\omega}$ is given by:

$$\boldsymbol{\omega} = \nabla_\wedge \mathbf{u} \tag{9.1-4}$$

Thus Eq. (9.1-2) yields

$$\rho \frac{\partial \mathbf{u}}{\partial t} + \frac{1}{2} \rho \nabla(|\mathbf{u}|^2) - \rho \, \mathbf{u}_\wedge \boldsymbol{\omega} = \rho \mathbf{g} - \nabla p \tag{9.1-5}$$

where the speed $|\mathbf{u}|$ is given by

$$|\mathbf{u}| = \sqrt{\mathbf{u} \cdot \mathbf{u}} \qquad (9.1\text{-}6)$$

We now suppose that the flow is irrotational (see Section 2.5), that is,

$$\boldsymbol{\omega} = \mathbf{0} \qquad (9.1\text{-}7)$$

Then $\mathbf{u}$ can be expressed in the form

$$\mathbf{u} = \nabla\phi \qquad (9.1\text{-}8)$$

where ϕ is a *scalar potential* field. Since $\mathbf{g} = \nabla(\mathbf{g} \cdot \mathbf{x})$ where $\mathbf{x}$ denotes position, Eq. (9.1-5) yields

$$\rho \frac{\partial}{\partial t} \nabla\phi + \frac{1}{2} \rho \nabla(|\nabla\phi|^2) = \rho \nabla(\mathbf{g} \cdot \mathbf{x}) - \nabla p \qquad (9.1\text{-}9)$$

or, since the operators $\partial/\partial t$ and ∇ commute (i.e., $\partial\nabla/\partial t = \nabla \, \partial/\partial t$),

$$\nabla\left(\frac{\partial\phi}{\partial t} + \frac{1}{2}|\nabla\phi|^2 - \mathbf{g} \cdot \mathbf{x} + \frac{1}{\rho}p\right) = \mathbf{0} \qquad (9.1\text{-}10)$$

which can be integrated to yield

$$\frac{\partial\phi}{\partial t} + \frac{1}{2}|\nabla\phi|^2 - \mathbf{g} \cdot \mathbf{x} + \frac{1}{\rho}p = \zeta(t) \qquad (9.1\text{-}11)$$

where ζ is a function of time t alone. This is *Bernoulli's equation*. If the flow is steady, it can be simplified to

$$\frac{1}{2}|\nabla\phi|^2 - \mathbf{g} \cdot \mathbf{x} + \frac{1}{\rho}p = \zeta \qquad (9.1\text{-}12)$$

or, since $|\mathbf{u}| = |\nabla\phi|$

$$\frac{1}{2}|\mathbf{u}|^2 - \mathbf{g} \cdot \mathbf{x} + \frac{1}{\rho}p = \zeta \qquad (9.1\text{-}13)$$

$$\uparrow \qquad \uparrow \quad \uparrow$$
$$(a) \qquad (b) \quad (c)$$

where ζ is a constant. The terms in Eq. (9.1-13) may be identified as kinetic energy (a), potential energy (b), and pressure energy (c) terms. Accordingly, Eq. (9.1–13) may be interpreted physically as stating that, for steady flow of an incompressible, inviscid fluid, the sum of the kinetic energy per unit mass $(\frac{1}{2}|\mathbf{u}|^2)$, the potential energy per unit mass $(-\mathbf{g} \cdot \mathbf{x})$ and the pressure energy per unit mass $[(1/\rho)p]$ is constant. Note, however, that Eq. (9.1–13) is not derived by energy conservation: it is thus analogous to the conservation equation for mechanical energy derived in Section 4.5.

Clearly, if we know ϕ, we can use Eq. (9.1-8) to determine $\mathbf{u}$ and then use Eq. (9.1–11) (or a simplified version of it) to determine p. It follows from Eqs. (9.1-1) and (9.1-8) that

$$\nabla^2 \phi = 0 \qquad (9.1\text{-}14)$$

that is, ϕ is determined by a scalar Laplace equation. It is possible to show that, if we know ϕ or $\nabla\phi \cdot \mathbf{n}$ on the boundary of a singly connected region, where $\mathbf{n}$ denotes the unit outer normal to an element of the boundary, Eq. (9.1-14) completely determines ϕ (to within an arbitrary function of time t which is irrelevant to the velocity $\mathbf{u}$; recall that $\mathbf{u}$ is obtained by taking spatial derivatives of ϕ). In a multiply connected region, additional conditions on circulation are also required (see Section 8.3).

We have seen that potential flows (i.e., flows for which the velocity field $\mathbf{u}$ is given as a gradient of a scalar potential field ϕ) satisfy Eqs. (9.1-1) and (9.1-2) which govern flows of incompressible, inviscid fluids. The question then arises: are all flows of incompressible, inviscid fluids also potential flows? The answer is as follows.

Suppose that $\mathbf{u}$ is a solution of Eqs. (9.1-1) and (9.1-2) and given by the gradient of a scalar potential ϕ at some (any) fixed time t. Then *Lagrange's theorem* states that $\mathbf{u}$ is given by the gradient of a scalar potential ϕ for all values of t.

Any flow of an incompressible, inviscid fluid that is a potential flow at some instant of time is always a potential flow.

More specifically,

Any flow of an incompressible, inviscid fluid that starts as a potential flow always remains a potential flow.

In particular, a flow of an incompressible, inviscid fluid that starts from rest (i.e., $\mathbf{u} = \mathbf{0}$ at $t = 0$, a degenerate potential flow for which the scalar potential ϕ is independent of position $\mathbf{x}$) always remains a potential flow.

9.2 FLOW AROUND A SOLID SPHERE

We consider a stationary, solid sphere of radius R in an unbounded, incompressible Newtonian fluid of density ρ and viscosity μ. We suppose that the flow at infinity is uniform with speed U_∞ and assume that

$$\text{Re} = \frac{\rho U_\infty 2R}{\mu} >> 1 \qquad (9.2\text{-}1)$$

Because Re $>> 1$, we expect that viscous effects are negligible and hence that there is inviscid flow about the sphere. Because of the obvious geometry, we use spherical polar coordinates (r, θ, α) to analyze the problem, with origin at the center of the sphere and the r direction aligned with the direction of the flow at infinity when $\theta = 0$ (see Fig. 9.1). For convenience, we also use rectangular

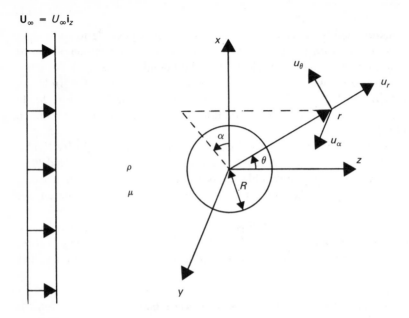

Figure 9.1. Geometry of flow past a solid sphere.

coordinates (x, y, z) with the same origin and the z direction aligned with the direction of the flow at infinity.

We expect that the flow round the sphere is axisymmetric and swirl-free. Thus we expect $\mathbf{u}$ to have no α component and we expect that the r and θ components of $\mathbf{u}$ (u_r and u_θ, respectively) are independent of α; we also expect that p is independent of α. Thus Eq. (9.1-8) becomes, using Eq. (A1.3-5) of Appendix A,

$$\mathbf{u} = \frac{\partial \phi}{\partial r} \mathbf{i}_r + \frac{1}{r} \frac{\partial \phi}{\partial \theta} \mathbf{i}_\theta \qquad (9.2\text{-}2)$$

and so, using Eq. (A1.3-18) of appendix A, Eq. (9.1-14) yields

$$\frac{1}{r^2} \frac{\partial}{\partial r} \left(r^2 \frac{\partial \phi}{\partial r} \right) + \frac{1}{r^2 \sin \theta} \frac{\partial}{\partial \theta} \left(\sin \theta \frac{\partial \phi}{\partial \theta} \right) = 0 \qquad (9.2\text{-}3)$$

which is a second-order partial differential equation for ϕ. The boundary conditions on $\mathbf{u}$ are no flow-through at the surface of the sphere, that is,

$$u_r = 0 \qquad \text{at } r = R \qquad (9.2\text{-}4)$$

and uniform flow at infinity, that is,

$$u_r \rightarrow U_\infty \cos \theta \qquad \text{as } r \rightarrow \infty \qquad (9.2\text{-}5)$$

Note that no conditions are specified on u_θ (1) because the fluid is taken to be inviscid, no conditions can be set on u_θ; and (2) because the partial differential equation to be solved is only second order [and not fourth order like Eq. (8.2-6) for slow flow round a sphere], no conditions need be set on u_θ (see also Section 6.2). It follows from Eqs. (9.2-4) and (9.2-5) that the boundary conditions on ϕ are

$$\frac{\partial \phi}{\partial r} = 0 \quad \text{at } r = 0 \tag{9.2-6}$$

$$\frac{\partial \phi}{\partial r} \rightarrow U_\infty \cos \theta \qquad \text{as } r \rightarrow \infty \tag{9.2-7}$$

We note that these are conditions on the normal gradient of ϕ, that is, on $\nabla \phi \cdot \mathbf{n}$, where the unit outer normal $\mathbf{n}$ is given by $-\mathbf{i}_r$ at the sphere and $\mathbf{i}_r$ at infinity. Because of this, and because the region occupied by the fluid is singly connected, we are guaranteed (see Section 9.1) that Eq. (9.2-3) has a solution ϕ which in turn determines $\mathbf{u}$.

We now proceed to solve Eq. (9.2-3) subject to boundary conditions (9.2-6) and (9.2-7). Motivated by the form of the boundary conditions, we follow the argument given in Section 8.2 and assume that

$$\phi(r, \theta) = f(r) \cos \theta \tag{9.2-8}$$

Then Eqs. (9.2-3), (9.2-6), and (9.2-7) yield

$$\left(\frac{d^2}{dr^2} + \frac{2}{r} \frac{d}{dr} - \frac{2}{r^2} \right) f = 0 \tag{9.2-9}$$

$$\frac{df}{dr} = 0 \qquad \text{at } r = R \tag{9.2-10}$$

$$\frac{df}{dr} \rightarrow U_\infty \qquad \text{as } r \rightarrow \infty \tag{9.2-11}$$

Clearly, our assumption of the validity of Eq. (9.2-8) was justified; Eqs. (9.2-9), (9.2-10), and (9.2-11) do not involve θ in any way. A trial solution

$$f = c_n r^n \tag{9.2-12}$$

where c_n is a constant reveals that

$$n = -2 \text{ or } 1 \tag{9.2-13}$$

Hence, using boundary conditions (9.2-10) and (9.2-11),

$$\phi = U_\infty \cos \theta \left(r + \frac{R^3}{2r^2} \right) \tag{9.2-14}$$

Thus,

$$\mathbf{u} = U_\infty \cos \theta \left(1 - \frac{R^3}{r^3} \right) \mathbf{i}_r - U_\infty \sin \theta \left(1 + \frac{R^3}{2r^3} \right) \mathbf{i}_\theta \tag{9.2-15}$$

We note that, at $r = R$,

$$u_\theta = -\frac{3}{2} U_\infty \sin \theta \neq 0 \qquad (9.2\text{-}16)$$

confirming that there is slip at the surface of the sphere. We also note that, as $r \to \infty$,

$$u_\theta \to -U_\infty \sin \theta \qquad (9.2\text{-}17)$$

so that both u_r and u_θ match with the imposed uniform flow at infinity. Although it might appear surprising that u_θ matches in this way, it is easy to see that if $\partial\phi/\partial r \to U_\infty \cos \theta$ as $r \to \infty$, then it follows that $\phi \to r\, U_\infty \cos \theta + g(\theta)$ [where $g(\theta)$ is an arbitrary function of θ] and hence that

$$\frac{1}{r}\frac{\partial\phi}{\partial\theta} \to -U_\infty \sin \theta \qquad \text{as } r \to \infty$$

that is, the behavior of u_r at infinity implies the behavior of ϕ and hence of u_θ there.

Having determined the velocity field $\mathbf{u}$, we now proceed to determine the pressure field p. We do this by using Bernoulli's equation (9.1-13) and incorporating any gravitational effects into p [which should thus strictly be the modified pressure $\bar{p}$; see Eq. (4.3-13)]:

$$\frac{1}{2}|\mathbf{u}|^2 + \frac{1}{\rho}p = \zeta = \frac{1}{2}U_\infty^2 + \frac{1}{\rho}p_\infty \qquad (9.2\text{-}18)$$

where the speed $|\mathbf{u}| = \sqrt{\mathbf{u}\cdot\mathbf{u}}$, p_∞ is the pressure at infinity (hence the subscript ∞) and ζ is a constant. Thus,

$$p = p_\infty + \frac{1}{2}\rho U_\infty^2 \left[\cos^2 \theta \left(\frac{2R^3}{r^3} - \frac{R^6}{r^6} \right) - \sin^2 \theta \left(\frac{R^3}{r^3} + \frac{R^6}{4r^6} \right) \right] \qquad (9.2\text{-}19)$$

We now calculate the force $\mathbf{F}$ exerted on the sphere by the fluid, which is given by

$$\mathbf{F} = \int_S -\mathbf{n}\cdot(-p\mathbf{I} + \boldsymbol{\tau})\, dS \qquad (9.2\text{-}20)$$

where $\mathbf{n}$ is the unit outer normal to an element dS of the surface S of the sphere (i.e., $\mathbf{n} = -\mathbf{i}_r$). Because the fluid is inviscid,

$$\boldsymbol{\tau} = \mathbf{0} \qquad (9.2\text{-}21)$$

A symmetry argument enables us to put

$$\mathbf{F} = F\, \mathbf{i}_z \qquad (9.2\text{-}22)$$

that is, $\mathbf{F}$ is a drag force, and

$$F = \int_0^\pi (-p \cos \theta) \, 2\pi R \sin \theta \, R \, d\theta \big|_{r=R} \qquad (9.2\text{-}23)$$

whence, using Eq. (9.2-19), we obtain

$$F = \int_0^\pi \{[-p_\infty + \tfrac{1}{2} \rho U_\infty^2 (-\cos^2 \theta + \tfrac{5}{4} \sin^2 \theta)] \cos \theta\} \, 2\pi R^2 \sin \theta \, d\theta \qquad (9.2\text{-}24)$$

On integration, we obtain

$$F = 0 \qquad (9.2\text{-}25)$$

that is,

An incompressible, inviscid fluid moving at constant velocity around a stationary sphere exerts no drag force on the sphere.

This is normally referred to as *d'Alembert's paradox*. In fact, we might have anticipated Eq. (9.2-25) without integration. We note from Eq. (9.2-19) that the pressure field is symmetric fore and aft (see Fig. 9.2), that is, the pressure at any point (r, θ) is precisely the same as the pressure at the point $(r, \pi - \theta)$. Thus, the z component of any force induced by pressure ahead of the sphere is exactly balanced by the z component of an equal and opposite force induced by pressure behind the sphere. Because the flow is assumed to be inviscid, there are no viscous forces on the sphere. Hence there can be no drag force on the sphere. Experimental evidence shows that, for flow of a real fluid, the viscosity of which never vanishes, Eq. (9.2-25) is false, no matter how large Re may be. The reason for this is that Re $>>$ 1 implies that flows are effectively inviscid and hence approximately irrotational almost, but not quite, everywhere. Viscous effects, which lead to drag forces, are always important near and downstream of walls (in boundary layers, which we will discuss in Chapter 11, and wakes, respectively). This is because, if $\mu \neq 0$, there is no slip at rigid, impermeable walls (see Section 6.2), however small μ may be; slip is permissible only if $\mu = 0$.

There is an essential difference between flows for which the Reynolds number is very large and flows for which the Reynolds number is infinite.

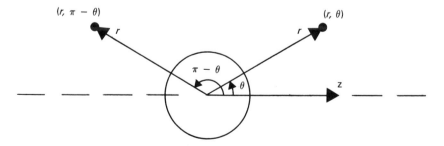

Figure 9.2. Equivalent positions fore and aft of the midplane of a sphere.

The question then arises: if assuming that a fluid is inviscid leads to an apparent paradox, that is, it leads to conclusions at variance with experimental observations, is there any value in studying flows of such fluids? We will see in the remaining sections of this chapter that there is indeed value is such studies. Fortunately, paradoxes are relatively scarce.

We conclude by considering a sphere moving at velocity $-U_\infty \, \mathbf{i}_z$ in a fluid stationary at infinity, instead of flow at infinity with velocity $U_\infty \, \mathbf{i}_z$ past a stationary sphere. Because we are making a *Galilean transformation* of reference frame (see Section 8.2), if we subtract the velocity field $U_\infty \cos \theta \, \mathbf{i}_r - U_\infty \sin \theta \, \mathbf{i}_\theta$ from $\mathbf{u}$ given by Eq. (9.2-15), we obtain the resultant velocity field $\mathbf{u}$ given by

$$\mathbf{u} = -U_\infty \cos \theta \, \frac{R^3}{r^3} \, \mathbf{i}_r - U_\infty \sin \theta \, \frac{R^3}{2r^3} \, \mathbf{i}_\theta \tag{9.2-26}$$

The kinetic energy K of the flow is given by

$$K = \int_V \tfrac{1}{2}\rho |\mathbf{u}|^2 \, dV \tag{9.2-27}$$

where V denotes the volume occupied by the fluid. The volume dV of an elementary annular region about the z axis through the sphere is given by the product of its perimeter ($2\pi r \sin \theta$), width ($r \, d\theta$) and depth (dr). Thus,

$$K = \int_{\theta=0}^{\pi} \int_{r=R}^{\infty} \tfrac{1}{2} \rho \, (u_r^2 + u_\theta^2) \, 2\pi r \sin \theta \, r \, d\theta \, dr \tag{9.2-28}$$

and so, using Eq. (9.2-26),

$$K = \pi\rho U_\infty^2 \, R^6 \int_R^{\infty} \frac{1}{r^4} \, dr \left[\int_0^{\pi} \left(\cos^2 \theta + \frac{1}{4} \sin^2 \theta \right) \sin \theta \, d\theta \right] \tag{9.2-29}$$

whence

$$K = \tfrac{1}{3} \, \pi\rho U_\infty^2 \, R^3 \tag{9.2-30}$$

If we now put

$$K = \tfrac{1}{2} \, mU_\infty^2 \tag{9.2-31}$$

then m denotes the mass of fluid moving with a uniform speed U_∞ which has the same kinetic energy as the unbounded fluid around the sphere. Clearly

$$m = \tfrac{2}{3} \, \pi R^3 \rho \tag{9.2-32}$$

which we note is precisely half the mass of the fluid displaced by the sphere. The mass m is called the *inertial added mass* (as in Section 8.2; see also Section 8.4) or the *added mass* or *virtual mass*. It is the effective mass of the fluid that must be added to that of the sphere in unsteady flows in which the sphere accelerates relative to the flow at infinity and thus where the total inertia is important.

In a steady flow, in order that fluid moves from the region in front of the sphere to the region behind it, the fluid must accelerate in front of the sphere and then decelerate behind it. There is, however, no net acceleration of the fluid because there is no net acceleration of the sphere relative to the fluid at infinity. In an unsteady flow, in contrast, there is a net acceleration of the fluid. The force required to produce this acceleration in the fluid is given by the product of the inertial added mass and the acceleration of the sphere relative to the fluid at infinity.

To see this more clearly, we consider the initial motion from rest of a rigid sphere of density $\rho\dagger$ in a fluid of density ρ. The buoyancy force on the sphere, which is aligned with the gravitational acceleration $\mathbf{g}$, is given by $\frac{4}{3}\pi R^3(\rho\dagger - \rho)$ $\mathbf{g}$. The total effective mass of the sphere and the fluid around it is $\frac{4}{3}\pi R^3(\rho\dagger + \frac{1}{2}\rho)$. At the instant at which the sphere starts to move, the velocity of the sphere is zero. Thus there can be no viscous force on the sphere initially, whether or not the fluid is inviscid. More generally:

Any flow of any fluid which starts from rest is effectively inviscid at the instant at which it starts.

It follows, therefore, that the product of the total effective mass and the initial acceleration $\mathbf{a}$ (in the direction of $\mathbf{g}$) of the sphere and the fluid around it just balances the buoyancy force, whence

$$\mathbf{a} = \mathbf{g}\,\frac{\rho\dagger/\rho - 1}{\rho\dagger/\rho + 1/2} \qquad (9.2\text{-}33)$$

It now follows that the initial acceleration of a spherical gas bubble into a liquid is given by

$$\mathbf{a} \simeq -2\mathbf{g} \qquad (9.2\text{-}34)$$

since the density $\rho\dagger$ of a gas is much less than the density ρ of a liquid. The fact that the sphere comprises a gas and not a solid is, of course, irrelevant provided the sphere is rigid. Thus a spherical gas bubble initially rises with approximately twice the gravitational acceleration.

9.3 FLOW AROUND A LARGE BUBBLE

We consider a bubble of a gas rising because of gravitational (buoyancy) effects, through a much denser, immiscible liquid. If the bubble is small, we expect it to be spherical, or nearly so, even when moving. This is because when the bubble is small, the gravitational and inertial forces that tend to distort it are small compared with interfacial tension and viscous forces which tend to prevent distortion of it. To see this, we note that the order of magnitude of the gravitational force F_g is given by

$$F_g \sim \rho g l_c^3 \qquad (9.3\text{-}1)$$

where g denotes the magnitude of the gravitational acceleration, l_c denotes a characteristic length of the bubble, and ρ denotes the density of the liquid (relative to which the density of the gas is negligible). Similarly, the order of magnitude of the inertial force F_i is given by

$$F_i \sim \rho u_c^2 l_c^2 \tag{9.3-2}$$

where u_c denotes a characteristic velocity (strictly, speed), say that of the bubble relative to the liquid. The order of magnitude of the interfacial (or surface) force F_s is given by

$$F_s \sim \sigma l_c \tag{9.3-3}$$

where σ denotes the interfacial tension. The order of magnitude of the viscous force F_v is given by

$$F_v \sim \mu u_c l_c \tag{9.3-4}$$

where μ denotes the viscosity of the liquid (relative to which the viscosity of the gas is negligible). Thus we expect that the bubble is nearly spherical (and that, for example, l_c might be identified with the radius R of the bubble) if $F_g << F_s$ and if $F_i << F_v$, that is, if the Eotvos number

$$Eo = \frac{\rho g l_c^2}{\sigma} \tag{9.3-5}$$

if such that Eo $<< 1$ and if

$$Re = \frac{\rho u_c l_c}{\mu} \tag{9.3-6}$$

is such that Re $<< 1$. In fact, as we saw in Section 8.2, provided Re $<< 1$, the value of Eo is irrelevant to the shape of the bubble but does affect the details of the flow. For present purposes, however, we note that Eo varies as l_c^2 and Re varies as l_c. Thus, if a bubble is small enough, we expect that Eo $<< 1$ and Re $<< 1$ and hence that the bubble is spherical. If, on the other hand, a bubble is large, we expect that Eo $>> 1$ and Re $>> 1$. We would *not* expect a large bubble to be spherical and, without solving the flow equations, we have no way of determining its shape theoretically. Instead of solving these equations, which is a difficult task, we appeal to experimental evidence which shows that,

Large bubbles in motion all have the same shape, which is approximately that of a slice of a sphere.

Thus we now consider inviscid flow (because Re $>> 1$) past such a spherical cap bubble. We assume that the bubble is stationary (thus fixing our reference frame) and that there is uniform flow at infinity toward the bubble with speed U_∞ (which we might identify with the characteristic velocity u_c). Because of the obvious geometry, we use spherical polar coordinates (r, θ, α) to analyze the problem, with origin at the center of the sphere of which the bubble is a segment and the r direction aligned with the direction of the flow at infinity when $\theta = 0$ (see Fig. 9.3). We also use rectangular coordinates (x, y, z) with the same origin and the z axis aligned with the direction of the flow at infinity, and assume that the

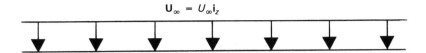

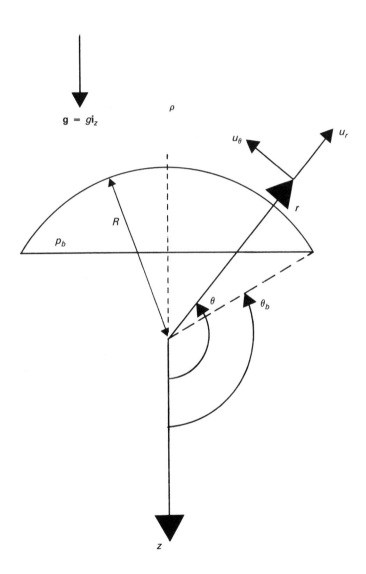

Figure 9.3. Geometry of flow past a spherical cap bubble.

flow is axisymmetric about the z axis. We let g denote the magnitude of the gravitational acceleration $\mathbf{g}$, ρ the density of the liquid, R the radius of the sphere of which the bubble is a segment (which we assume to be independent of time and which we might identify with the characteristic dimension l_c), θ_b the half-angle of the bubble [measured from the back of (i.e., below) the bubble], and p_b the pressure inside the bubble. We note, incidentally, that p_b is uniform within the bubble because the gas density is too low for there to be significant hydrostatic effects and significant pressure variations as a result of inertial effects. Thus we do not need to solve the flow equations for the gas within the bubble. We take the flow in the liquid round the bubble to be irrotational, that is, we take it to be a potential flow.

Around the front or nose of the bubble (i.e., above the bubble where $\theta_b \leq \theta \leq \pi$), we assume that the flow is the same as it would be if the bubble were a complete sphere, and not just a segment of a sphere. The justification for this assumption does not come from rigorous mathematical or physical arguments, though plausible ones can be made. Instead, it comes from the fact that the assumption will lead to a self-consistent result which is corroborated experimentally. It follows from Eq. (9.2-15) that the velocity field $\mathbf{u}$ is given by

$$\mathbf{u} = U_\infty \cos\theta \left(1 - \frac{R^3}{r^3}\right) \mathbf{i}_r - U_\infty \sin\theta \left(1 + \frac{R^3}{2r^3}\right) \mathbf{i}_\theta \qquad (9.3\text{-}7)$$

whence the speed $|\mathbf{u}(r = R, \theta)|$ at the surface of the bubble is given by

$$|\mathbf{u}(r = R, \theta)| = \tfrac{3}{2} U_\infty \sin\theta \qquad (9.3\text{-}8)$$

Substitution of Eq. (9.3-8) into Bernoulli's equation (9.1-13) yields

$$p(r = R, \theta) + \tfrac{9}{8}\rho U_\infty^2 \sin^2\theta - \rho g R \cos\theta = p(r = R, \theta = \pi) + \rho g R \qquad (9.3\text{-}9)$$

since $|\mathbf{u}|$ vanishes at the nose of the bubble. Because interfacial tension and viscous effects are negligible, it follows from Eq. (6.3-6) that, at the surface of the bubble,

$$p(r = R, \theta) = p_b \qquad \text{for } \theta_b \leq \theta < \pi \qquad (9.3\text{-}10)$$

so Eq. (9.3-9) yields

$$p_b + \tfrac{9}{8}\rho U_\infty^2 \sin^2\theta - \rho g R \cos\theta = p_b + \rho g R \qquad (9.3\text{-}11)$$

and so,

$$\frac{9}{8}\frac{U_\infty^2}{gR} = \frac{1 + \cos\theta}{\sin^2\theta} \qquad (9.3\text{-}12)$$

Clearly, Eq. (9.3-12) cannot hold for all values of θ, since the left-hand side is independent of θ while the right-hand side is not. It can, however, hold for small values of $\pi - \theta$, because

$$\frac{1 + \cos\theta}{\sin^2\theta} \simeq \frac{1}{2} \qquad \text{for } \pi - \theta \ll 1 \qquad (9.3\text{-}13)$$

Provided $\pi - \theta \ll 1$, which is the case in practice if $\pi > \theta_b > \pi/2$ or so, Eq. (9.3-12) yields

$$U_\infty \simeq \tfrac{2}{3} \sqrt{gR} \tag{9.3-14}$$

or, in terms of the Froude number [see Eq. (7.2-8)],

$$\text{Fr} = \frac{U_\infty^2}{gR} = \frac{4}{9} \tag{9.3-15}$$

For practical purposes, the radius R is a somewhat inconvenient quantity: a more convenient one is the volume V of the bubble. We now identify the characteristic velocity u_c with U_∞ and the characteristic length l_c with the diameter $(6V/\pi)^{1/3}$ of the sphere of the same volume as the bubble. Then, because it is observed experimentally that for $\text{Re} > 150$ or so, θ_b is approximately constant and given by $\theta_b \simeq 2.269$ (radians), it follows by elementary trigonometry that

$$U_\infty \simeq 0.792 \, g^{1/2} \, V^{1/6} \tag{9.3-16}$$

This expression for U_∞ is found experimentally to be valid for $\text{Eo} > 40$ or so (in order for the bubble to be large enough to form a spherical cap) and, as already noted, $\text{Re} > 150$ or so, where Eo and Re are given by Eqs. (9.3-5) and (9.3-6), respectively, with $u_c = U_\infty$ and $l_c = (6V/\pi)^{1/3}$.

More generally, it can be shown that the speed of rise U_∞ of a spherical cap of one fluid of density $\rho\dagger$ in another immiscible fluid of density ρ is given by

$$U_\infty \simeq 0.792 \, g^{1/2} \, V^{1/6} \left| 1 - \frac{\rho\dagger}{\rho} \right|^{1/2} \frac{\rho - \rho\dagger}{|\rho - \rho\dagger|} \tag{9.3-17}$$

Clearly, the spherical cap rises (i.e., $U_\infty > 0$) if $\rho > \rho\dagger$ and falls (i.e., $U_\infty < 0$) if $\rho < \rho\dagger$. It is found experimentally that the conditions for the validity of Eq. (9.3-17) are the same as those for that of Eq. (9.3-16), except that Eo is now given by

$$\text{Eo} = \frac{|\rho - \rho\dagger| \, g l_c^2}{\sigma} \tag{9.3-18}$$

instead of by Eq. (9.3-5). Note, however, that if the viscosity $\mu\dagger$ of the fluid comprising the bubble or drop is greater than about half the viscosity μ of the fluid around it, it is found experimentally that the bubble or drop is usually unstable and breaks up if its volume V exceeds a maximum volume V_{max} given by

$$V_{\text{max}} \simeq \frac{32}{3} \pi \left(\frac{\sigma}{g \, |\rho - \rho\dagger|} \right)^{3/2} \tag{9.3-19}$$

Thus, if $\mu\dagger > \tfrac{1}{2} \mu$ or so, which generally arises when a liquid drop moves through an immiscible liquid or falls through a gas, there is a maximum size the drop can attain (which has an obvious and fortunate implication in the case of raindrops in air). It also follows from Eq. (9.3-19) that, if $\mu\dagger > \tfrac{1}{2} \mu$ or so, there is a maximum value of Eo given by

$$\mathrm{Eo}_{\max} \simeq 16 \qquad (9.3\text{-}20)$$

As we have already noted, spherical cap drops (and bubbles) exist only for Eo $>$ 40 or so: for smaller values of Eo, they are spherical or ellipsoidal in shape. Thus it follows that, if $\mu\dagger > \frac{1}{2} \mu$ or so, spherical cap drops are usually unstable and do not, therefore, exist.

9.4 SURFACE WAVES

We consider an infinite expanse of an incompressible Newtonian liquid of density ρ, viscosity μ, and depth h on a flat, horizontal, rigid surface. Above the liquid, we suppose that there is a gas of negligible density and viscosity and uniform pressure p_a. The interfacial tension of the liquid-gas system is σ. We suppose that the liquid is caused to move in such a way that a planar flow is established in which there are traveling two-dimensional waves of amplitude a, wavelength λ, and frequency ω on its otherwise flat, level surface. Because of the obvious geometry, we use rectangular coordinates (x, y, z) to analyze the problem, with origin on the undisturbed surface of the liquid, the x direction aligned horizontally normal to the crests and troughs of the waves, and the y direction aligned vertically upward, that is, in a direction opposite to the gravitational acceleration $\mathbf{g}$ (see Fig. 9.4).

We can ignore motion in the gas because it is at uniform pressure. In the liquid, the flow equations are mass conservation

$$\nabla \cdot \mathbf{u} = 0 \qquad (9.4\text{-}1)$$

and linear momentum conservation, that is, the Navier-Stokes equations

$$\rho \frac{\partial \mathbf{u}}{\partial t} + \rho \mathbf{u} \cdot \nabla \mathbf{u} = \rho \mathbf{g} - \nabla p + \mu \Delta \mathbf{u} \qquad (9.4\text{-}2)$$

We now note that, to an order of magnitude, $|\mathbf{u}| \sim a\omega$, $\partial/\partial t \sim \omega$, $|\nabla| \sim 1/l_c$, and $|\Delta| \sim 1/l_c^2$. Here the characteristic length l_c is the lesser of the wavelength λ and the depth h. The reason for this is that, for a quantity that varies by a given amount, the magnitudes of its spatial derivatives are larger over shorter distances than over longer ones. Thus,

$$\left| \rho \frac{\partial \mathbf{u}}{\partial t} \right| \sim \rho a \omega^2 \qquad |\rho \mathbf{u} \cdot \nabla \mathbf{u}| \sim \frac{\rho a^2 \omega^2}{l_c} \qquad |\mu \, \Delta \mathbf{u}| \sim \frac{\mu a \omega}{l_c^2} \qquad (9.4\text{-}3)$$

It follows that the convection term is negligible compared with the transient term, that is, nonlinear inertial effects are negligible, if $|\rho \, \mathbf{u} \cdot \nabla \mathbf{u}| \ll |\rho \, \partial \mathbf{u}/\partial t|$, that is, if

$$\frac{l_c}{a} \gg 1 \qquad (9.4\text{-}4)$$

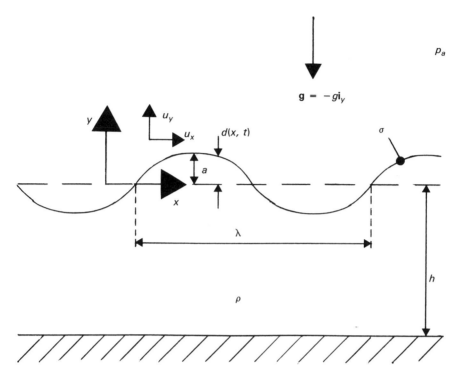

Figure 9.4. Geometry of two-dimensional surface waves.

Similarly, the viscous term is negligible compared with the transient term, that is, viscous effects are negligible and the flow may be assumed to be inviscid, if $|\mu\,\Delta\mathbf{u}| \ll |\rho\,\partial\mathbf{u}/\partial t|$, that is, if

$$\frac{\rho\omega l_c^2}{\mu} \gg 1 \qquad (9.4\text{-}5)$$

Note that, because ωl_c is a characteristic velocity u_c, we might interpret $\rho\omega l_c^2/\mu$ as a Reynolds number. Henceforth, we assume that conditions (9.4-4) and (9.4-5) hold, so that Eq. (9.4-2) becomes

$$\rho\frac{\partial\mathbf{u}}{\partial t} = \rho\mathbf{g} - \nabla p \qquad (9.4\text{-}6)$$

Because the flow is inviscid, we put

$$\mathbf{u} = \nabla\phi \qquad (9.4\text{-}7)$$

whence, using Eq. (9.4-1),

$$\nabla^2\phi = 0 \qquad (9.4\text{-}8)$$

We are considering two-dimensional waves, so we seek a solution of the form

$$\phi = f(y) \sin (\kappa x - \omega t) \tag{9.4-9}$$

Note that we are *assuming* the form of the solution: it remains to be seen whether or not such a form is consistent with the equations governing the flow. The parameter κ in Eq. (9.4-9) is the wave number given by

$$\kappa = \frac{2\pi}{\lambda} \tag{9.4-10}$$

Substitution of Eq. (9.4-9) into Eq. (9.4-8) and using Eq. (A1.3-16) of Appendix A yields

$$-\kappa^2 f \sin (\kappa x - \omega t) + \frac{d^2 f}{dy^2} \sin (\kappa x - \omega t) = 0 \tag{9.4-11}$$

that is,

$$\frac{d^2 f}{dy^2} - \kappa^2 f = 0 \tag{9.4-12}$$

the solution of which is

$$f = c_1 e^{\kappa y} + c_2 e^{-\kappa y} \tag{9.4-13}$$

or

$$f = (c_1 + c_2) \cosh (\kappa y) + (c_1 - c_2) \sinh (\kappa y) \tag{9.4-14}$$

where c_1 and c_2 are constants. Hence,

$$\phi = [(c_1 + c_2) \cosh (\kappa y) + (c_1 - c_2) \sinh (\kappa y)] \sin (\kappa x - \omega t) \tag{9.4-15}$$

Boundary conditions must be imposed at the free surface $y = d$, where d is a function of horizontal position x and time t, and at the base of the liquid $y = -h$. At the free surface, the vertical component of velocity u_y is given by

$$u_y = \frac{Dd}{Dt} = \frac{\partial d}{\partial t} + u_x \frac{\partial d}{\partial x} \approx \frac{\partial d}{\partial t} \tag{9.4-16}$$

since it follows from Eq. (9.4-4) that $|u_x \, \partial d/\partial x|/|\partial d/\partial t| \sim a/l_c \ll 1$. Thus, since $u_y = \partial \phi/\partial y$ [see Eq. (A1.3-1) of Appendix A],

$$\frac{\partial \phi}{\partial y} = \frac{\partial d}{\partial t} \tag{9.4-17}$$

It follows from Eq. (6.3-6) that

$$-p + p_a = -\sigma \left(\frac{1}{R_\mathrm{I}} + \frac{1}{R_\mathrm{II}} \right) \tag{9.4-18}$$

where the principal radii of curvature of the surface of the liquid, R_I and R_II, can be shown to be given by

$$R_\mathrm{I} = -\frac{[1 + (\partial d/\partial x)^2]^{3/2}}{\partial^2 d/\partial x^2} \qquad R_\mathrm{II} \to \infty \qquad (9.4\text{-}19)$$

It follows from Eq. (9.4-4) that $|\partial d/\partial x| \sim a/\lambda \ll 1$ and so,

$$R_\mathrm{I} \simeq -\frac{1}{\partial^2 d/\partial x^2} \qquad R_\mathrm{II} \to \infty \qquad (9.4\text{-}20)$$

Thus Eq. (9.4-18) yields

$$-p + p_a = \sigma \frac{\partial^2 d}{\partial x^2} \qquad (9.4\text{-}21)$$

We now note that it follows from Eqs. (9.4-6) and (9.4-7) that

$$\nabla\left(\frac{\partial \phi}{\partial t}\right) = -\nabla\left(\frac{p}{\rho}\right) + \nabla(\mathbf{g} \cdot \mathbf{x}) \qquad (9.4\text{-}22)$$

which may be integrated to yield

$$\frac{\partial \phi}{\partial t} = -\frac{p}{\rho} + \mathbf{g} \cdot \mathbf{x} + \zeta(t) \qquad (9.4\text{-}23)$$

which is the linearized form of Bernoulli's equation (9.1-11). Because it is the linearized form, we now set the function $\zeta(t)$ arbitrarily to zero, since the addition of any function just of time t to ϕ does not alter $\mathbf{u}$. Thus,

$$p = -\rho \frac{\partial \phi}{\partial t} + \rho \mathbf{g} \cdot \mathbf{x} \qquad (9.4\text{-}24)$$

or, since $\mathbf{g} \cdot \mathbf{x} = -gd$ on the free surface, where $g = |\mathbf{g}|$, combining Eqs. (9.4-21) and (9.2-24), we obtain

$$\rho \frac{\partial \phi}{\partial t} + \rho g d + p_a = \sigma \frac{\partial^2 d}{\partial x^2} \qquad (9.4\text{-}25)$$

Differentiation with respect to time t yields

$$\rho \frac{\partial^2 \phi}{\partial t^2} + \rho g \frac{\partial d}{\partial t} = \sigma \frac{\partial^3 d}{\partial t \partial x^2} \qquad (9.4\text{-}26)$$

It thus follows that the boundary condition to be applied at the free surface is

$$-\rho \frac{\partial^2 \phi}{\partial t^2} - \rho g \frac{\partial \phi}{\partial y} + \sigma \frac{\partial^3 \sigma}{\partial y \partial x^2} = 0 \qquad \text{at } y = d \qquad (9.4\text{-}27)$$

which is obtained by combining Eqs. (9.4-17) and (9.4-26). Strictly, this condition should be applied at $y = d$. In fact, however, since $|d| \leq a$, it follows from Eq. (9.4-4) that $|d|/l_c \ll 1$ and so the condition may be applied at $y = 0$ instead without significant error [i.e., the error made is no larger than those others made as a result of using Eq. (9.4-4)]. The boundary condition to be applied at the base of the liquid is much more straightforward than that at the free surface: no flow-through (see Section 6.2) means that $u_y = 0$ there and hence that

$$\frac{\partial \phi}{\partial y} = 0 \qquad \text{at } y = -h \tag{9.4-28}$$

Imposition of boundary conditions (9.4-27) and (9.4-28) on Eq. (9.4-15) yields

$$\frac{c_1 - c_2}{c_1 + c_2} = \tanh (\kappa h) \tag{9.4-29}$$

$$\omega^2 = \frac{c_1 - c_2}{c_1 + c_2}\left(g\kappa + \frac{\kappa^3 \sigma}{\rho}\right) \tag{9.4-30}$$

Hence,

$$\omega^2 = \tanh (\kappa h)\left(g\kappa + \frac{\kappa^3 \sigma}{\rho}\right) \tag{9.4-31}$$

Equation (9.4-31) relates frequency ω to wave number κ and hence to wavelength λ. We note that (1) if $h >> \lambda = 2\pi/\kappa$, which we refer to as the *short wave* (or *deep water*) asymptote, then $\tanh (\kappa h) \simeq 1$, and so,

$$\omega \simeq \left(g\kappa + \frac{\kappa^3 \sigma}{\rho}\right)^{1/2} \tag{9.4-32}$$

while (2) if $h << \lambda = 2\pi/\kappa$, which we refer to as the *long wave* (or *shallow water*) asymptote, then $\tanh (\kappa h) \simeq \kappa h$ and so

$$\omega \simeq \left(g\kappa^2 h + \frac{\kappa^4 \sigma h}{\rho}\right)^{1/2} \tag{9.4-33}$$

The phase velocity c is defined by

$$c = \frac{\omega}{\kappa} = \frac{\omega\lambda}{2\pi} \tag{9.4-34}$$

Thus, for short waves,

$$c \simeq \left(\frac{g}{\kappa} + \frac{\sigma\kappa}{\rho}\right)^{1/2} = \left(\frac{g\lambda}{2\pi} + \frac{2\pi\sigma}{\rho\lambda}\right)^{1/2} \tag{9.4-35}$$

while, for long waves,

$$c \simeq \left(gh + \frac{\sigma\kappa^2 h}{\rho}\right)^{1/2} = \left(gh + \frac{4\pi^2\sigma h}{\rho\lambda^2}\right)^{1/2} \tag{9.4-36}$$

We note that it follows from Eq. (9.4-31) that the relative importance of gravitational and interfacial tension effects is given by the relative sizes of $g\kappa$ and $\kappa^3\sigma/\rho$. Thus, gravitational effects are negligible if

$$\frac{\rho g}{\sigma \kappa^2} \sim \frac{\rho g \lambda^2}{\sigma} << 1 \qquad (9.4\text{-}37)$$

and the waves are called *capillary waves* or *ripples*. For short capillary waves,

$$c \simeq \left(\frac{2\pi\sigma}{\rho\lambda}\right)^{1/2} \qquad (9.4\text{-}38)$$

and for long capillary waves;

$$c \simeq \left(\frac{4\pi^2 \sigma h}{\rho\lambda^2}\right)^{1/2} \qquad (9.4\text{-}39)$$

Interfacial tension effects are negligible if

$$\frac{\rho g}{\sigma \kappa^2} \sim \frac{\rho g \lambda^2}{\sigma} >> 1 \qquad (9.4\text{-}40)$$

and the waves are called *gravity waves*. For short gravity waves,

$$c \simeq \left(\frac{g\lambda}{2\pi}\right)^{1/2} \qquad (9.4\text{-}41)$$

and for long gravity waves,

$$c \simeq (gh)^{1/2} \qquad (9.4\text{-}42)$$

These results for the phase velocity c as a function of wavelength λ are summarized schematically in Fig. 9.5. We note immediately that, with one exception,

Waves of different wavelengths or frequencies travel at different speeds.

This is referred to as the *dispersive* property of waves. The exception is long gravity waves, all of which travel at the same speed $(gh)^{1/2}$. We should note, incidentally, that the speed at which wave energy is propagated can be shown to be given not by the phase velocity c but instead by the group velocity C:

$$C = \frac{d\omega}{d\kappa} \qquad (9.4\text{-}43)$$

It follows from Eq. (9.4-31) that

$$\frac{C}{c} = \frac{\kappa}{\omega}\frac{d\omega}{d\kappa} = \frac{1}{2}\left[\frac{1 + (3\sigma\kappa^2/\rho g)}{1 + (\sigma\kappa^2/\rho g)} + \frac{2\kappa h}{\sinh(2\kappa h)}\right] \qquad (9.4\text{-}44)$$

from which it follows that

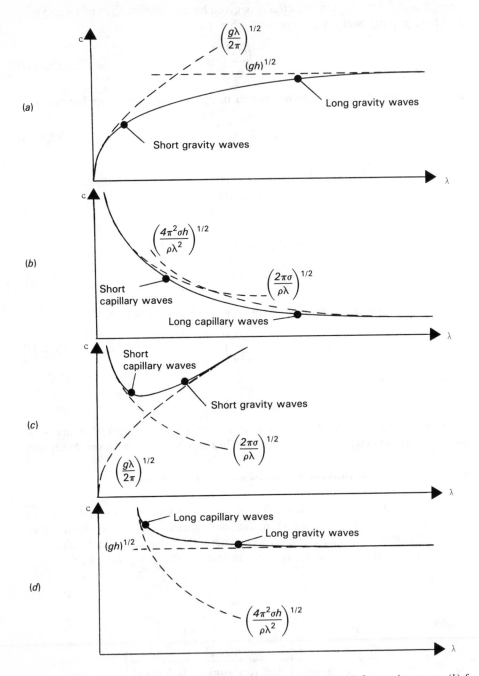

Figure 9.5. Schematic variation of phase velocity with wavelength: (*a*) for gravity waves, (*b*) for capillary waves, (*c*) for short waves, and (*d*) for long waves.

$$\frac{C}{c} \simeq \begin{cases} 1 & \text{for long gravity waves} \\ \frac{1}{2} & \text{for short gravity waves} \\ 2 & \text{for long capillary waves} \\ \frac{3}{2} & \text{for short capillary waves} \end{cases} \qquad (9.4\text{-}45)$$

Note that $C \neq c$ (other than for long gravity waves) because of the dispersive property of waves.

We conclude by noting that we have implicitly confirmed that Eq. (9.4-9) is consistent with the equations governing the flow, since no inconsistency arose. Also, more general wave motions than the *simple harmonic motion* implied by Eq. (9.4-9) can be analyzed by superposition, since Eq. (9.4-8) and its boundary conditions are linear in ϕ.

9.5 STABILITY OF A JET

We consider a liquid jet of circular cross section moving initially at constant speed U and constant radius R through a gas of negligible density and viscosity and uniform pressure p_a. The liquid is of density ρ; the interfacial tension of the liquid-gas system is σ. We assume that viscous and gravitational effects are negligible compared with inertial and interfacial tension effects in the jet. Because of the obvious geometry, we use cylindrical polar coordinates (r, θ, z) to analyze the problem, with origin on, and z direction aligned with, the centerline of the jet (see Fig. 9.6).

Such jets are observed in practice to be unstable and break up to produce drops. Our aim is to obtain a criterion for instability of the jet to very small, in fact to infinitesimal, disturbances. Any real jet, indeed any real flow, always contains some disturbances, such as roughness on walls: disturbances can never be entirely eliminated from any real flow.

We confine our attention to axisymmetric disturbances, since it can be shown (though we will not do so here) that the jet is stable if the disturbances are non-axisymmetric. Thus the jet always has a circular cross section; its local radius R^* is, however, a function of axial position z (and time t).

We choose, for convenience, a reference frame which is moving, initially with the jet, at (constant) velocity $U\mathbf{i}_z$ (note that we are implicitly making a Galilean transformation of reference frame; see Section 8.2). As a result, the speed U no longer enters our analysis so the stability or otherwise of the jet is unaffected by its initial velocity. Because the jet is initially at rest in our chosen reference frame, it follows from Lagrange's theorem (see Section 9.1) that there exists a scalar potential ϕ such that the velocity field $\mathbf{u}$ is given by:

$$\mathbf{u} = \nabla\phi \qquad (9.5\text{-}1)$$

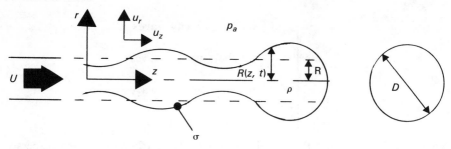

Figure 9.6. Geometry of breakup of a jet.

Because the jet comprises a liquid, which we may assume to be incompressible, mass conservation implies that $\nabla \cdot \mathbf{u} = 0$ and hence that

$$\nabla^2 \phi = 0 \tag{9.5-2}$$

Using Equation (A1.3-17) of Appendix A, and assuming that the flow is axisymmetric and swirl-free, Eq. (9.5-2) yields

$$\frac{1}{r}\frac{\partial}{\partial r}\left(r\frac{\partial \phi}{\partial r}\right) + \frac{\partial^2 \phi}{\partial z^2} = 0 \tag{9.5-3}$$

We now assume that ϕ is a separable function of z, r, and t, that is,

$$\phi(z, r, t) = f(z)\, g(r)\, h(t) \tag{9.5-4}$$

(This assumption is justified implicitly in what follows because no inconsistency arises from making it.) Thus Eq. (9.5-3) becomes

$$\frac{1}{rg}\frac{d}{dr}\left(r\frac{dg}{dr}\right) = -\frac{1}{f}\frac{d^2 f}{dz^2} \tag{9.5-5}$$

The left-hand side of Eq. (9.5-5) is independent of z and t; the right-hand side is independent of r and t. Hence both sides must be constant, that is,

$$\frac{1}{f}\frac{d^2 f}{dz^2} = -\kappa^2 \tag{9.5-6}$$

where the wave number κ is a constant; we may, without loss of generality, assume that $\kappa > 0$. The solution of the ordinary differential Eq. (9.5-6) is

$$f = c_1 \cos(\kappa z) + c_2 \sin(\kappa z) \tag{9.5-7}$$

where c_1 and c_2 are constants. It then follows that

$$\frac{1}{rg}\frac{d}{dr}\left(r\frac{dg}{dr}\right) = \kappa^2 \tag{9.5-8}$$

We now put

$$\eta = \kappa r \tag{9.5-9}$$

so that Eq. (9.5-8) becomes

$$\eta^2 \frac{d^2 g}{d\eta^2} + \eta \frac{dg}{d\eta} - \eta^2 g = 0 \qquad (9.5\text{-}10)$$

which is a modified form of Bessel's equation. Its solution is

$$g = c_3 I_0(\eta) + c_4 K_0(\eta) \qquad (9.5\text{-}11)$$

where c_3 and c_4 are constants and $I_0(\ )$ and $K_0(\ )$ denote the zero-order modified Bessel functions of the first and second kinds, respectively. Hence,

$$\phi = [c_1 \cos(\kappa z) + c_2 \sin(\kappa z)][c_3 I_0(\kappa r) + c_4 K_0(\kappa r)] h(t) \qquad (9.5\text{-}12)$$

Boundary conditions must be imposed at the centerline ($r = 0$) of the jet and at its free surface ($r = R^*$). At the centerline of the jet, we require that the velocity be finite, which [since $K_0(\eta)$ and its derivatives become unbounded as $\eta \to 0$] implies that

$$c_4 = 0 \qquad (9.5\text{-}13)$$

We may now, without loss of generality, put

$$c_3 = 1 \qquad (9.5\text{-}14)$$

Hence,

$$\phi = [c_1 \cos(\kappa z) + c_2 \sin(\kappa z)] I_0(\kappa r) h(t) \qquad (9.5\text{-}15)$$

At the free surface of the jet, the radial component of velocity u_r is given by

$$u_r = \frac{DR^*}{Dt} = \frac{\partial R^*}{\partial t} + u_z \frac{\partial R^*}{\partial z} \qquad (9.5\text{-}16)$$

If the amplitude of the perturbation to the jet radius is small, that is, if

$$\left| \frac{R^*}{R} - 1 \right| \ll 1 \qquad (9.5\text{-}17)$$

then $|u_z\, \partial R^*/\partial z| \ll |\partial R^*/\partial t|$ and so, since $u_r = \partial\phi/\partial r$ [see Eq. (A1.3-3) of Appendix A],

$$\frac{\partial\phi}{\partial r} \simeq \frac{\partial R^*}{\partial t} \qquad \text{at } r = R^* \qquad (9.5\text{-}18)$$

Strictly, this condition should be applied at $r = R^*$. In fact, however, provided Eq. (9.5-17) holds, the condition may be applied at $r = R$ instead without significant error. Combining Eqs. (9.5-15) and (9.5-18), we obtain

$$R^* = R \{1 + [c_1 \cos(\kappa z) + c_2 \sin(\kappa z)] H(t)\} \qquad (9.5\text{-}19)$$

and

$$\frac{dH}{dt} = \frac{\kappa}{R} I_1(\kappa R) h \tag{9.5-20}$$

where $I_1(\ )$ denotes the first-order modified Bessel function of the first kind. Hence

$$\phi = [c_1 \cos(\kappa z) + c_2 \sin(\kappa z)] \frac{I_0(\kappa r)}{I_1(\kappa R)} \frac{R}{\kappa} \frac{dH}{dt} \tag{9.5-21}$$

It follows from Eq. (6.3-6) that

$$-p + p_a = -\sigma \left(\frac{1}{R_I} + \frac{1}{R_{II}} \right) \qquad \text{at } r = R^* \tag{9.5-22}$$

Provided Eq. (9.5-17) holds, we may apply condition (9.5-22) at $r = R$ without significant error; we may, moreover, show that

$$\frac{1}{R_I} + \frac{1}{R_{II}} \simeq \frac{1}{R^*} - \frac{\partial^2 R^*}{\partial z^2} \tag{9.5-23}$$

and hence that

$$\frac{1}{R_I} + \frac{1}{R_{II}} \simeq \frac{1}{R} \{1 - H [c_1 \cos(\kappa z) + c_2 \sin(\kappa z)](1 - \kappa^2 R^2)\} \tag{9.5-24}$$

Using Eq. (9.5-17), it now follows from Bernoulli's equation (9.1-11), neglecting the nonlinear inertial term $\frac{1}{2} |\nabla\phi|^2$ and the gravitational term $-\mathbf{g} \cdot \mathbf{x}$, that

$$p + \rho \frac{\partial\phi}{\partial t} = \rho\zeta(t) \tag{9.5-25}$$

We may arbitrarily set

$$\rho\zeta(t) = p_a + \frac{\sigma}{R} \tag{9.5-26}$$

which is, in fact, the (constant) pressure in the initial undisturbed jet, since the addition of an arbitrary function of time t to ϕ does not alter the velocity field $\mathbf{u}$. Hence Eq. (9.5-22) yields

$$-p_a - \frac{\sigma}{R} + \frac{\rho R}{\kappa} [c_1 \cos(\kappa z) + c_2 \sin(\kappa z)] \frac{I_0(\kappa R)}{I_1(\kappa R)} \frac{d^2 H}{dt^2} + p_a \tag{9.5-27}$$

$$= -\frac{\sigma}{R} \{1 - H [c_1 \cos(\kappa z) + c_2 \sin(\kappa z)](1 - \kappa^2 R^2)\}$$

Thus it follows that

$$\frac{d^2 H}{dt^2} = \beta^2 H \tag{9.5-28}$$

where
$$\beta^2 = \frac{\sigma \kappa}{\rho R^2} (1 - \kappa^2 R^2) \frac{I_1(\kappa R)}{I_0(\kappa R)} \qquad (9.5\text{-}29)$$

The solution of Eq. (9.5-28) is

$$H = H_1 e^{\beta t} + H_2 e^{-\beta t} \qquad (9.5\text{-}30)$$

The constants H_1 and H_2 are determined by the form of the disturbance that is assumed to exist at time $t = 0$. The precise form is irrelevant to our consideration of stability.

The relevant issue is whether or not the disturbance grows: if it does, the jet is unstable.

Clearly,

- If $\beta^2 = 0$, $(R^* - R)$ is independent of time and the jet is said to be stationary.
- If $\beta^2 > 0$, so that β is real, $|R^* - R|$ grows exponentially with time t and the jet is said to be unstable.
- If $\beta^2 < 0$, so that β is imaginary, $|R^* - R|$ varies sinusoidally with time t and the jet is said to be oscillatory.

In order to clarify these ideas, consider a heavy, frictionless ball (see Fig. 9.7).

- If it is placed on a flat, horizontal surface, it has no tendency to move if it is displaced from its original position: it corresponds to a stationary jet.
- If it is placed on top of a hill, any displacement down the hill will cause it to roll down ever further away from its original position: it corresponds to an unstable jet.
- If it is placed at the bottom of a valley, any displacement up one side of the valley will cause it to roll down, through its original position, up the opposite side of the valley and then back again endlessly: it corresponds to an oscillatory jet.

It follows that the jet is unstable if

$$\kappa R (1 - \kappa^2 R^2) \frac{I_1(\kappa R)}{I_0(\kappa R)} > 0 \qquad (9.5\text{-}31)$$

that is,
$$0 < \kappa R < 1 \qquad (9.5\text{-}32)$$

since $I_0(\eta)$ and $I_1(\eta)$ are positive for $\eta > 0$. The jet is, therefore, unstable if the wavelength λ given by

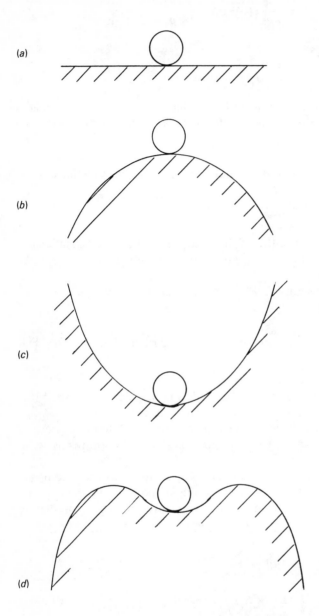

Figure 9.7. A. heavy frictionless ball has no tendency to move if it is displaced on a flat, horizontal surface (*a*); it tends to roll down if it is displaced from the top of a hill (*b*); it tends to oscillate if it is displaced from the bottom of a valley (*c*); it tends to oscillate for small displacements and roll down for large displacements from the bottom of a hollow on top of a hill (*d*).

$$\lambda = \frac{2\pi}{\kappa}$$

(9.5-33)

exceeds the initial circumference ($2\pi R$) of the jet. The most unstable wavenumber $\kappa^\S$ (or wavelength $\lambda^\S$) corresponds to the maximum (real) value of β, which we denote $\beta^\S$. This is because disturbances of such a frequency will grow faster than disturbances of any other frequency and eventually completely dominate them: $\kappa^\S$, $\lambda^\S$, and $\beta^\S$ are given by (see Table 9.1):

$$\kappa^\S = \frac{2\pi}{\lambda^\S} \simeq \frac{0.697}{R}$$

(9.5-34)

$$\beta^\S \simeq 0.3433 \sqrt{\frac{\sigma}{\rho R^3}}$$

(9.5-35)

We now estimate the diameter of the drops that form on breakup of the jet at the most unstable wavelength $\lambda^\S$. We assume that the drops are all of the same diameter D and that they have the same volume as a cylinder of length $\lambda^\S$ and radius R, that is,

$$\frac{\pi}{6} D^3 = \pi R^2 \lambda^\S$$

(9.5-36)

whence
$$D \simeq 3.78R$$
(9.5-37)

Equation (9.5-37) is approximately corroborated experimentally for the breakup of many liquid jets in gases, such as water jets in air, provided the Reynolds number $Re = \rho U 2R / \mu \sim 1000$ where μ denotes the viscosity of the liquid. It is, however, observed that additional small satellite drops tend to form in practice between drops of approximate diameter D.

We conclude by noting that our analysis gives a criterion only for instability with respect to small, in fact to infinitesimal, disturbances [see Eq. (9.5-17)]. A jet that is stable with respect to infinitesimal disturbances may be unstable with respect to sufficiently large finite disturbances. To see this, we consider a heavy, frictionless ball placed in a hollow on top of a hill (see Fig. 9.7). Clearly, it will not move down the hill if it is given a small enough displacement but will do so if given a large enough displacement. Thus our analysis only gives a *sufficient* criterion for instability. We note, incidentally, that because instability of the jet means that $|R^* - R|$ grows exponentially with time, it follows that after a short period of time $|(R^*/R) - 1| \lll 1$ and eventually (when drops are about to form) $|(R^*/R) - 1| \sim 1$. Clearly, when this happens, Eq. (9.5-17) is no longer satisfied, and our analysis, which assumes small disturbances, no longer holds and so the growth of $|R^* - R|$, which starts off by being exponential with time, soon becomes nonexponential and beyond the scope of our analysis here. We note finally that we chose, in Eq. (9.5-12), a particular form or mode of disturbance: this choice

Table 9.1 Values of $\beta\sqrt{\rho R^3/\sigma}$ as a function of κR

κR	$\beta\sqrt{\rho R^3/\sigma}$
0	0
0.1	0.0703
0.2	0.1382
0.3	0.2012
0.4	0.2567
0.5	0.3016
0.6	0.3321
0.697	0.3433
0.7	0.3433
0.8	0.3269
0.9	0.2647
1.0	0

may be justified by noting that (1) instability to *any* particular mode is *sufficient* for a system to be unstable because all possible modes of infinitesimal disturbance may be assumed to exist; (2) Eq. (9.5-3) and its boundary conditions are linear in ϕ, so that more general (axisymmetric) forms of disturbance can be analyzed by superposition of modes; and (3) the choice yields results that are, in certain circumstances, corroborated experimentally.

9.6 RUPTURE OF A DAM

We consider a semi-infinite volume of an incompressible Newtonian liquid of density ρ and viscosity μ on a smooth, flat, horizontal surface of infinite extent. Initially, for time $t \leq 0$, the liquid is of depth h_0 and constrained by a vertical, planar barrier of infinite length. Above the liquid, we suppose that there is a gas of negligible density and viscosity and uniform pressure p_a; the interfacial tension of the liquid-gas system is σ. We might think of the volume of liquid as a reservoir, the barrier as a dam, and the horizontal surface as the ground. At time $t = 0$, the dam is instantaneously removed or ruptured: we analyze the subsequent flow of liquid out of the reservoir and over the ground. This comprises what is called the dam break problem. We use rectangular coordinates (x, y, z) to analyze the problem with origin at the intersection of the plane comprising the ground and the plane comprising the initial position of the dam. The x direction is aligned horizontally normal to the plane comprising the initial position of the dam and away from the region initially comprising the reservoir. The y direction is aligned vertically upward, that is, in the opposite direction to the gravitational acceleration $\mathbf{g}$ (see Fig. 9.8).

We assume that viscous and interfacial tension effects are negligible compared with inertial and gravitational effects in the liquid. Symmetry means that the velocity field $\mathbf{u}$ is planar, that is, it has no z component and its x and y components (u_x and u_y, respectively) are, like the pressure field p, independent of z. Thus, since a liquid is effectively incompressible, mass conservation yields

$$\frac{\partial u_x}{\partial x} + \frac{\partial u_y}{\partial y} = 0 \qquad (9.6\text{-}1)$$

[see Eq. (A2.2-2) of Appendix B] and linear momentum conservation yields Euler's equations:

$$\frac{\partial u_x}{\partial t} + u_x \frac{\partial u_x}{\partial x} + u_y \frac{\partial u_x}{\partial y} = -\frac{1}{\rho}\frac{\partial p}{\partial x} \qquad (9.6\text{-}2)$$

$$\frac{\partial u_y}{\partial t} + u_x \frac{\partial u_y}{\partial x} + u_y \frac{\partial u_y}{\partial y} = -\frac{1}{\rho}\frac{\partial p}{\partial y} - g \qquad (9.6\text{-}3)$$

[see Eqs. (A2.3-2) of Appendix B, which have been simplified by omission of the viscous terms] where $g = |\mathbf{g}|$. It follows by integration of Eq. (9.6-1) that

$$\int_0^h \frac{\partial u_y}{\partial y}\, dy = u_y|_{y=h} - u_y|_{y=0} = -\int_0^h \frac{\partial u_x}{\partial x}\, dy \qquad (9.6\text{-}4)$$

where $h(x, t)$ denotes the height of the liquid above the plane $y = 0$ (see Fig. 9.8). Because $u_y|_{y=h} = Dh/Dt$ (where D/Dt denotes the substantial derivative; see Section 2.3) and $u_y|_{y=0} = 0$ (there is no flow-through at the base of the liquid, that is, the ground is impermeable), it follows that

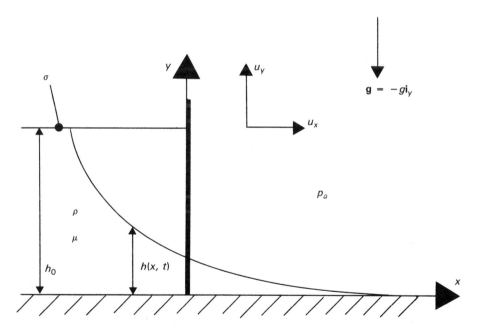

Figure 9.8. Geometry of rupture of a dam.

$$\frac{\partial h}{\partial t} + u_x \frac{\partial h}{\partial x} = -\int_0^h \frac{\partial u_x}{\partial x} \, dy \qquad (9.6\text{-}5)$$

We now assume that $|Du_y/Dt| \ll g$, but note that this is not a good assumption for a short period of time immediately after the dam is removed when, for liquid near $x = 0$ and $y = h$, the initial motion is almost entirely vertically downward with acceleration $\mathbf{g}$. Then the vertical linear momentum conservation equation (9.6-3) yields

$$0 = -\frac{1}{\rho}\frac{\partial p}{\partial y} - g \qquad (9.6\text{-}6)$$

which may be integrated immediately to yield

$$p = p_a + \rho g(h - y) \qquad (9.6\text{-}7)$$

since it follows from Eq. (6.3-6) that $p = p_a$ at $y = h$ (recall that interfacial tension and viscous effects are assumed to be negligible). Hence,

$$\frac{\partial p}{\partial x} = \rho g \frac{\partial h}{\partial x} \qquad (9.6\text{-}8)$$

and so the horizontal linear momentum conservation equation ((9.6-2) yields

$$\frac{Du_x}{Dt} = \frac{\partial u_x}{\partial t} + u_x \frac{\partial u_x}{\partial x} + u_y \frac{\partial u_x}{\partial y} = -g \frac{\partial h}{\partial x} \qquad (9.6\text{-}9)$$

Clearly, $g\,\partial h/\partial x$ is independent of y, so Du_x/Dt is independent of y. Initially, for time $t \leq 0$, the liquid is at rest so that u_x is initially independent of y. Thus u_x remains independent of y at all subsequent times $t > 0$. Hence $\partial u_x/\partial y = 0$ and so Eq. (9.6-9) becomes

$$\frac{\partial u_x}{\partial t} + u_x \frac{\partial u_x}{\partial x} + g \frac{\partial h}{\partial x} = 0 \qquad (9.6\text{-}10)$$

while Eq. (9.6-5) yield

$$\frac{\partial h}{\partial t} + u_x \frac{\partial h}{\partial x} + h \frac{\partial u_x}{\partial x} = 0 \qquad (9.6\text{-}11)$$

For convenience, we now put

$$c^2 = gh \qquad (9.6\text{-}12)$$

where c is the *phase velocity* of long gravity waves of very small amplitude on the surface of the liquid [see Eq. (9.4-42)]. Because the long wave asymptote is also the shallow water asymptote (see Section 9.4). Eqs. (9.6-10) and (9.6-11) are called the *shallow water equations*. We can assume, without loss of generality, that $c > 0$. The reason for introducing the single variable c is that it replaces the two variables g and h. Then, because $2c\,dc = g\,dh$, Eqs. (9.6-10) and (9.6-11) yield

$$\frac{\partial u_x}{\partial t} + u_x \frac{\partial u_x}{\partial x} + 2c \frac{\partial c}{\partial x} = 0 \qquad (9.6\text{-}13)$$

$$2 \frac{\partial c}{\partial t} + 2u_x \frac{\partial c}{\partial x} + c \frac{\partial u_x}{\partial x} = 0 \qquad (9.6\text{-}14)$$

These equations must be solved subject to the initial conditions

$$c = \begin{cases} c_0 = \sqrt{gh_0} & \text{at } t = 0 \text{ if } x \le 0 \\ 0 & \text{at } t = 0 \text{ if } x > 0 \end{cases} \qquad (9.6\text{-}15)$$

$$u_x = 0 \qquad \text{at } t = 0 \qquad (9.6\text{-}16)$$

To obtain a solution of Eqs. (9.6-13) and (9.6-14) subject to initial conditions (9.6-15) and (9.6-16), we note that

- c_0 is a characteristic velocity u_c with which u_x and c can be made dimensionless.
- There is no characteristic length l_c with which x can be made dimensionless (h_0 is not a suitable characteristic length, at least it is not a suitable characteristic horizontal length, though it is a suitable vertical one, as we will see).
- There is no characteristic time t_c with which t can be made dimensionless (though, were a characteristic length l_c to exist, a characteristic time t_c would be given by l_c/c_0).

As a result, we seek (as we noted in Section 7.2) a *similarity solution*. We start by defining in an obvious way the dimensionless variables α and β given by

$$\alpha = \frac{u_x}{c_0} \qquad \beta = \frac{c}{c_0} \qquad (9.6\text{-}17)$$

Then Eqs. (9.6-13) and (9.6-14) become

$$\frac{1}{c_0} \frac{\partial \alpha}{\partial t} + \alpha \frac{\partial \alpha}{\partial x} + 2\beta \frac{\partial \beta}{\partial x} = 0 \qquad (9.6\text{-}18)$$

$$\frac{2}{c_0} \frac{\partial \beta}{\partial t} + 2\alpha \frac{\partial \beta}{\partial x} + \beta \frac{\partial \alpha}{\partial x} = 0 \qquad (9.6\text{-}19)$$

We now define the similarity variable η;

$$\eta = c_1 t^{c_2} x \qquad (9.6\text{-}20)$$

where c_1 and c_2 are constants yet to be determined. In fact, c_1 will be determined by considerations of convenience and c_2 by the requirement that α and β be functions of η alone. Note that we are assuming the form of the functional dependence of the similarity variable η on the independent variables (here, t and x). In some problems, different functional forms are necessary. For the other similarity solutions that we consider here (in Sections 10.6, 11.2, and 11.4), however, the similarity variable always has this functional form, though the independent variables and associated constants are not always the same, of course. It follows from Eq. (9.6-20) that

$$\frac{\partial}{\partial t} = \frac{d}{d\eta} c_1 c_2 t^{c_2 - 1} x \quad \text{and} \quad \frac{\partial}{\partial x} = \frac{d}{d\eta} c_1 t^{c_2}$$

Thus Eqs. (9.6-18) and (9.6-19) become

$$\frac{1}{c_0} \frac{d\alpha}{d\eta} c_1 c_2 t^{c_2 - 1} x + \alpha \frac{d\alpha}{d\eta} c_1 t^{c_2} + 2\beta \frac{d\beta}{d\eta} c_1 t^{c_2} = 0 \qquad (9.6\text{-}21)$$

$$\frac{2}{c_0} \frac{d\beta}{d\eta} c_1 c_2 t^{c_2 - 1} x + 2\alpha \frac{d\beta}{d\eta} c_1 t^{c_2} + \beta \frac{d\alpha}{d\eta} c_1 t^{c_2} = 0 \qquad (9.6\text{-}22)$$

or

$$\frac{1}{c_0} \frac{d\alpha}{d\eta} c_2 \frac{x}{t} + \alpha \frac{d\alpha}{d\eta} + 2\beta \frac{d\beta}{d\eta} = 0 \qquad (9.6\text{-}23)$$

$$\frac{2}{c_0} \frac{d\beta}{d\eta} c_2 \frac{x}{t} + 2\alpha \frac{d\beta}{d\eta} + \beta \frac{d\alpha}{d\eta} = 0 \qquad (9.6\text{-}24)$$

Because, for a similarity solution, we require that α and β are functions of η only, it is clearly necessary that Eqs. (9.6-23) and (9.6-24) contain no terms involving x and t. Thus it is necessary either that $c_2 = 0$ (which implies a trivial steady solution) or that

$$c_2 = -1 \qquad (9.6\text{-}25)$$

so that

$$\frac{1}{c_0} c_2 \frac{x}{t} = -\frac{1}{c_0 c_1} \eta$$

It is also convenient to make the similarity variable η dimensionless. Accordingly, we choose

$$c_1 = \frac{1}{c_0} \qquad (9.6\text{-}26)$$

so that Eq. (9.6-20) becomes

$$\eta = \frac{x}{c_0 t} \qquad (9.6\text{-}27)$$

Note that η, which is dimensionless, is defined in terms of x and t. The only way in which the independent variable x can be made dimensionless is with respect to the independent variable t (and with c_0, of course), and not with respect to any characteristic length associated with the parameters of the problem. More generally,

A problem generally admits a similarity solution if the only way in which independent variables can be made dimensionless is with respect to one another.

Thus the *partial* differential equations (9.6-13) and (9.6-14) become

$$\frac{d\alpha}{d\eta} (\alpha - \eta) + 2\beta \frac{d\beta}{d\eta} = 0 \qquad (9.6\text{-}28)$$

$$2 \frac{d\beta}{d\eta} (\alpha - \eta) + \beta \frac{d\alpha}{d\eta} = 0 \qquad (9.6\text{-}29)$$

which are *ordinary* differential equations. The reduction from partial to ordinary differential equations occurs because use of a similarity variable reduces the number of independent variables by one, in this case from two (and hence partial differential equations) to one (and hence ordinary differential equations).

Multiplying the terms in Eq. (9.6-28) by $2\, d\beta/d\eta$ and those in Eq. (9.6-29) by $d\alpha/d\eta$ and subtracting one set of terms from the other yields

$$\beta \left[4 \left(\frac{d\beta}{d\eta} \right)^2 - \left(\frac{d\alpha}{d\eta} \right)^2 \right] = 0 \qquad (9.6\text{-}30)$$

Because $\beta \neq 0$ (except at the moving liquid front), it follows that

$$\frac{d\beta}{d\eta} = \pm \frac{1}{2} \frac{d\alpha}{d\eta} \qquad (9.6\text{-}31)$$

Because we expect that $\partial u_x/\partial x > 0$ and $\partial c/\partial x < 0$ in the moving liquid, it follows that $d\alpha/d\eta > 0$ and $d\beta/d\eta < 0$ and so

$$\frac{d\beta}{d\eta} = -\frac{1}{2} \frac{d\alpha}{d\eta} \qquad (9.6\text{-}32)$$

Thus Eq. (9.6-28) yields

$$\frac{d\alpha}{d\eta} (\alpha - \eta - \beta) = 0 \qquad (9.6\text{-}33)$$

Because $d\alpha/d\eta \neq 0$ in the moving liquid, it follows that

$$\alpha - \eta - \beta = 0 \qquad (9.6\text{-}34)$$

and hence that

$$\frac{d\alpha}{d\eta} - 1 - \frac{d\beta}{d\eta} = 0 \tag{9.6-35}$$

Combining Eqs. (9.6-32) and (9.6-35) yields

$$\frac{d\alpha}{d\eta} = \frac{2}{3} \qquad \frac{d\beta}{d\eta} = -\frac{1}{3} \tag{9.6-36}$$

Thus,
$$\alpha = c_3 + \tfrac{2}{3}\eta \qquad \beta = c_3 - \tfrac{1}{3}\eta \tag{9.6-37}$$

and hence
$$u_x = c_3 c_0 + \frac{2x}{3t} \qquad c = c_3 c_0 - \frac{x}{3t} \tag{9.6-38}$$

where c_3 is a constant yet to be determined.

It follows from Eqs. (9.6-13) and (9.6-14) that

$$\frac{\partial}{\partial t}(u_x \pm 2c) + (u_x \pm c)\frac{\partial}{\partial x}(u_x \pm 2c) = 0 \tag{9.6-39}$$

Hence the quantities $(u_x + 2c)$ and $(u_x - 2c)$, which are called the *Riemann invariants,* are constant for points moving with (horizontal) velocities $(u_x + c)$ and $(u_x - c)$, respectively, that is, they are constant along trajectories which are called *characteristics* given by $dx/dt = u_x + c$ and $dx/dt = u_x - c$, respectively. To see this more clearly, consider a flow variable f which is a function of horizontal position x and time t. Then

$$df = \frac{\partial f}{\partial t}dt + \frac{\partial f}{\partial x}dx \tag{9.6-40}$$

Suppose that f is constant along a trajectory given by $x = \int F\,dt$, that is, along a trajectory given by $dx/dt = F$. Then, along this trajectory, df vanishes and so

$$\frac{\partial f}{\partial t}dt + \frac{\partial f}{\partial x}dx = \left(\frac{\partial f}{\partial t} + F\frac{\partial f}{\partial x}\right)dt = 0 \tag{9.6-41}$$

that is,
$$\frac{\partial f}{\partial t} + F\frac{\partial f}{\partial x} = 0 \tag{9.6-42}$$

Conversely, it follows that, if Eq. (9.6-42) holds, f is constant along a trajectory given by $dx/dt = F$. Thus, for example, if the substantial derivative $Df/Dt = \partial f/\partial t + \mathbf{u} \cdot \nabla f = 0$, it follows that f is constant along a trajectory given by $dx/dt = \mathbf{u}$, where $\mathbf{x}$ and $\mathbf{u}$ denote position and velocity, respectively (i.e., f is constant for each material point X in the flow). By identifying f with $(u_x \pm 2c)$ and F with $(u_x \pm c)$, we thus see that it follows from Eq. (9.6-39) that $(u_x \pm 2c)$ is constant along a trajectory given by $dx/dt = u_x \pm c$. We now deduce, using Eqs. (9.6-38), (1) that the Riemann invariant $3c_3c_0$ is constant along the characteristic

$$\frac{dx_f}{dt} = 2c_3c_0 + \frac{x_f}{3t} \qquad (9.6\text{-}43)$$

which we call the forward characteristic, hence the subscript f (of course, $3c_3c_0$ is in fact constant everywhere, not just along the forward characteristic; this Riemann invariant is of a particularly simple form); and (2) that the Riemann invariant $(4x/3t) - c_3c_0$ is constant along the characteristic

$$\frac{dx_b}{dt} = \frac{x_b}{t} \qquad (9.6\text{-}44)$$

which we call the backward characteristic, hence the subscript b.

It follows from initial conditions (9.6-15) and (9.6-16) that $c = c_0$ and $u_x = 0$ at the space-time origin $(x = 0, t = 0)$. Hence,

1. On the forward characteristic through $(x = 0, t = 0)$, $(u_x + 2c) = 2c_0$. So, using Eqs. (9.6-38),

$$c_3 = \tfrac{2}{3} \qquad (9.6\text{-}45)$$

and $$u_x = \frac{2}{3}\left(c_0 + \frac{x}{t}\right) \qquad c = \frac{1}{3}\left(2c_0 - \frac{x}{t}\right) \qquad (9.6\text{-}46)$$

We see that

$$u_x = \tfrac{2}{3}c_0 \qquad c = \tfrac{2}{3}c_0 \qquad \text{at } x = 0 \text{ for all } t > 0 \qquad (9.6\text{-}47)$$

which means that the height h of liquid at the initial position of the dam $x = 0$ is always $\tfrac{4}{9}h_0$ for $t > 0$. It follows from Eqs. (9.6-43) and (9.6-45) that

$$\frac{dx_f}{dt} = \frac{4}{3}c_0 + \frac{x_f}{3t} \qquad (9.6\text{-}48)$$

which may be integrated to yield

$$x_f = c_4 t^{1/3} + 2c_0 t \qquad (9.6\text{-}49)$$

where c_4 is a constant yet to be determined. Substitution of equation (9.6-49) into Eqs. (9.6-46) yields

$$u_x = 2c_0 + \tfrac{2}{3}c_4 t^{-2/3} \qquad c = -\tfrac{1}{3}c_4 t^{-2/3} \qquad (9.6\text{-}50)$$

from which we deduce that

$$c_4 = 0 \qquad (9.6\text{-}51)$$

for otherwise u_x and c would become unbounded as $t \to 0$. Thus Eq. (9.6-49) yields

$$x_f = 2c_0 t \qquad (9.6\text{-}52)$$

and Eqs. (9.6-46) yield

$$u_x|_{x=x_f} = 2c_0 \qquad c|_{x=x_f} = 0 \qquad (9.6\text{-}53)$$

Because $h = 0$ and, hence, $c = 0$ at the moving liquid front, it follows from Eqs. (9.6-53) that the forward characteristic through $(x = 0, t = 0)$ is the trajectory of the moving liquid front and, moreover, that the front moves at a constant velocity $2c_0$.

2. On the backward characteristic through $(x = 0, t = 0)$, $u_x - 2c = -2c_0$. So, using Eqs. (9.6-38) and (9.6-45),

$$x_b = -c_0 t \qquad (9.6\text{-}54)$$

and

$$u_x|_{x=x_b} = 0 \qquad c|_{x=x_b} = c_0 \qquad (9.6\text{-}55)$$

Because $h = h_0$ and so $c = c_0$ initially (for time $t \le 0$) and also subsequently (for $t > 0$) if the liquid is stationary (i.e., it is not yet moving), it follows from Eqs. (9.6-55) that the backward characteristic through $(x = 0, t = 0)$ is the trajectory of a boundary between a region $x < x_b$ of stationary liquid and a region $x > x_b$ of moving liquid. Thus, the forward and backward characteristics through $(x = 0, t = 0)$ define a region of space-time in which the effect of the rupture of the dam is felt inside but not outside (see Fig. 9.9).

The forward characteristic through $(x = 0, t = 0)$ given by Eq. (9.6-52) is,

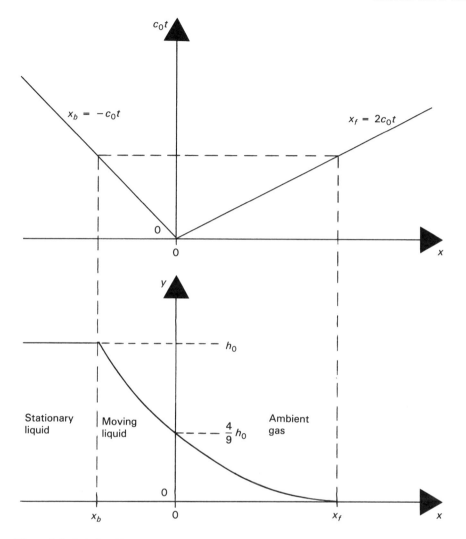

Figure 9.9. Relationship between characteristics and flow regions after rupture of a dam.

in fact, just one of an infinite family of nonintersecting forward characteristics given by Eqs. (9.6-49) where, for those forward characteristics within the moving liquid, the constant c_4 satisfies the inequality

$$c_4 \leq 0 \tag{9.6-56}$$

Similarly, the backward characteristic through $(x = 0, t = 0)$ given by Eq. (9.6-54) is, in fact, just one of an infinite family of backward characteristics, all of which pass through the space-time origin $(x = 0, t = 0)$, given from Eq. (9.6-44) by

$$x_b = c_5 t \tag{9.6-57}$$

where, for those characteristics within the moving liquid, the constant c_5 satisfies the inequality

$$-c_0 \le c_5 \le 2c_0 \tag{9.6-58}$$

The two families of characteristics are shown schematically in Fig. 9.10.

In our analysis, we assume that viscous and interfacial tension effects are negligible compared with inertial and gravitational effects in the liquid. We note first that it follows from Eqs. (9.6-46) that $u_x \sim c_0$. Thus the characteristic velocity of the liquid u_c may justifiably be taken to be $c_0 = \sqrt{gh_0}$. It also follows from all the preceding equations [such as Eqs. (9.6-46)] that $c_0 t$ is a characteristic horizontal length; h_0 is a characteristic vertical length. If, for the sake of argument, we now identify the characteristic length l_c with h_0, then it follows that the Froude number [see Eq. (7.2-8)] is given by

$$\text{Fr} = \frac{c_0^2}{gh_0} \tag{9.6-59}$$

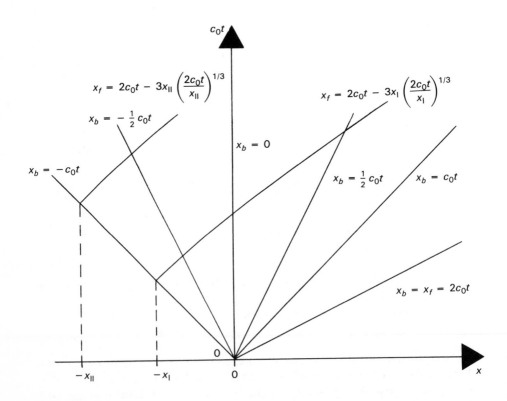

Figure 9.10. Families of characteristics after rupture of a dam.

and so

$$Fr = 1 \tag{9.6-60}$$

(i.e., inertial and gravitational effects balance, as is obvious physically); the Reynolds and Weber numbers [see Eqs. (7.2-9) and (7.2-13)] are given by

$$Re = \frac{\rho g^{1/2} h_0^{3/2}}{\mu} \qquad We = \frac{\rho g h_0^2}{\sigma} \tag{9.6-61}$$

so that viscous and interfacial tension effects are negligible if $Re \gg 1$ and $We \gg 1$, respectively. If in fact these effects are not negligible, their main consequence is to slow the flow down and to blunt its leading edge.

We conclude by extending our analysis to the case when there is a vertical, planar wall of infinite length parallel to, and at a distance x_w from, the initial position of the dam (see Fig. 9.11). We assume that the wall is so high that *no* flow occurs over its top. Because the liquid front moves with velocity $2c_0$, it follows that the wall has no effect on the flow for $t < x_w/2c_0$. For $t \geq x_w/2c_0$, liquid starts to accumulate at the wall and, as a result, the liquid level rises there, forming a bore or a hydraulic jump. At the leading edge of the bore, at a distance x_s from the initial position of the dam, there is an almost discontinuous change in liquid height called a shock, by analogy with the shock wave which forms when a supersonic gas flow decelerates to become subsonic. With relatively small error, we may assume that there is in fact a discontinuous change in liquid height at the shock. Let u_x^- and h^- denote the horizontal velocity and height of liquid entering the shock, u_x^+ and h^+ the horizontal velocity and height of liquid leaving it and U the horizontal velocity of the shock itself, so that $U = dx_s/dt$ (see Fig. 9.12). (Note that we expect that the shock forms initially at the wall and then moves away from it; thus we expect that $U < 0$.) Then, relative to a reference frame moving with the shock, that is, one moving with a velocity U in the x direction, conservation of mass means that liquid must flow into the shock at the same rate at which it leaves it (otherwise mass would have to be created or destroyed within the shock). Thus,

$$\rho(u_x^- - U)h^- = \rho(u_x^+ - U)h^+ \tag{9.6-62}$$

Similarly, conservation of linear momentum means that the difference between the rate at which horizontal linear momentum flows out of the shock and the rate at which it enters it is just the force due to pressure acting horizontally on a vertical surface through the shock. Thus, using Eq. (9.6-7),

$$\rho(u_x^+ - U)^2 h^+ - \rho(u_x^- - U)^2 h^- = \int_0^{h^-} [p_a + \rho g(h^- - y)]dy$$

$$+ p_a(h^+ - h^-) - \int_0^{h^+} [(p_a + \rho g(h^+ - y)]dy \tag{9.6-63}$$

that is,

$$\rho(u_x^+ - U)^2 h^+ - \rho(u_x^- - U)^2 h^- = \tfrac{1}{2} \rho g(h^{-2} - h^{+2}) \tag{9.6-64}$$

Equations (9.6-62) and (9.6-64) together yield

$$U = u_x^- - \sqrt{\frac{1}{2} gh^+ \left(1 + \frac{h^+}{h^-} \right)} \qquad (9.6\text{-}65)$$

and

$$u_x^+ = u_x^- - \left(1 - \frac{h^-}{h^+} \right) \sqrt{\frac{1}{2} gh^+ \left(1 + \frac{h^+}{h^-} \right)} \qquad (9.6\text{-}66)$$

We now determine the rate E' at which mechanical (i.e., pressure, potential, and

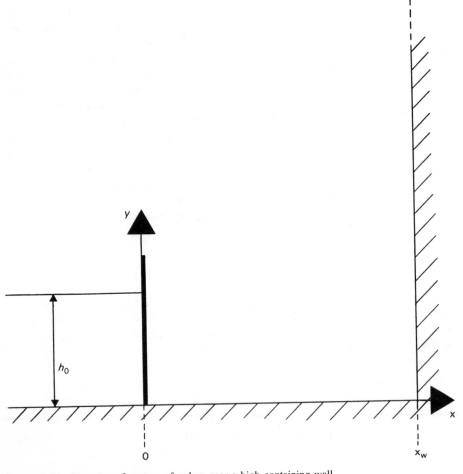

Figure 9.11. Geometry of rupture of a dam near a high containing wall.

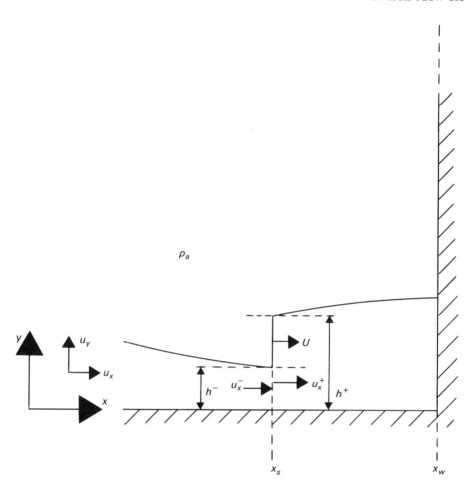

Figure 9.12. Geometry of flow in the vicinity of a shock.

kinetic) energy is dissipated in the shock, per unit width of the shock. Relative to the shock,

- Pressure energy flows at a rate given by $\int_0^{h^\S} [p_a + \rho g(h^\S - y)](u_x^\S - U)dy$.
- Potential energy flows at a rate given by $\frac{1}{2} \rho g h^\S (u_x^\S - U)h^\S$ (since the average potential energy per unit volume is $\frac{1}{2} \rho g h^\S$).
- Kinetic energy flows at a rate given by $\frac{1}{2} \rho (u_x^\S - U)^2 (u_x^\S - U)h^\S$. Here $u_x^\S$ and $h^\S$ denote u_x^- or u_x^+ and h^- or h^+, respectively.

Thus,

$$E' = \int_0^{h^-} [p_a + \rho g(h^- - y)](u_x^- - U)dy$$
$$+ \tfrac{1}{2}\rho g h^-(u_x^- - U)h^- + \tfrac{1}{2}\rho(u_x^- - U)^2(u_x^- - U)h^-$$
$$- \int_0^{h^+} [p_a + \rho g(h^+ - y)](u_x^+ - U)dy$$
$$- \tfrac{1}{2}\rho g h^+(u_x^+ - U)h^+ - \tfrac{1}{2}\rho(u_x^+ - U)^2(u_x^+ - U)h^+ \qquad (9.6\text{-}67)$$

whence, using Eqs. (9.6-62) and (9.6-64);

$$E' = \rho g h^- \frac{(u_x^- - U)(h^+ - h^-)^3}{4h^+ h^-} \qquad (9.6\text{-}68)$$

It follows from Eqs. (9.6-65) that $u_x^- - U > 0$ so $E' > 0$ if $h^+ > h^-$, and vice versa. Physically, mechanical energy cannot be generated in the shock, though it can be dissipated. Excluding the trivial case $E' = 0$, which implies that $h^+ = h^-$, we see that the shock must be such that $E' > 0$ and hence $h^+ > h^-$. Thus, the shock must involve liquid flowing from a shallower (and hence a faster-moving) region to a deeper (and hence a slower-moving) region, which is precisely the circumstance in which we earlier asserted that the bore forms.

The way in which mechanical energy is dissipated is found experimentally to depend on the strength of the shock. Some energy is lost by the formation of waves beyond the shock (i.e., for $x > x_s$). Indeed, most is lost in this way for weak shocks (i.e., for shocks for which h^+/h^- is not too large). For stronger shocks, most energy is lost by turbulence at the shock, which can cause significant entrainment of the gas initially above the liquid, and hence foaming, at the shock.

TEN

(NEARLY) NONACCELERATING FLOW

10.1 STOKES' EQUATIONS

The presence of the convection term $\rho\mathbf{u}\cdot\nabla\mathbf{u}$ in the Navier-Stokes equations

$$\rho\frac{\partial\mathbf{u}}{\partial t} + \rho\mathbf{u}\cdot\nabla\mathbf{u} = \rho\mathbf{g} - \nabla p + \mu\Delta\mathbf{u} \qquad (10.1\text{-}1)$$

makes them nonlinear and hence very difficult to solve as a rule. It happens, however, that in some common flows $|\rho\mathbf{u}\cdot\nabla\mathbf{u}|$ either vanishes or is at most small compared with $|\rho\mathbf{g} - \nabla p + \mu\Delta\mathbf{u}|$, even though the Reynolds number is not small; and, moreover, that the flow is steady, or nearly steady, so that $|\rho\,\partial\mathbf{u}/\partial t|$ either vanishes or is not significantly larger, and is often much smaller, than $|\rho\mathbf{u}\cdot\nabla\mathbf{u}|$.

We recall from Section 7.3 that the transient term $\rho\,\partial\mathbf{u}/\partial t$ and the convection term $\rho\mathbf{u}\cdot\nabla\mathbf{u}$ together comprise the inertial terms in the Navier-Stokes equations. Indeed, they are also called the linear and nonlinear inertial terms, respectively. A flow in which the inertial terms either vanish or are negligible compared with the gravitational, pressure, and viscous terms is a flow in which there is no, or at most very little, acceleration. We call such a flow nonaccelerating if the inertial terms vanish and nearly nonaccelerating if they are negligible. In general, we call it a (nearly) nonaccelerating flow.

In a (nearly) nonaccelerating flow of an incompressible Newtonian fluid, mass conservation means that

$$\nabla\cdot\mathbf{u} = 0 \qquad (10.1\text{-}2)$$

and linear momentum conservation means that the Navier-Stokes equations (10.1-1) reduce to Stokes' equations:

$$-\nabla p + \mu\,\Delta\mathbf{u} = \mathbf{0} \qquad\qquad (10.1\text{-}3)$$

where gravitational effects, if any, are incorporated in the pressure term [so that p should strictly be replaced by $\bar{p}$; see Eq. (4.3-13)]. We note, however, that Eq. (10.1-1) reduces to Eq. (10.1-3) for a (nearly) nonaccelerating flow only if the Reynolds number is not so large that the flow becomes turbulent. Turbulence is, as we will see in Chapter 12, an advanced manifestation of instability to disturbances in a flow.

All (nearly) nonaccelerating flows are unstable if the Reynolds number, or some other essentially equivalent dimensionless number, becomes sufficiently large.

There is, therefore, always a limit to the applicability of the analyses that follow in this chapter.

10.2 PRESSURE–INDUCED FLOW IN A PIPE

We consider the flow of an incompressible Newtonian fluid of density ρ and viscosity μ induced by a constant positive pressure difference or pressure drop $(p_0 - p_L)$ in a pipe of length L and radius $R \ll L$. (By a pipe, we always mean a right circular cylindrical duct, that is, a duct with a circular cross section normal to its axis or generator.) Because of the obvious geometry, we use cylindrical polar coordinates (r, θ, z) to analyze the problem, with origin on the centerline of the pipe entrance and z direction aligned with the centerline (see Fig. 10.1).

- Because of the obvious symmetry, we expect that the flow is (1) *swirl-free,* so that the only nonzero components of the velocity field $\mathbf{u}$ are the radial component u_r and the axial component u_z (i.e., we expect that the angular com-

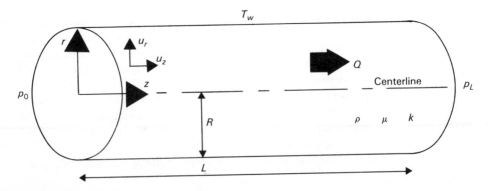

Figure 10.1. Geometry of flow in a pipe.

ponent u_θ vanishes); and (2) *axisymmetric*, so that u_z and u_r are independent of θ, as is the pressure field p.

- Because the pressure drop $(p_0 - p_L)$ is constant, we expect that the flow is *steady*, so we expect that the transient or linear inertial term $\rho\, \partial \mathbf{u}/\partial t$ in the Navier-Stokes equations (10.1-1) vanishes.
- Because the pipe is long, we expect that the flow is *fully developed*, that is, we expect that the velocity field $\mathbf{u}$ is independent of axial position z everywhere [except near the entrance $(z = 0)$ and exit $(z = L)$ of the pipe].

Thus we expect, in particular, that u_z is independent of z, that is,

$$\frac{\partial u_z}{\partial z} = 0 \qquad (10.2\text{-}1)$$

Because the flow is swirl-free, the mass conservation Eq. (A2.2-3) of Appendix B yields

$$\frac{1}{r}\frac{\partial}{\partial r}(ru_r) + \frac{\partial u_z}{\partial z} = 0 \qquad (10.2\text{-}2)$$

whence, using Eq. (10.2-1),

$$\frac{1}{r}\frac{\partial}{\partial r}(ru_r) = 0 \qquad (10.2\text{-}3)$$

This may be integrated to yield

$$ru_r = 0 \qquad (10.2\text{-}4)$$

since no flow-through (see Section 6.2) at the pipe wall $(r = R)$ means that u_r vanishes there. Hence,

$$u_r = 0 \qquad (10.2\text{-}5)$$

It then follows that

$$\mathbf{u} = u_z(r)\mathbf{i}_z \qquad (10.2\text{-}6)$$

that is, $\mathbf{u}$ has only one nonzero component which varies only in one direction normal to that component. As a result, the flow is a *shear flow* with *shear rate* γ given by

$$\gamma = \left|\frac{du_z}{dr}\right| \qquad (10.2\text{-}7)$$

[see Eqs. (5.4-32) and (5.4-33)]. Also, using the expression for $\mathbf{u}\cdot\nabla\mathbf{u}$ given from those parts of the left-hand sides of Eqs. (A2.3-3) of Appendix B that do not involve a time derivative, we deduce that the convection or nonlinear inertial term $\rho\mathbf{u}\cdot\nabla\mathbf{u}$ in the Navier-Stokes equations (10.1-1) vanishes. The reason for this is that the velocity of each material point is constant since the only variation in velocity is in a direction normal to the direction of motion. Because both the linear and the nonlinear inertial terms vanish, it follows that the flow in the pipe is a nonaccelerating flow. In fact, this is strictly true everywhere except near $z = 0$ and $z = L$ where entrance and exit effects, respectively, are important. These effects are not, however, significant to the overall flow if the pipe is long enough.

Because the flow is nonaccelerating, the Navier-Stokes equations (10.1-1) reduce to Stokes' equations (10.1-3). We have already used the mass conservation equation (10.1-2) to obtain Eq. (10.2-6). In order to determine the functional dependence of the axial component of velocity u_z on radial position r, we must use linear momentum conservation, that is, we must solve Eq. (10.1-3). To do this, we use Eqs. (A2.3-3) of Appendix B and omit the transient and convection terms (i.e., we set their left-hand sides to zero). We also incorporate the gravitational term in the pressure term [so that, instead of p, we should strictly use $\bar{p}$; see Eq. (4.3-13)]. Then, using Eq. (10.2-6), we obtain

$$-\frac{\partial p}{\partial r} = 0 \qquad (10.2\text{-}8)$$

$$-\frac{\partial p}{\partial z} + \mu \frac{1}{r}\frac{d}{dr}\left(r\frac{du_z}{dr}\right) = 0 \qquad (10.2\text{-}9)$$

Because p is independent of r by Eq. (10.2-8) and of θ by axisymmetry, it follows that p is a function of z alone. Hence, since u_z is independent of z, it follows from Eq. (10.2-9) that $\partial p/\partial z$, which may now write as dp/dz, is a constant. Because there is a pressure drop $(p_0 - p_L)$ over the length L of pipe, it thus follows that Eq. (10.2-9) becomes

$$\frac{1}{r}\frac{d}{dr}\left(r\frac{du_z}{dr}\right) = -\frac{p_0 - p_L}{\mu L} \qquad (10.2\text{-}10)$$

The boundary conditions on u_z are no slip at the pipe wall, that is,

$$u_z = 0 \qquad \text{at } r = R \qquad (10.2\text{-}11)$$

and symmetry about the centerline of the pipe:

$$\frac{du_z}{dr} = 0 \qquad \text{at } r = 0 \qquad (10.2\text{-}12)$$

(see Section 6.5). We note, incidentally, that we could replace this condition by the condition that u_z is finite at the centerline (i.e., $|u_z| < \infty$ at $r = 0$); this would not alter the analysis that follows. We can integrate Eq. (10.2-10) with respect to r to yield

$$r\frac{du_z}{dr} = -\frac{p_0 - p_L}{\mu L}\frac{r^2}{2} + c_1 \qquad (10.2\text{-}13)$$

where c_1 is a constant. Boundary condition (10.2-12) implies that

$$c_1 = 0 \qquad (10.2\text{-}14)$$

Thus, integrating Eq. (10.2-13) with respect to r, we obtain

$$u_z = -\frac{p_0 - p_L}{\mu L} \frac{r^2}{4} + c_2 \tag{10.2-15}$$

where c_2 is a constant. Boundary condition (10.2-11) implies that

$$c_2 = \frac{p_0 - p_L}{\mu L} \frac{R^2}{4} \tag{10.2-16}$$

so that

$$u_z = \frac{p_0 - p_L}{4 \mu L} (R^2 - r^2) \tag{10.2-17}$$

It follows from Eq. (10.2-17) that the axial velocity profile is parabolic (see Fig. 10.2). The maximum axial velocity $u_{z_{max}}$ occurs at the centerline and is given by

$$u_{z_{max}} = \frac{p_0 - p_L}{4 \mu L} R^2 \tag{10.2-18}$$

whereas the mean axial velocity $\bar{u}_z$ is given by

$$\bar{u}_z = \frac{1}{\pi R^2} \int_0^R u_z 2 \pi r \, dr \tag{10.2-19}$$

whence

$$\bar{u}_z = \frac{p_0 - p_L}{8 \mu L} R^2 \tag{10.2-20}$$

from which we see that $\bar{u}_z$ is just $\frac{1}{2} u_{z_{max}}$. The *volumetric flow rate Q* through the pipe is given by

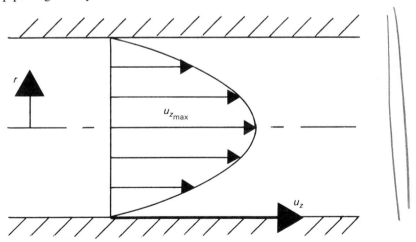

Figure 10.2. Axial velocity profile for flow in a pipe of a Newtonian fluid.

$$Q = \bar{u}_z \pi R^2 \qquad (10.2\text{-}21)$$

whence
$$Q = \frac{p_0 - p_L}{8\mu L} \pi R^4 \qquad (10.2\text{-}22)$$

This is the *Hagen-Poiseuille equation*. Experimentally, it is found to be corroborated provided that (1) the Reynolds number given by

$$\text{Re} = \frac{\rho \bar{u}_z 2R}{\mu} \qquad (10.2\text{-}23)$$

is less than some critical value $\text{Re}_c \simeq 2000$, so that the flow is not turbulent
(though it should be noted that Re_c appears to be very much larger than 2000 if
particular care is taken to minimize disturbances that might cause instabilities);
and (2) $L/R \gg c\,\text{Re}$ where $c \simeq 0.1$ (see Section 11.3), so that the pipe is long
enough for entrance and exit effects to be negligible and hence for **u** to be fully
developed.

We now show how, by use of the energy conservation equation (7.1-3), we
can determine the temperature field T in the fluid in the pipe. We suppose that
the pipe is so long that T is, like **u**, fully developed. Experimentally, this is found
to be the case if $L/R \gg c\,\text{Pe}$ where $c \simeq 0.1$ and the Péclet number [see Eq.
(7.2-10)] is given by

$$\text{Pe} = \frac{\rho c \bar{u}_z 2R}{k} \qquad (10.2\text{-}24)$$

Here c denotes the specific heat of the fluid and k its thermal conductivity. We
also assume that T is steady and axisymmetric so that

$$T = T(r) \qquad (10.2\text{-}25)$$

Then Eq. (7.1-3) becomes

$$k\nabla^2 T + \tfrac{1}{2}\mu\mathbf{e}:\mathbf{e} = 0 \qquad (10.2\text{-}26)$$

where **e** denotes the rate of strain. Using Eqs. (10.2-6) and (10.2-25) and Eq.
(A2.5-3) of Appendix B, Eq. (10.2-26) yields

$$\frac{1}{r}\frac{d}{dr}\left(r\frac{dT}{dr}\right) = -\frac{\mu}{k}\left(\frac{du_z}{dr}\right)^2 \qquad (10.2\text{-}27)$$

We suppose that the pipe wall temperature T_w is specified, that is,

$$T = T_w \qquad \text{at } r = R \qquad (10.2\text{-}28)$$

while symmetry about the centerline means that

$$\frac{dT}{dr} = 0 \qquad \text{at } r = 0 \qquad (10.2\text{-}29)$$

We can integrate Eq. (10.2-27) twice with respect to r. Imposition of the boundary conditions (10.2-28) and (10.2-29) then yields

$$T = T_w + \frac{\mu}{k}\left(\frac{p_0 - p_L}{8\mu L}\right)^2 (R^4 - r^4) \qquad (10.2\text{-}30)$$

or, from Eq. (10.2-20),

$$T = T_w + \frac{\mu \bar{u}_z^2}{k}\left(1 - \frac{r^4}{R^4}\right) \qquad (10.2\text{-}31)$$

Clearly, $T > T_w$ for $0 \le r < R$ (and $T = T_w$ for $r = R$). From this, we deduce that heat generation by viscous dissipation raises the temperature everywhere within the fluid. Note, incidentally, that if we define a dimensionless temperature θ as

$$\theta = \frac{T}{T_w} \qquad (10.2\text{-}32)$$

then Eq. (10.2-31) becomes

$$\theta = 1 + \text{Br}\,(1 - \sigma^4) \qquad (10.2\text{-}33)$$

where the dimensionless radial coordinate σ is given by

$$\sigma = \frac{r}{R} \qquad (10.2\text{-}34)$$

and the Brinkman number Br [see Eq. (7.2-11)] is given by

$$\text{Br} = \frac{\mu \bar{u}_z^2}{k T_w} \qquad (10.2\text{-}35)$$

We now extend our analysis to the flow of non-Newtonian fluid. In particular, we consider the flow of a *power-law* fluid, the viscosity μ of which varies algebraically with shear-rate γ [see Eq. (5.4-34)]. For fully developed flow in a pipe, γ is given by Eq. (10.2-7). Hence μ is given by

$$\mu = \mu_0\left(-\frac{du_z}{dr}\right)^{(1/m)-1} \qquad (10.2\text{-}36)$$

where μ_0 denotes the consistency and m the shear-rate exponent. Of course, m is related to n in Eq. (5.4-34): $m = 1/(1 - 2n)$; we use m rather than n for later convenience. For a *Newtonian* fluid, $m = 1$. Note that $|du_z/dr| = (-du_z/dr)$. This is because u_z is a minimum (it vanishes) at the pipe wall by the no-slip condition and a maximum at the centerline of the pipe by symmetry. Thus u_z decreases monotonically with increasing r. The analysis up to and including Eq. (10.2-8) is unaltered by changing the constitutive equation of the fluid. Setting to zero the left-hand sides of equations (A2.4-3) of Appendix B and using Eq. (10.2-6), we obtain

$$-\frac{\partial p}{\partial z} + \frac{1}{r}\frac{d}{dr}\left(\mu r \frac{du_z}{dr}\right) = 0 \qquad (10.2\text{-}37)$$

Thus Eq. (10.2-10) is replaced by

$$\frac{1}{r}\frac{d}{dr}\left[r\left(-\frac{du_z}{dr}\right)^{1/m}\right] = \frac{p_0 - p_L}{\mu_0 L} \tag{10.2-38}$$

The boundary conditions on u_z are still given by Eqs. (10.2-11) and (10.2-12). Integration of Eq. (10.2-38) twice with respect to r and imposition of the boundary conditions yields

$$u_z = \left(\frac{p_0 - p_L}{2\mu_0 L}\right)^m \frac{R^{m+1} - r^{m+1}}{m+1} \tag{10.2-39}$$

It follows from Eq. (10.2-39) that the axial velocity profile is sharper than parabolic when $m < 1$ and blunter than parabolic when (as is more usually the case in practice) $m > 1$; see Fig. 10.3. The volumetric flow rate Q through the pipe is given from Eqs. (10.2-19) and (10.2-21) by

$$Q = \left(\frac{p_0 - p_L}{2\mu_0 L}\right)^m \frac{\pi R^{m+3}}{(m+3)} \tag{10.2-40}$$

which is the analogue of the Hagen-Poiseuille equation (10.2-22) for power-law fluids.

We conclude by extending our analysis to the flow in a long pipe of a power-law fluid, the consistency μ_0 of which is not constant but varies with temperature T. For most liquids, adequate agreement with experimental results is often achieved by assuming that the consistency decreases exponentially with increasing temperature

$$\mu_0 = \mu_1 e^{-\zeta T} \tag{10.2-41}$$

where μ_1 denotes the consistency at zero temperature and ζ the temperature exponent. If, as before, we assume that the pipe is long enough for the velocity field $\mathbf{u}$ and temperature field T to be fully developed so that Eqs. (10.2-6) and

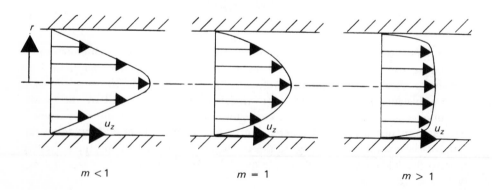

$m < 1$ $m = 1$ $m > 1$

Figure 10.3. Axial velocity profiles for flow in a pipe of power-law fluids with different shear-rate exponents.

(10.2-25) hold, the axial linear momentum conservation equation (10.2-10) is replaced by

$$\frac{1}{r}\frac{d}{dr}\left(re^{-\zeta T}\left(-\frac{du_z}{dr}\right)^{1/m}\right) = \frac{p_0 - p_L}{\mu_1 L} \tag{10.2-42}$$

and the energy conservation equation (10.2-27) is replaced by

$$\frac{1}{r}\frac{d}{dr}\left(r\frac{dT}{dr}\right) = -\frac{\mu_1}{k}e^{-\zeta T}\left(-\frac{du_z}{dr}\right)^{1+1/m} \tag{10.2-43}$$

The boundary conditions on u_z and T are given by Eqs. (10.2-11), (10.2-12), (10.2-28), and (10.2-29). The method of obtaining the solution of Eqs. (10.2-42) and (10.2-43) subject to these boundary conditions is too lengthy, if nevertheless straightforward, to give here. It is, however, easy to verify that the solution, which may most be conveniently be given in dimensionless form, is

$$Y = \frac{Na}{\Pi}\int_\sigma^1 \sigma^m\left[\frac{2/\Pi}{(2/\Pi) - 1 + \sigma^{m+3}}\right]^2 d\sigma \tag{10.2-44}$$

$$\theta = 2\ln\frac{2/\Pi}{2/\Pi - 1 + \sigma^{m+3}} \tag{10.2-45}$$

where the dimensionless axial velocity Y and temperature θ are given by

$$Y = \frac{u_z \pi R^2}{(m+3)Q} \qquad \theta = m\zeta(T - T_w) \tag{10.2-46}$$

The dimensionless radial coordinate σ is given by Eq. (10.2-34) and the dimensionless power Π is given by

$$\Pi = \frac{m\zeta Q(p_0 - p_L)}{(m+3)2\pi kL} \tag{10.2-47}$$

Note that, for an incompressible fluid, the power required to maintain a volumetric flow rate Q when there is a pressure drop $(p_0 - p_L)$ is given by $Q(p_0 - p_L)$. The *Nahme number* Na is given by

$$Na = \frac{m\zeta\mu_1 R^2 e^{m\zeta T_w}}{(m+3)^2 k}\left[\frac{R(p_0 - p_L)}{2\mu_1 L}\right]^{m+1} \tag{10.2-48}$$

The Nahme number may be interpreted as the ratio of the temperature rise as a result of heat generation by viscous dissipation to the temperature change required significantly to alter the consistency (and hence viscosity). It can be shown that Π is determined by the quadratic equation

$$Na = \Pi - \tfrac{1}{2}\Pi^2 \tag{10.2-49}$$

(see Fig. 10.4). We note immediately that $0 \le \Pi \le 2$ since Na > 0. We also note that there is *no* value of Π corresponding to any Na $> \frac{1}{2}$, *one* value of Π

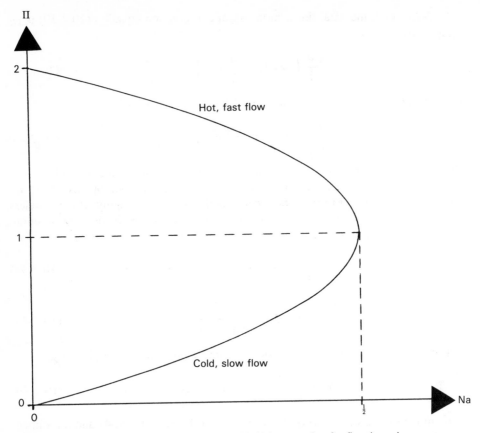

Figure 10.4. Variation of dimensionless power with Nahme number for flow in a pipe.

corresponding to $Na = \frac{1}{2}$ and *two* values of Π corresponding to $Na < \frac{1}{2}$. It then follows that there exists a critical pressure drop $(p_0 - p_L)_c$ given from Eq. (10.2-48) by:

$$(p_0 - p_L)_c = \left[\frac{(m + 3)^2 k}{2m\zeta\mu_1 R^2 e^{m\zeta T_w}} \right]^{1/(m+1)} \frac{2\mu_1 L}{R} \tag{10.2-50}$$

- For a specified pressure drop $(p_0 - p_L)$ greater than $(p_0 - p_L)_c$, there corresponds *no* value of volumetric flow rate Q and it is not clear what would occur in practice, since we may in principle specify any value of $(p_0 - p_L)$: possibly, the flow would not be steady.
- For $(p_0 - p_L)$ equal to $(p_0 - p_L)_c$, there corresponds *one* (i.e., a unique) value of Q.
- For $(p_0 - p_L)$ less than $(p_0 - p_L)_c$, there correspond *two* (i.e., nonunique) values of Q and it is not at all clear which of them would be realized in practice; presumably, the way in which the flow is established would determine the appropriate value of Q (but see also the discussion of instability which follows).

For a specified volumetric flow rate Q, however, there always corresponds one (i.e., a unique) value of pressure drop $(p_0 - p_L)$. That there may, in certain circumstances, be two values of Q corresponding to a specified value of $(p_0 - p_L)$ can be explained as follows,

1. If $\zeta = 0$, so that the consistency is independent of temperature, $(p_0 - p_L)$ increases monotonically as Q increases, in accordance with Eq. (10.2-40) [see curve (i) in Fig. 10.5].
2. If $\zeta > 0$, so that the consistency depends on temperature, $(p_0 - p_L)$ increases and then decreases as Q increases [see curve (ii) in Fig. 10.5]. The reason for this is that $(p_0 - p_L)$ increases with increasing Q, all other variables being held constant, and with increasing consistency μ_0 and hence viscosity μ, again all other variables being held constant; a low value of Q means that little heat is generated by viscous dissipation and so the temperature is low; μ_0 is, therefore, high; similarly, a high value of Q means that μ_0 is low.

Thus, of the two values of Q corresponding to a specified value of $(p_0 - p_L)$

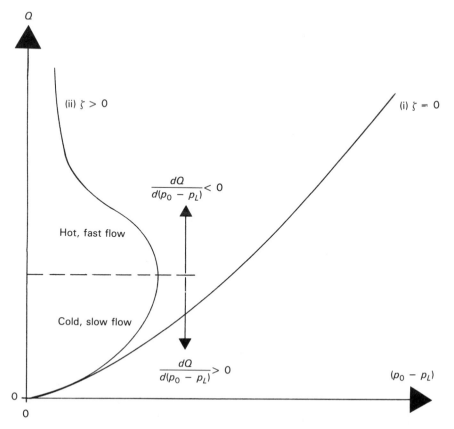

Figure 10.5. Variation of volumetric flow rate with pressure drop for flow in a pipe.

less than $(p_0 - p_L)_c$, one corresponds to a hot, fast flow and the other to a cold, slow flow. We note, however, that for the former,

$$\frac{dQ}{d(p_0 - p_L)} < 0 \qquad (10.2\text{-}51)$$

which, it may be shown, occurs when $d\text{Na}/d\Pi < 0$ (see Fig. 10.4). It is, therefore, possible (but by no means certain) that the hot, fast flow is unstable since, for such a flow, Eq. (10.2-51) implies that the volumetric flow rate decreases as the pressure drop increases, which might be expected to be unrealizable in practice.

To see why this might be the case, we note that if, for a specified pressure drop $(p_0 - p_L)$ in the pipe, the radius R of the pipe is made smaller, all other variables (except the volumetric flow rate Q) being held constant, then Na decreases [see Eq. (10.2-48)]. For a cold, slow flow, a decrease in Na leads to a decrease in Π (see Fig. 10.4) and hence [see Eq. (10.2-47)] a decrease in Q. For a hot, fast flow, in contrast, a decrease in Na leads to an increase in Π and hence an increase in Q; the question then arises: how can such a flow be stopped?

10.3 DRAG–INDUCED FLOW BETWEEN PARALLEL FLAT PLATES

We consider the flow of an incompressible Newtonian fluid of density ρ and viscosity μ between two parallel flat plates a constant distance H apart. The lower plate is of length $L \gg H$ and infinite width; it is fixed (relative to our chosen reference frame). The upper plate is of infinite length and infinite width; it moves at constant speed U in the direction of the length of the lower plate. We suppose that there is no difference in pressure between the ends of the lower plate, that is, the pressure difference $(p_0 - p_L)$ vanishes. Because of the obvious geometry, we use rectangular coordinates (x, y, z) to analyze the problem, with origin on the leading edge of the lower plate. The x direction is aligned with the direction of the length of the lower plate and the y direction is aligned vertically upward (see Fig. 10.6).

Because the upper plate moves at *constant* velocity, we expect that the drag-induced flow between the plates is *steady*, and hence that the linear inertial term $\rho \, \partial \mathbf{u}/\partial t$ in the Navier-Stokes equations (10.1-1) vanishes. Because $L \gg H$, we expect that the flow is *fully developed*, that is, we expect that the velocity field $\mathbf{u}$ is independent of axial position x; we also expect that we may neglect what happens near the ends of the fixed plate (at $x = 0$ and $x = L$). The question of how much larger L has to be than H for this to be the case is addressed in Section 10.4. Because the plates are infinitely wide, we expect that the flow is planar, that is, that $\mathbf{u}$ and the pressure field p are independent of the coordinate z and that the z component of velocity u_z vanishes, so that the only nonzero components of $\mathbf{u}$ are the transverse component u_y and the axial component u_x. Thus, in particular, we expect that u_x is independent of x, that is,

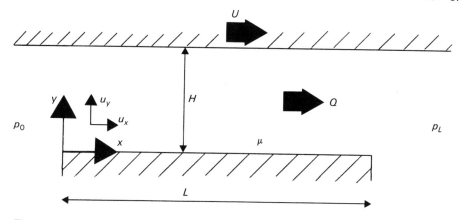

Figure 10.6. Geometry of flow between parallel flat plates.

$$\frac{\partial u_x}{\partial x} = 0 \qquad (10.3\text{-}1)$$

Because $u_z = 0$, Eq. (A2.2-2) of Appendix B yields

$$\frac{\partial u_x}{\partial x} + \frac{\partial u_y}{\partial y} = 0 \qquad (10.3\text{-}2)$$

whence, using Eq. (10.3-1),

$$\frac{\partial u_y}{\partial y} = 0 \qquad (10.3\text{-}3)$$

which may be integrated to yield

$$u_y = 0 \qquad (10.3\text{-}4)$$

since no flow-through at both plates means that u_y vanishes there. It then follows that

$$\mathbf{u} = u_x(y)\mathbf{i}_x \qquad (10.3\text{-}5)$$

(i.e., $\mathbf{u}$ has only one nonzero component which varies only in one direction normal to that component). As a result, the flow is a *shear flow*, just like the flow in a pipe that we analyzed in Section 10.2. Also, using the left-hand sides of Eqs. (A2.3-2) of Appendix B, we deduce that the nonlinear inertial term $\rho\mathbf{u}\cdot\nabla\mathbf{u}$ in the Navier-Stokes equations (10.1-1) vanishes. Because both the linear and nonlinear inertial terms vanish, it follows that the flow between the plates is a nonaccelerating flow (except near $x = 0$ and $x = L$).

The mass conservation equation (A2.2-2) of Appendix B is automatically satisfied by the velocity field given by Eq. (10.3-5) because the latter is derived from the former. The linear momentum conservation equations (A2.3-2) of Appendix B (with the transient and convection terms omitted and the gravitational term incorporated in the pressure term) and Eq. (10.3-5) yield

$$-\frac{\partial p}{\partial y} = 0 \qquad (10.3\text{-}6)$$

$$-\frac{\partial p}{\partial x} + \mu \frac{d^2 u_x}{dy^2} = 0 \qquad (10.3\text{-}7)$$

Because p is independent of y [by Eq. (10.3-6)] and of z since the flow is planar, it follows that p is a function of x alone. Since u_x is independent of x, it follows from Eq. (10.3-7) that $\partial p/\partial x$ is a constant. The pressure p_0 at $x = 0$ is the same as the pressure p_L at $x = L$. Thus it follows that pressure p is a constant and hence that

$$\frac{\partial p}{\partial x} = 0 \qquad (10.3\text{-}8)$$

Hence Eq. (10.3-7) becomes

$$\frac{d^2 u_x}{dy^2} = 0 \qquad (10.3\text{-}9)$$

The boundary conditions on u_x are no slip at the lower and upper plates, that is,

$$u_x = \begin{cases} U & \text{at } y = H \\ 0 & \text{at } y = 0 \end{cases} \qquad (10.3\text{-}10)$$

Equation (10.3-9) may be integrated twice with respect to y to yield, on substitution of the boundary conditions (10.3-10),

$$u_x = \frac{Uy}{H} \qquad (10.3\text{-}11)$$

It follows that the axial velocity profile is linear (see Fig. 10.7). The mean axial velocity $\bar{u}_x$ is given by

$$\bar{u}_x = \frac{1}{H} \int_0^H u_x \, dy \qquad (10.3\text{-}12)$$

whence
$$\bar{u}_x = \tfrac{1}{2} U \qquad (10.3\text{-}13)$$

The *volumetric flow rate per unit width* Q' between the plates is given by

$$Q' = \bar{u}_x H \qquad (10.3\text{-}14)$$

whence
$$Q' = \tfrac{1}{2} U H \qquad (10.3\text{-}15)$$

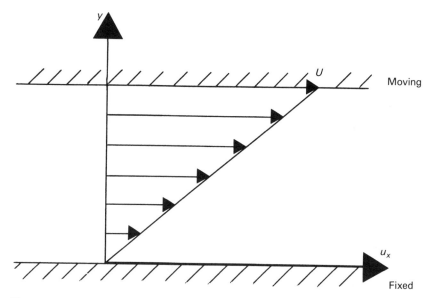

Figure 10.7. Axial velocity profile between parallel flat plates.

10.4 FLOW BETWEEN INCLINED FLAT PLATES

We consider combined pressure-induced and drag-induced flow of an incompressible Newtonian fluid of density ρ and viscosity μ between two flat plates. We consider first flow between parallel plates, which is an extension of the problem discussed in Section 10.3 to the case when there is a nonzero pressure difference $(p_0 - p_L)$ between the ends of the lower plate (see Fig. 10.6). Using precisely the same arguments as in Section 10.3, we immediately deduce that the analysis up to and including Eq. (10.3-7) still holds. Equations (10.3-8) and (10.3-9) do not now hold, but boundary conditions (10.3-10) do. It then follows that

$$
u_x = U\frac{y}{H} - \frac{H^2}{2\mu}\frac{dp}{dx}\frac{y}{H}\left(1 - \frac{y}{H}\right)
$$
$$
\quad\quad\uparrow \quad\quad\quad\quad\quad \uparrow
$$
$$
\quad\quad (a) \quad\quad\quad\quad\quad (b)
$$
$$
\tag{10.4-1}
$$

We see that the axial velocity u_x comprises a drag-induced component (a) and a pressure-induced component (b). The drag-induced component is identical to that given in Eq. (10.3-11). In the pressure-induced component, $dp/dx = -[(p_0 - p_L)/L]$. Because the flow equations and boundary conditions are linear, the axial velocity profile is just the sum of the linear profile arising from component (a) and the parabolic profile arising from component (b) (see Fig. 10.8).

We now consider combined pressure and drag induced flow between two inclined plates (see Fig. 10.9). The lower, fixed plate is a distance H_0 from the

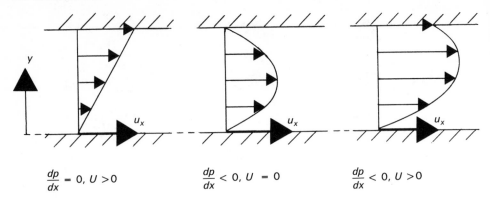

$$\frac{dp}{dx} = 0, U > 0 \qquad\qquad \frac{dp}{dx} < 0, U = 0 \qquad\qquad \frac{dp}{dx} < 0, U > 0$$

Figure 10.8. Axial velocity profiles for different combinations of pressure-induced and drag-induced flow between parallel flat plates.

upper, moving plate at its leading edge ($x = 0$) and a distance H_L at its trailing edge ($x = L$). We suppose that $H_L < H_0 << L$ and that the plates are only slightly inclined with respect to each other, that is,

$$\left| \frac{dH}{dx} \right| = \frac{H_0 - H_L}{L} << 1 \qquad (10.4\text{-}2)$$

As a result, we assume that combined pressure-induced and drag-induced flow between inclined plates is locally the same as it would be between parallel plates, that is, we assume that Eq. (10.4-1) holds *locally*. We justify this assumption by the following order-of-magnitude analysis.

The mass conservation equation (A2.2-2) of Appendix B yields

$$\frac{\partial u_x}{\partial x} + \frac{\partial u_y}{\partial y} = 0 \qquad (10.4\text{-}3)$$

To an order of magnitude (see Section 7.2), $u_x \sim U$, $x \sim L$, and $y \sim H$, so that $\partial/\partial x \sim 1/L$ and $\partial/\partial y \sim 1/H$. Hence it follows from Eq. (10.4-3) that $u_y \sim UH/L$ (i.e., $u_y/u_x << 1$). The axial and transverse linear momentum conservation equations (A2.3-2) of Appendix B yield, respectively,

$$u_x \frac{\partial u_x}{\partial x} + u_y \frac{\partial u_x}{\partial y} = -\frac{1}{\rho} \frac{\partial p}{\partial x} + \frac{\mu}{\rho} \left(\frac{\partial^2 u_x}{\partial x^2} + \frac{\partial^2 u_x}{\partial y^2} \right) \qquad (10.4\text{-}4)$$

and

$$u_x \frac{\partial u_y}{\partial x} + u_y \frac{\partial u_y}{\partial y} = -\frac{1}{\rho} \frac{\partial p}{\partial y} + \frac{\mu}{\rho} \left(\frac{\partial^2 u_y}{\partial x^2} + \frac{\partial^2 u_y}{\partial y^2} \right) \qquad (10.4\text{-}5)$$

We consider first Eq. (10.4-4). To an order of magnitude, the inertial (strictly, the convection) terms

$$u_x \frac{\partial u_x}{\partial x} + u_y \frac{\partial u_x}{\partial y} \sim \frac{U^2}{L}$$

and, since $H \ll L$, the viscous terms

$$\frac{\mu}{\rho} \left(\frac{\partial^2 u_x}{\partial x^2} + \frac{\partial^2 u_x}{\partial y^2} \right) \sim \frac{\mu}{\rho} \left(\frac{U}{L^2} + \frac{U}{H^2} \right) \sim \frac{\mu U}{\rho H^2}$$

Thus, to an order of magnitude, the ratio of the inertial terms to the viscous terms is Re H/L where

$$\text{Re} = \frac{\rho U H}{\mu} \tag{10.4-6}$$

Note that Re is a local Reynolds number since it depends on the local quantity H. We deduce that, even if Re $\ll 1$, provided Re $H/L \ll 1$, inertial effects may be neglected compared with viscous effects, that is, although the flow accelerates, the acceleration may be neglected.

We now consider Eq. (10.4-5). To an order of magnitude, the ratio of the inertial terms $u_x (\partial u_y/\partial x) + u_y (\partial u_y/\partial y)$ to the viscous terms $\mu/\rho (\partial^2 u_y/\partial x^2 + \partial^2 u_y/\partial y^2)$ is Re H/L. Thus the order-of-magnitude analyses of Eqs. (10.4-4) and (10.4-5) are consistent and, provided Re $H/L \ll 1$, they yield

$$\frac{\partial p}{\partial x} \simeq \mu \frac{\partial^2 u_x}{\partial y^2} \tag{10.4-7}$$

$$\frac{\partial p}{\partial y} \simeq \mu \frac{\partial^2 u_y}{\partial y^2} \tag{10.4-8}$$

We see that $\partial p/\partial x \sim \mu U/H^2$ and $\partial p/\partial y \sim \mu U/LH$. [Note, incidentally, that we have not made these order-of-magnitude estimates of $\partial p/\partial x$ and $\partial p/\partial y$ because

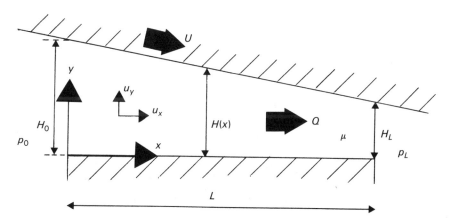

Figure 10.9. Geometry of flow between inclined flat plates.

of the difficulty of choosing in advance a characteristic pressure (see Section 7.2). Instead, the estimates emerged from those of the other terms in the flow equations.] It follows that, to an order of magnitude, the ratio of the transverse and axial pressure gradients $(\partial p/\partial y)/(\partial p/\partial x)$ is H/L. This means that $\partial p/\partial y$ is much smaller than $\partial p/\partial x$. Thus $\partial p/\partial y$ can be assumed to vanish approximately [recall that it vanishes exactly for flow between parallel plates; see Eq. (10.3–6)]. It also means that $\partial p/\partial x$ can be replaced by dp/dx and that Eq. (10.4-7) can then be integrated to yield Eq. (10.4-1) on imposition of the boundary conditions (10.3-10).

We now return to the main analysis and note that the volumetric flow rate per unit width Q' between the plates is given by

$$Q' = \int_0^H u_x \, dy \qquad (10.4-9)$$

whence, using Eq. (10.4-1),

$$Q' = \frac{1}{2} UH - \frac{H^3}{12\mu} \frac{dp}{dx}$$
$$\uparrow \qquad \uparrow \qquad (10.4-10)$$
$$(a) \qquad (b)$$

We note that Q', like u_x, comprises a drag-induced component (a) [which is identical to that given in Eq. (10.3-15)] and a pressure-induced component (b). Because the flow is incompressible, Q' is independent of axial position x. Clearly, U and μ are also independent of x. But the distance H between the plates is a function of x. It then follows from Eq. (10.4-10) that dp/dx is a function of x, so that, in particular, $dp/dx \neq -[(p_0 - p_L)/L]$. Instead,

$$\frac{dp}{dx} = 12\mu \left(\frac{U}{2H^2} - \frac{Q'}{H^3} \right) \qquad (10.4-11)$$

But
$$H = H_0 + (H_L - H_0) \frac{x}{L} \qquad (10.4-12)$$

so, integrating Eq. (10.4-11) with respect to x,

$$\int_{p_0}^{p_L} dp = \int_0^L 12\mu \left(\frac{U}{2H^2} - \frac{Q'}{H^3} \right) dx \qquad (10.4-13)$$

that is,

$$p_L - p_0 = \frac{6\mu L}{H_L - H_0} \left[-U \left(\frac{1}{H_L} - \frac{1}{H_0} \right) + Q' \left(\frac{1}{H_L^2} - \frac{1}{H_0^2} \right) \right] \qquad (10.4-14)$$

We now suppose that there is no overall pressure drop, that is,

$$p_0 = p_L \qquad (10.4-15)$$

Then it follows that

$$Q' = \frac{UH_0 H_L}{H_0 + H_L} \qquad (10.4\text{-}16)$$

and also that

$$p = p_0 + \frac{6\mu UL}{H_0 - H_L} \frac{(H_0 - H)(H - H_L)}{H^2(H_0 + H_L)} \qquad (10.4\text{-}17)$$

The distribution of pressure p with axial position x is shown schematically in Fig. 10.10. We note the asymmetry in the distribution: the maximum pressure $p_{\max}$ occurs at $x = H_0 L/(H_0 + H_L) > \frac{1}{2} L$.

The magnitude of the *normal force per unit width* F'_n exerted by the fluid on each plate is given by

$$F'_n = \int_0^L (p - p_0)\,dx \qquad (10.4\text{-}18)$$

Note that there is no normal stress contribution to F'_n. Recall from Section 8.2 that all normal stress components vanish at a rigid impermeable wall in an incompressible Newtonian fluid. Hence,

$$F'_n = \frac{6\mu UL^2}{(H_0 - H_L)^2} \left[\ln\frac{H_0}{H_L} - 2\frac{H_0 - H_L}{H_0 + H_L} \right] \qquad (10.4\text{-}19)$$

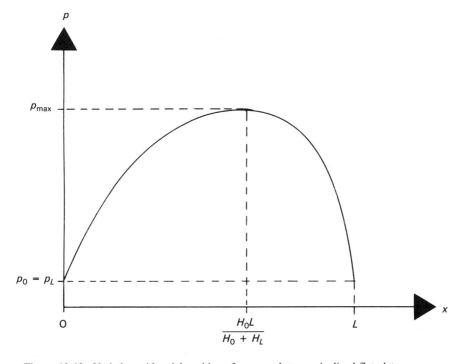

Figure 10.10. Variation with axial position of pressure between inclined flat plates.

The magnitude of the *tangential force per unit width* F'_{t_l} exerted on the lower (fixed) plate is given by

$$F'_{t_l} = \int_0^L \tau_{yx}|_{y=0}\, dx \qquad (10.4\text{-}20)$$

Because u_y vanishes at $y = 0$, the shear stress component $\tau_{yx} = \mu(\partial u_x/\partial y + \partial u_y/\partial x)$ is given by $\mu\, \partial u_x/\partial y$ at $y = 0$. Hence,

$$F'_{t_l} = \frac{2\mu UL}{H_0 - H_L}\left(3\,\frac{H_0 - H_L}{H_0 + H_L} - \ln\frac{H_0}{H_L}\right) \qquad (10.4\text{-}21)$$

The magnitude of the *tangential force per unit width* F'_{t_u} exerted on the upper (moving) plate is given by

$$F'_{t_u} = \int_0^L \tau_{yx}|_{y=H(x)}\, dx \qquad (10.4\text{-}22)$$

Because u_y vanishes at $y = H$, $\tau_{yx} = \mu\, \partial u_x/\partial y$ at $y = H$. Hence,

$$F'_{t_u} = -\frac{2\mu UL}{H_0 - H_L}\left(3\,\frac{H_0 - H_L}{H_0 + H_L} - 2\ln\frac{H_0}{H_L}\right) \qquad (10.4\text{-}23)$$

Note that $F'_{t_l} \neq F'_{t_u}$ because the two plates are not quite parallel to each other. If $H_0/H_L - 1 \sim 1$, it follows that

$$\frac{F'_t}{F'_n} \sim \frac{H_0 - H_L}{L} \qquad (10.4\text{-}24)$$

where F'_t denotes either F'_{t_l} or F'_{t_u}. We might interpret F'_t/F'_n as a *coefficient of friction*. Because $H_0 - H_L \ll L$, it follows that $F'_t/F'_n \ll 1$: the tangential force is much smaller than the normal force. This is the basis of *hydrodynamic lubrication* in which high solid-solid friction (here, between the two plates) is replaced by low solid-fluid friction. The large normal force generated by the pressure between the plates prevents fluid lubricant from being squeezed out of the gap.

In many flow problems, the flow region comprises a relatively long, narrow duct with slowly varying transverse dimensions. The length over which the transverse dimensions vary is very large compared with the transverse dimensions and hence with any change in the those dimensions. For flows in such ducts, it can be shown by an order-of-magnitude analysis that the flow is locally fully developed (i.e., the flow is locally not accelerating). This is called the *lubrication approximation* after the lubrication problem that we have just analyzed. It involves the assumption that viscous forces (dominated by large gradients of the principal velocity component in directions normal to that component) are much larger than inertial forces (dominated by changes of velocity components over large distances in the directions of those components or, for unsteady problems, over long times). When the lubrication approximation can be made, the Navier-Stokes equations

(10.1-1) simplify to Stokes' equations (10.1-3): the flow is a nearly nonaccelerating flow.

10.5 GRAVITATIONALLY INDUCED FLOW IN A LAYER ON AN INCLINED FLAT PLATE

We consider an incompressible Newtonian liquid of density ρ and viscosity μ in a layer of thickness or depth h on a flat plate of infinite extent inclined at an angle α to the horizontal. The liquid flows down the plate as a result of the gravitational acceleration **g**. Above the liquid is a gas of negligible density and viscosity and uniform pressure p_a. Because of the obvious geometry, we use rectangular coordinates (x, y, z), with origin on the plate, x direction aligned with the direction of the flow down the plate, and y direction aligned normal to the plate and toward the liquid (see Fig. 10.11).

Provided $|dh/dx| \ll 1$, so that h changes only slowly (if at all) with x, we

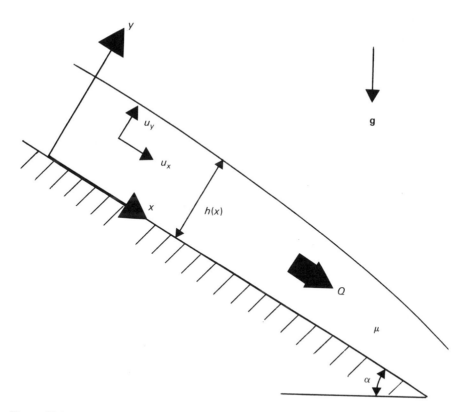

Figure 10.11. Geometry of flow in a layer down an inclined flat plate.

expect that the velocity field **u** is approximately fully developed. Hence, by the same arguments as in Sections 10.2 and 10.3, we expect that

$$\mathbf{u} \simeq u_x(y)\mathbf{i}_x \tag{10.5-1}$$

assuming that the flow is steady. The approximate equality in this equation would be replaced by an exact equality if dh/dx were to vanish. Then, again by the same arguments as in Sections 10.2 and 10.3, it follows that the linear and nonlinear inertial terms in the Navier-Stokes equations vanish, though only approximately so unless again dh/dx were to vanish. Thus the flow down the plane is a nearly nonaccelerating flow.

The mass conservation equation (A2.2-2) of Appendix B is automatically satisfied by the velocity field given by Eq. (10.5-1). The linear momentum conservation equation (10.1-1) reduces to

$$\rho\mathbf{g} - \nabla p + \mu\Delta\mathbf{u} = 0 \tag{10.5-2}$$

We use this form in preference to that in Eq. (10.1-3) because the flow down the plane is gravitationally induced, so it is convenient to make gravitational effects explicit. Equations (A2.3-2) of Appendix B (omitting the transient and convection terms) and Eq. (10.5-1) together yield

$$-\frac{\partial p}{\partial y} - \rho g \cos \alpha = 0 \tag{10.5-3}$$

$$-\frac{\partial p}{\partial x} + \rho g \sin \alpha + \mu \frac{d^2 u_x}{dy^2} = 0 \tag{10.5-4}$$

where

$$\mathbf{g} = g \sin \alpha \mathbf{i}_x - g \cos \alpha \mathbf{i}_y \qquad g = |\mathbf{g}| \tag{10.5-5}$$

The boundary conditions on u_x and p are

$$u_x = 0 \qquad \text{at } y = 0 \tag{10.5-6}$$

and

$$\frac{du_x}{dy} = 0 \qquad p = p_a \qquad \text{at } y = h \tag{10.5-7}$$

Boundary conditions (10.5-7) follow from Eq. (6.3-6), assuming that interfacial tension effects are negligible. The principal radii of curvature of the free surface, R_I and R_II, are given by

$$R_\mathrm{I} \simeq \frac{-1}{d^2h/dx^2} \qquad R_\mathrm{II} \rightarrow \infty \tag{10.5-8}$$

[see Eqs. (9.4-20)]. Thus neglect of interfacial tension effects is justifiable if $|d^2h/$

dx^2| is small enough. Note that du_x/dy vanishes at the free surface because the shear stress τ_{yx} vanishes there.

We can integrate Eq. (10.5-3) with respect to y and impose the boundary condition on p at h [see Eqs. (10.5-7)] to yield

$$p + \rho gy \cos \alpha = p_a + \rho gh \cos \alpha \qquad (10.5\text{-}9)$$

It now follows that, provided $|dh/dx| \ll 1$,

$$\frac{\partial p}{\partial x} = 0 \qquad (10.5\text{-}10)$$

and hence that Eq. (10.5-4) becomes

$$\frac{d^2 u_x}{dy^2} = -\frac{\rho g}{\mu} \sin \alpha \qquad (10.5\text{-}11)$$

which, on integration twice with respect to y and imposition of boundary conditions (10.5-6) and (10.5-7), yields

$$u_x = \frac{\rho gh^2}{\mu} \sin \alpha \left(\frac{y}{h} - \frac{y^2}{2h^2} \right) \qquad (10.5\text{-}12)$$

It follows that the axial velocity profile is parabolic (see Fig. 10.12).

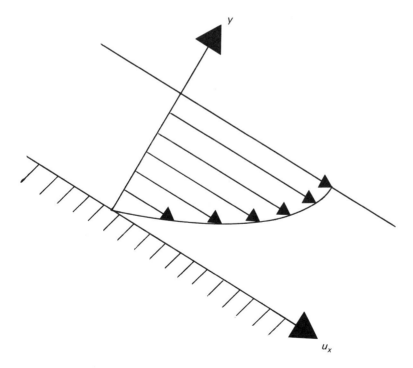

Figure 10.12. Axial velocity profile in a layer on an inclined flat plate.

The volumetric flow rate per unit width Q' is given by

$$Q' = \int_0^h u_x \, dy \tag{10.5-13}$$

whence

$$Q' = \frac{\rho g h^3}{3\mu} \sin \alpha \tag{10.5-14}$$

It follows that the thickness h has a weak dependence on the parameters Q', g, ρ, μ, and α which define the flow—thus it is difficult in practice to alter h significantly—and that the thickness h is constant down the inclined plate (i.e., $dh/dx = 0$) because ρ, g, μ, and α are constant and, for a steady flow, Q' is clearly constant.

Because dh/dx vanishes, it follows that the flow is in fact nonaccelerating, and not just nearly nonaccelerating, though we had no reason to assume this prior to solving the flow problem. It also follows that d^2h/dx^2 vanishes and hence that there are no interfacial tension effects. Thus boundary conditions (10.5-7) in fact hold exactly.

We note that this flow is unstable for large enough values of Re given by

$$\text{Re} = \frac{\rho Q'}{\mu} \tag{10.5-15}$$

The form that the instability takes is waves on the free surface. There is experimental evidence that such waves occur in practice at rather small values of Re, perhaps even as Re $\to$ 0 (in which case the flow is unconditionally unstable).

10.6 FLOW PAST AN IMPULSIVELY MOVED FLAT PLATE

We consider an infinite volume of an incompressible Newtonian fluid of density ρ and viscosity μ above an infinite flat plate. The plate is at rest initially (i.e., for time $t < 0$). Subsequently, for $t \geq 0$, it is moved impulsively with constant speed U in a fixed direction in its own plane. Because of the obvious geometry, we use rectangular coordinates (x, y, z) to analyze the problem, with origin on the plate, x direction aligned with the direction of motion of the plate and y direction vertically upward (see Fig. 10.13).

The velocity of the plate is given by $UH(t)\mathbf{i}_x$ where $H(t)$ denotes the Heaviside or unit step function:

$$H(t) = \begin{cases} 0 & \text{if } t < 0 \\ 1 & \text{if } t \geq 0 \end{cases} \tag{10.6-1}$$

Clearly, the impulsive motion of the plate means that the fluid above it is brought into an unsteady, drag-induced motion. Symmetry means that there is no x or z dependence of flow variables and also that the fluid velocity $\mathbf{u}$ has no z component. Thus $\mathbf{u}$ has just two components, $u_x(y, t)$ in the x direction and $u_y(y, t)$ in

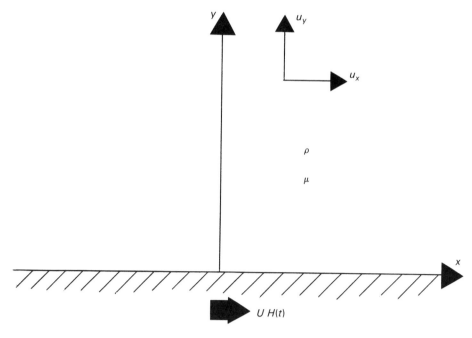

Figure 10.13. Geometry of flow past an impulsively moved plate.

the y direction. As a result, the mass conservation equation (A2.2-2) of Appendix B yields

$$\frac{\partial u_y}{\partial y} = 0 \qquad (10.6\text{-}2)$$

This equation can be integrated to yield

$$u_y = 0 \qquad (10.6\text{-}3)$$

since no flow-through at the plate means that u_y vanishes there.

Hence,
$$\mathbf{u} = u_x(y, t)\mathbf{i}_x \qquad (10.6\text{-}4)$$

from which it follows that the convection or nonlinear inertial term $\rho\mathbf{u}\cdot\nabla\mathbf{u}$ in the Navier-Stokes equations (10.1-1) vanishes. The flow is, however, unsteady so that the transient or linear inertial term $\rho\,\partial\mathbf{u}/\partial t$ in the Navier-Stokes equations does not vanish. Thus the flow is not a nonaccelerating flow, though it is what we might call a convectionless flow for which the Navier-Stokes equations (10.1-1) simplify to

$$\rho\frac{\partial\mathbf{u}}{\partial t} = -\nabla p + \mu\,\Delta\mathbf{u} \qquad (10.6\text{-}5)$$

where the gravitational term is incorporated in the pressure term.

Equations (A2.3-2) of Appendix B and Eq. (10.6-4) together yield

$$0 = -\frac{\partial p}{\partial y} \tag{10.6-6}$$

from which it follows immediately that p is at most a function of time t, and

$$\rho \frac{\partial u_x}{\partial t} = \mu \frac{\partial^2 u_x}{\partial y^2} \tag{10.6-7}$$

The boundary conditions are no slip at the plate, that is,

$$u_x = U \qquad \text{at } y = 0 \text{ for } t \geq 0 \tag{10.6-8}$$

and, in order that the kinetic energy of the fluid is bounded, vanishing of the velocity far from the plate (at infinity; see Section 6.4), that is,

$$u_x \to 0 \qquad \text{as } y \to \infty \tag{10.6-9}$$

The initial condition, which is required because the flow is unsteady, is

$$u_x = 0 \qquad \text{at } t = 0 \text{ for } y > 0 \tag{10.6-10}$$

In order to solve Eq. (10.6-7) subject to conditions (10.6-8), (10.6-9), and (10.6-10), we note that there is no characteristic length with which y can be made dimensionless. Similarly, there is no characteristic time with which t can be made dimensionless. As a result, we seek (as we noted in Section 7.2) a similarity solution. We start by defining the dimensionless axial velocity α in an obvious way:

$$\alpha = \frac{u_x}{U} \tag{10.6-11}$$

Then Eqs. (10.6-7) and boundary and initial conditions (10.6-8), (10.6-9), and (10.6-10) become

$$\rho \frac{\partial \alpha}{\partial t} = \mu \frac{\partial^2 \alpha}{\partial y^2} \tag{10.6-12}$$

$$\alpha = 1 \quad \text{at } y = 0 \qquad \alpha \to 0 \quad \text{as } y \to \infty \qquad \alpha = 0 \quad \text{at } t = 0 \tag{10.6-13}$$

We now define the similarity variable η:

$$\eta = c_1 t^{c_2} y \tag{10.6-14}$$

where c_1 and c_2 are constants yet to be determined. In fact, c_1 will be determined by considerations of convenience and c_2 by the requirement that α be a function of η alone. It follows from Eq. (10.6-14) that

$$\frac{\partial \alpha}{\partial t} = \frac{d\alpha}{d\eta} \frac{\partial \eta}{\partial t} = \frac{d\alpha}{d\eta} c_1 c_2 t^{c_2-1} y \tag{10.6-15}$$

$$\frac{\partial \alpha}{\partial y} = \frac{d\alpha}{d\eta} \frac{\partial \eta}{\partial y} = \frac{d\alpha}{d\eta} c_1 t^{c_2} \tag{10.6-16}$$

$$\frac{\partial^2 \alpha}{\partial y^2} = \frac{\partial}{\partial y}\left(\frac{d\alpha}{d\eta}\right) c_1 t^{c_2} = \frac{d}{d\eta}\left(\frac{\partial \alpha}{\partial y}\right) c_1 t^{c_2} = \frac{d^2\alpha}{d\eta^2} c_1^2 t^{2c_2} \tag{10.6-17}$$

Thus Eq. (10.6-12) becomes

$$\frac{d^2\alpha}{d\eta^2} - \eta \frac{d\alpha}{d\eta} \frac{\rho c_2}{\mu c_1^2 t^{2c_2+1}} = 0 \tag{10.6-18}$$

Because, for a similarity solution, we require that α is a function of η only, it is necessary that Eq. (10.6-18) contains no terms involving t. Equivalently, it must contain no terms involving y, since any term involving η and t can be rewritten as a term involving η and y. Thus it is necessary either that c_2 vanishes, which implies a trivial steady solution, or that t^{2c_2+1} is a constant, that is, that

$$c_2 = -\frac{1}{2} \tag{10.6-19}$$

It is convenient to make the similarity variable η dimensionless. Accordingly, we choose

$$c_1 = \frac{1}{2}\sqrt{\frac{\rho}{\mu}} \tag{10.6-20}$$

(the factor of 2 is, of course, entirely arbitrary), so that Eq. (10.6-14) becomes

$$\eta = \frac{y}{2}\sqrt{\frac{\rho}{\mu t}} \tag{10.6-21}$$

We note that η, which is dimensionless, is defined in terms of both x and t. The only way in which x and t can be made dimensionless is with respect to each other (and, of course, with ρ and μ). As a result, Eq. (10.6-12) becomes

$$\frac{d^2\alpha}{d\eta^2} + 2\eta \frac{d\alpha}{d\eta} = 0 \tag{10.6-22}$$

The reason why the partial differential equation (10.6-12) becomes the ordinary differential equation (10.6-22) is discussed in Section 9.6. The boundary and initial conditions (10.6-13) become

$$\alpha = 1 \quad \text{at } \eta = 0 \qquad \alpha \to 0 \quad \text{as } \eta \to \infty \tag{10.6-23}$$

We note that the boundary condition on α as $y \to \infty$ and the initial condition on α at $t = 0$ are combined into a single condition on α as $\eta \to \infty$. Thus $y \to \infty$ is, in a sense, equivalent to $t = 0$. This is consistent with the fact that η varies *directly* as y and *inversely* as the square root of t. We also note that conditions (10.6-23) and Eq. (10.6-22) involve α, or its derivatives, and η alone.

It is an essential condition for the existence of a similarity solution that all equations and conditions can be expressed in terms only of the similarity variable (or variables) and functions of it (or them).

Equation (10.6-22) may be rearranged to give

$$\frac{d(d\alpha/d\eta)/d\eta}{d\alpha/d\eta} = -2\eta \qquad (10.6\text{-}24)$$

which may be integrated to give

$$\ln \frac{d\alpha}{d\eta} = -\eta^2 + \ln c_3 \qquad (10.6\text{-}25)$$

or

$$\frac{d\alpha}{d\eta} = c_3 e^{-\eta^2} \qquad (10.6\text{-}26)$$

where c_3 is a constant. Equation (10.6-26) may be integrated to yield

$$\alpha = c_3 \int^{\eta} e^{-\eta^{\#2}} d\eta^{\#} + c_4 \qquad (10.6\text{-}27)$$

where c_4 is another constant. Since

$$\int_0^{\infty} e^{-\eta^{\#2}} d\eta^{\#} = \frac{\sqrt{\pi}}{2} \qquad (10.6\text{-}28)$$

imposition of conditions (10.6-23) yields

$$\alpha = \frac{2}{\sqrt{\pi}} \int_{\eta}^{\infty} e^{-\eta^{\#2}} d\eta^{\#} \qquad (10.6\text{-}29)$$

or

$$\alpha = \text{erfc}(\eta) \qquad (10.6\text{-}30)$$

where erfc( ) denotes the complementary error function. Thus:

$$u_x = U \, \text{erfc}\left(\frac{y}{2}\sqrt{\frac{\rho}{\mu t}}\right) \qquad (10.6\text{-}31)$$

We conclude by determining the stress τ exerted on the plate by the fluid. Because $\tau = \mu e$ for a Newtonian fluid, where e denotes the rate of strain, it follows from Eq. (10.6-4) and Eq. (A2.1-2) of Appendix B that the only non-vanishing components of τ are the shear stress components τ_{xy} and τ_{yx}. At the plate, they are given by

$$\tau_{xy}|_{y=0} = \tau_{yx}|_{y=0} = \mu \frac{\partial u_x}{\partial y}\bigg|_{y=0} \qquad (10.6\text{-}32)$$

whence, using Eq. (10.6-31);

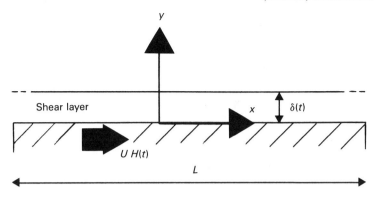

Figure 10.14. Shear layer by an impulsively moved plate.

$$\tau_{xy}\big|_{y=0} = \tau_{yx}\big|_{y=0} = -\sqrt{\frac{\rho U^2 \mu}{\pi t}} \tag{10.6-33}$$

We note that $\tau_{xy}\big|_{y=0}$ and $\tau_{yx}\big|_{y=0}$ vary as $1/\sqrt{t}$, that is, they decrease as time elapses. This decrease is related to the increase in the thickness δ of the shear layer as time elapses. The shear layer is the region by the plate in which $|\partial u_x/\partial y|$ is significant and is related to the boundary layer on a plate that we discuss in Section 11.2. It follows from Eq. (10.6-32) that $\tau_{xy}\big|_{y=0} = \tau_{yx}\big|_{y=0} \sim \mu U/\delta$ and hence from Eq. (10.6-33) that

$$\delta \sim \sqrt{\frac{\mu t}{\rho}} \tag{10.6-34}$$

so δ varies as $\sqrt{t}$ (i.e., it increases as time elapses). We might, therefore, expect our analysis to hold for a plate of finite length L and width W if $\delta << L$ and $\delta << W$ so that end and edge effects are negligible, that is,

$$t << \begin{cases} \dfrac{\rho L^2}{\mu} \\[2ex] \dfrac{\rho W^2}{\mu} \end{cases} \tag{10.6-35}$$

(see Fig. 10.14).

10.7 STRESS–INDUCED FLOW IN A ROTATING FLUID

We consider an incompressible Newtonian liquid of density ρ and viscosity μ and of infinite extent. We suppose that the fluid rotates as a whole with constant angular velocity Ω about a vertical axis. Above the liquid, we suppose that there is a gas which, in a manner which need not concern us, causes a stress $\mathbf{t}$ to be applied to, and act in the plane of, the flat, horizontal surface of the liquid (note that **t** is a stress vector; see Section 3.2). For convenience, we use a reference frame which rotates with the liquid. Because of the obvious geometry, we use rectangular coordinates (x_R, y_R, z_R) to analyze the problem, with origin at the intersection of the plane comprising the surface of the liquid and the axis of rotation; the subscript R denotes the rotating reference frame. The y_R direction is aligned with **t** and the z_R direction is aligned vertically upward along the axis of rotation (see Fig. 10.15).

We suppose that $\mathbf{\Omega} = \Omega \mathbf{i}_{z_R}$ so that, in a right-handed coordinate system (see Section 1.2), $\Omega > 0$ means that the rotation is clockwise when it is viewed in the direction of increasing z_R (i.e., from beneath). In order to determine the stress-induced flow in the fluid, we must first modify the Navier-Stokes equations (10.1-1) for a rotating (and hence noninertial) reference frame.

We let the subscripts I and R denote inertial and rotating reference frames, respectively. We suppose that the rotating frame has a constant angular velocity $\mathbf{\Omega}$ with respect to the inertial frame. Then the positions $\mathbf{x}_I$ and $\mathbf{x}_R$ of a material point X in the two reference frames are related thus:

$$\mathbf{x}_R = \mathbf{Q} \cdot \mathbf{x}_I \tag{10.7-1}$$

Here $\mathbf{Q}$ is a rotation tensor (i.e., it is an orthogonal tensor), so that $\mathbf{Q}^T \cdot \mathbf{Q} = \mathbf{Q} \cdot \mathbf{Q}^T = I$ where I denotes the unit tensor [see Eqs. (A1.2-17) and (A1.2-18) of Appendix A]. It is defined by

$$\mathbf{\Omega}_\wedge \mathbf{v} = \mathbf{v} \cdot \frac{D\mathbf{Q}}{Dt} \cdot \mathbf{Q}^T = -\frac{D\mathbf{Q}}{Dt} \cdot \mathbf{Q}^T \cdot \mathbf{v} \tag{10.7-2}$$

for any vector field **v** where D/Dt denotes the substantial derivative (see Section

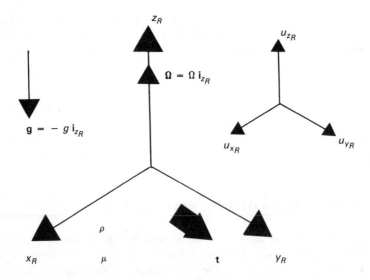

Figure 10.15. Geometry of flow in a rotating fluid.

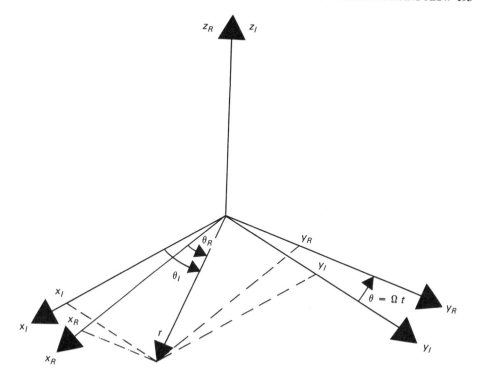

Figure 10.16. Relationship between coordinates in inertial and rotating reference frames.

2.3). (Note that $\boldsymbol{\Omega}$ is, therefore, the pseudovector formed from the antisymmetric tensor $(D\mathbf{Q}/Dt) \cdot \mathbf{Q}^T$ in the same way as the vorticity vector $\boldsymbol{\omega}$ is the pseudovector formed from the antisymmetric vorticity tensor $\mathbf{w}$; see Section 2.4.) To understand the role of $\mathbf{Q}$ more clearly, we note that if we consider rectangular coordinates (x_I, y_I, z_I) in the inertial reference frame and (x_R, y_R, z_R) in the rotating reference frame (see Fig. 10.16), then

$$x_I = r \cos \theta_I \qquad x_R = r \cos \theta_R \qquad (10.7\text{-}3)$$

$$y_I = r \sin \theta_I \qquad y_R = r \sin \theta_R \qquad (10.7\text{-}4)$$

$$z_I = z_R \qquad (10.7\text{-}5)$$

where

$$\theta_I - \theta_R = \theta = \Omega t \qquad (10.7\text{-}6)$$

Here, r denotes distance from the axis of rotation, θ_I and θ_R denote the angles between the x_I and r directions and the x_R and r directions, respectively, and t denotes time. It then follows by elementary trigonometry that

$$x_R = x_I \cos \theta + y_I \sin \theta \qquad (10.7\text{-}7)$$

$$y_R = -x_I \sin \theta + y_I \cos \theta \qquad (10.7\text{-}8)$$

$$z_R = z_I \tag{10.7-9}$$

that is, Eq. (10.7-1) holds with $\mathbf{Q}$ given by

$$\mathbf{Q} = \cos\theta \mathbf{i}_{x_R}\mathbf{i}_{x_I} - \sin\theta \mathbf{i}_{y_R}\mathbf{i}_{x_I} + \sin\theta \mathbf{i}_{x_R}\mathbf{i}_{y_I} + \cos\theta \mathbf{i}_{y_R}\mathbf{i}_{y_I} + \mathbf{i}_{z_R}\mathbf{i}_{z_I} \tag{10.7-10}$$

and θ given by Eq. (10.7-6) [note the similarity between Eq. (10.7-10) and Eq. (A1.2-32) of Appendix A]. It follows from Eq. (10.7-1) that

$$\frac{D\mathbf{x}_R}{Dt} = \frac{D\mathbf{Q}}{Dt} \cdot \mathbf{x}_I + \mathbf{Q} \cdot \frac{D\mathbf{x}_I}{Dt} = \frac{D\mathbf{Q}}{Dt} \cdot \mathbf{Q}^{-1} \cdot \mathbf{x}_R + \mathbf{Q} \cdot \frac{D\mathbf{x}_I}{Dt} \tag{10.7-11}$$

where $\mathbf{Q}^{-1}$ denotes the *inverse* of the tensor $\mathbf{Q}$ [see Eq. (A1.2-16) of Appendix A]. So, because $\mathbf{Q}$ is orthogonal (i.e., $\mathbf{Q}^T = \mathbf{Q}^{-1}$), use of Eq. (10.7-2) yields

$$\mathbf{u}_R = -\mathbf{\Omega}_{\wedge}\mathbf{x}_R + \mathbf{Q} \cdot \mathbf{u}_I \tag{10.7-12}$$

where $\mathbf{u}_R = D\mathbf{x}_R/Dt$ and $\mathbf{u}_I = D\mathbf{x}_I/Dt$ denote the velocities of the material point X in the rotating and inertial reference frames, respectively. Hence,

$$\frac{D\mathbf{u}_R}{Dt} = -\mathbf{\Omega}_{\wedge}\mathbf{u}_R + \frac{D\mathbf{Q}}{Dt} \cdot \mathbf{u}_I + \mathbf{Q} \cdot \frac{D\mathbf{u}_I}{Dt} \tag{10.7-13}$$

or

$$\mathbf{a}_R = -\mathbf{\Omega}_{\wedge}\mathbf{u}_R + \frac{D\mathbf{Q}}{Dt} \cdot \mathbf{Q}^{-1} \cdot \mathbf{u}_R + \frac{D\mathbf{Q}}{Dt} \cdot \mathbf{Q}^{-1} \cdot (\mathbf{\Omega}_{\wedge}\mathbf{x}_R) + \mathbf{Q} \cdot \mathbf{a}_I \tag{10.7-14}$$

where $\mathbf{a}_R = D\mathbf{u}_R/Dt$ and $\mathbf{a}_I = D\mathbf{u}_I/Dt$ denote the accelerations of the material point X in the rotating and inertial reference frames, respectively. Equation (10.7-14) may now be simplified, using Eq. (10.7-2), to yield

$$\mathbf{Q} \cdot \mathbf{a}_I = \mathbf{a}_R + 2\mathbf{\Omega}_{\wedge}\mathbf{u}_R + \mathbf{\Omega}_{\wedge}(\mathbf{\Omega}_{\wedge}\mathbf{x}_R) \tag{10.7-15}$$

Hence, in an inertial reference frame *instantaneously coincident* with the rotating reference frame, so that $\mathbf{Q} = \mathbf{I}$,

$$\mathbf{a}_I = \underset{\underset{(a)}{\uparrow}}{\mathbf{a}_R} + \underset{\underset{(b)}{\uparrow}}{2\,\mathbf{\Omega}_{\wedge}\mathbf{u}_R} + \underset{\underset{(c)}{\uparrow}}{\mathbf{\Omega}_{\wedge}(\mathbf{\Omega}_{\wedge}\mathbf{x}_R)} \tag{10.7-16}$$

Thus the rectilinear acceleration $\mathbf{a}_I$ in an inertial reference frame comprises three components in a steadily rotating reference frame: the rectilinear acceleration (a), the *Coriolis acceleration* (b), and the *centripetal acceleration* (c). The Coriolis acceleration acts in a direction normal to both $\mathbf{\Omega}$ and $\mathbf{u}_R$. The centripetal acceleration acts in a direction parallel to $\mathbf{x}_R$. It then follows that, in the rotating reference frame, the Navier-Stokes equations (10.1-1) become

$$\rho\frac{\partial\mathbf{u}_R}{\partial t} + \rho\mathbf{u}_R \cdot \nabla\mathbf{u}_R + 2\rho\mathbf{\Omega}_{\wedge}\mathbf{u}_R + \rho\mathbf{\Omega}_{\wedge}(\mathbf{\Omega}_{\wedge}\mathbf{x}_R) = \rho\mathbf{g} - \nabla p + \mu\Delta\mathbf{u}_R \tag{10.7-17}$$

It is easy to show that

$$\mathbf{\Omega}_\wedge(\mathbf{\Omega}_\wedge \mathbf{x}_R) = -\nabla(\tfrac{1}{2}|\mathbf{\Omega}_\wedge \mathbf{x}_R|^2) = -\nabla(\tfrac{1}{2}\,\Omega^2 r^2) \qquad (10.7\text{-}18)$$

where r denotes distance from the axis of the rotation (see Fig. 10.16). Because the fluid is incompressible, we can define a *modified pressure* $\tilde{p}$ as

$$\tilde{p} = p - \rho \mathbf{g} \cdot \mathbf{x}_R - \tfrac{1}{2}\rho\Omega^2 r^2 \qquad (10.7\text{-}19)$$

provided pressure does not appear explicitly in the boundary conditions. Clearly, $\tilde{p}$ is clearly a generalization of the modified pressure $\bar{p}$ defined in Eq. (4.3-13). By introducing $\tilde{p}$, we can eliminate both gravitational and centripetal effects from Eq. (10.7-17) which thus becomes

$$\rho\,\frac{\partial \mathbf{u}_R}{\partial t} + \rho \mathbf{u}_R \cdot \nabla \mathbf{u}_R + 2\rho\mathbf{\Omega}_\wedge \mathbf{u}_R = -\nabla\tilde{p} + \mu\Delta \mathbf{u}_R \qquad (10.7\text{-}20)$$

Having modified the Navier-Stokes equations for a steadily rotating reference frame, we can now proceed with the analysis of the stress-induced flow. The flow is steady, so $\partial \mathbf{u}_R/\partial t$ vanishes. By symmetry, the x_R, y_R, and z_R components of velocity $\mathbf{u}_R$ (u_{x_R}, u_{y_R}, and u_{z_R}, respectively) and the modified pressure $\tilde{p}$ are independent of the horizontal coordinates x_R and y_R. The mass conservation equation (A2.2-2) of Appendix B thus yields

$$\frac{\partial u_{z_R}}{\partial z_R} = 0 \qquad (10.7\text{-}21)$$

which can be integrated to yield

$$u_{z_R} = 0 \qquad (10.7\text{-}22)$$

since no flow-through at the surface of the fluid means that u_{z_R} vanishes there. It thus follows that $\mathbf{u}_R \cdot \nabla \mathbf{u}_R$ vanishes: together with the fact that $\partial \mathbf{u}_R/\partial t$ vanishes, this means that the flow might be regarded as a *nonaccelerating flow*. The presence of the Coriolis acceleration $2\,\mathbf{\Omega}_\wedge \mathbf{u}_R$, however, means that strictly speaking it is not. Using Eqs. (A1.1-19) of Appendix A and (A2.3-2) of Appendix B, it follows from Eq. (10.7-20) that

$$-2\rho\Omega u_{y_R} = \mu\,\frac{d^2 u_{x_R}}{dz_R^2} \qquad (10.7\text{-}23)$$

$$+ 2\rho\Omega u_{x_R} = \mu\,\frac{d^2 u_{y_R}}{dz_R^2} \qquad (10.7\text{-}24)$$

It also follows that $\partial\tilde{p}/\partial z_R$ vanishes, so that $\tilde{p}$ is independent of position. The first boundary condition that must be imposed is that the shear stress in the fluid matches with the applied shear stress at the surface of the fluid, that is,

$$\frac{du_{x_R}}{dz_R} = 0 \qquad \frac{du_{y_R}}{dz_R} = \frac{\tau}{\mu} \qquad \text{at } z_R = 0 \qquad (10.7\text{-}25)$$

where

$$\tau = |\mathbf{t}| \qquad (10.7\text{-}26)$$

The second boundary condition that must be imposed is that the velocity vanishes far below the surface (at infinity, so that the kinetic energy of the fluid is bounded), that is,

$$u_{x_R} \to 0 \qquad u_{y_R} \to 0 \qquad \text{as } z_R \to -\infty \qquad (10.7\text{-}27)$$

The solution of Eqs. (10.7-23) and (10.7-24) subject to boundary conditions (10.7-25) and (10.7-27) is easily shown to be

$$u_{x_R} = \frac{\tau Z}{\sqrt{2}\,\mu} e^{z_R/Z} \cos\left(\frac{z_R}{Z} + \frac{\pi}{4}\right) \qquad u_{y_R} = \frac{\tau Z}{\sqrt{2}\,\mu} e^{z_R/Z} \sin\left(\frac{z_R}{Z} + \frac{\pi}{4}\right) \qquad (10.7\text{-}28)$$

The parameter Z is a characteristic length l_c of the flow (more descriptively, it is a characteristic depth) given by

$$Z = \sqrt{\frac{\mu}{\Omega \rho}} \qquad (10.7\text{-}29)$$

We now let U denote the speed at the surface of the liquid, that is,

$$U = \sqrt{u_{x_R}^2 + u_{y_R}^2}\big|_{z_R=0} = \frac{\tau Z}{\sqrt{2}\,\mu} \qquad (10.7\text{-}30)$$

so that Eqs. (10.7-28) yield

$$\mathbf{u}_R = U e^{z_R/Z} \cos\left(\frac{z_R}{Z} + \frac{\pi}{4}\right)\mathbf{i}_{x_R} + U e^{z_R/Z} \sin\left(\frac{z_R}{Z} + \frac{\pi}{4}\right)\mathbf{i}_{y_R} \qquad (10.7\text{-}31)$$

The velocity profile is plotted schematically in Fig. 10.17. Note that $\mathbf{u}_R$ spirals inward and clockwise viewed from above (for $\Omega > 0$) as the depth increases, that is, as z_R decreases. This is the *Ekman spiral*. Note also that the velocity at the surface is not aligned with the surface stress but is at an angle $\pi/4$ to it.

We now calculate the *volumetric fluxes* Q'_{x_R} and Q'_{y_R} of fluid in the x_R and y_R directions, respectively, which are given by

$$Q'_{x_R} = \int_{-\infty}^{0} u_{x_R} dz_R \qquad Q'_{y_R} = \int_{-\infty}^{0} u_{y_R} dz_R \qquad (10.7\text{-}32)$$

It follows by substitution from Eq. (10.7-31) that

$$Q'_{x_R} = \frac{\tau}{2\Omega\rho} \qquad Q'_{y_R} = 0 \qquad (10.7\text{-}33)$$

Thus the net volumetric flux of fluid is in the x_R direction and hence normal to the direction of the surface stress. The force associated with the surface stress acts in the y_R direction.

A practical application of this problem is to wind-driven flows in oceans, where the Coriolis acceleration arises from the rotation of the earth. Provided the flow is over a sufficiently small region, there is negligible variation in the vertical

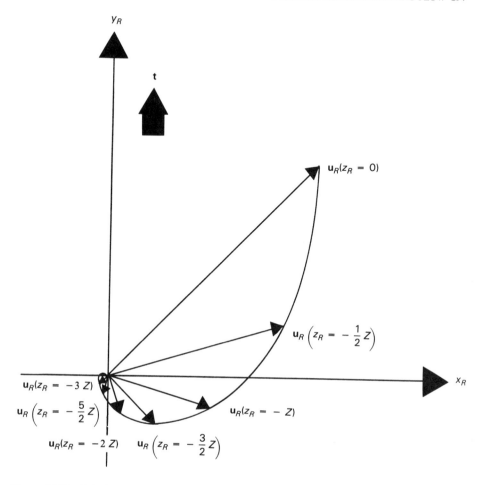

Figure 10.17. Velocity profile in a rotating fluid: the Ekman spiral.

component of the angular velocity of the earth. In fact, however, oceans are not of infinite depth but of a typical finite depth D and their surfaces are not flat but have waves of typical height H on them. Thus our analysis is reasonable only if

$$H << Z = \sqrt{\frac{\mu}{\Omega\rho}} << D \qquad (10.7\text{-}34)$$

Moreover, ocean flows are in fact turbulent, not laminar. If, however, the kinematic viscosity $v = \mu/\rho$ is replaced by an *eddy viscosity* v_t (see Chapter 12) where $v_t >> v$, fairly realistic results can be obtained.

ELEVEN

BOUNDARY LAYER FLOW

11.1 BOUNDARY LAYER APPROXIMATION

We noted in Section 9.2 that a high Reynolds number flow is effectively inviscid almost everywhere. Near a wall, however, in what is called a boundary layer, viscous effects are always important. The reason for this is that there is no slip of a viscous fluid at a rigid, impermeable wall, however small its viscosity may be: there is slip only if the viscosity vanishes (see Section 6.2). If we wish to analyze high Reynolds number flows in which there are walls, we can adopt one of two courses. We can solve the unsimplified Navier-Stokes equations, with all the attendant difficulties which that would imply. Alternatively, we can make approximations and assume that there are two distinct regions in the flow: (1) a region away from the walls in which viscous effects are negligible; and (2) a region immediately adjacent to the walls in which viscous effects are not negligible.

In the region away from the walls, the Navier-Stokes equations simplify to Euler's equations and the methods discussed in Chapter 9 may be used to solve them. In the boundary layer region adjacent to the walls, in which inertial effects and viscous effects are both significant, it would appear that we would need to solve the unsimplified Navier-Stokes equations, with all the attendant difficulties which that would again imply. Fortunately, however, we do not need to this because boundary layers are thin. Let us call the axial direction the one along the boundary layer aligned with the main flow and the transverse direction the one across the thin dimension of the boundary layer and normal to the main flow. Then, as we will see, we can simplify the Navier-Stokes equations by assuming (1) that the transverse velocity is much smaller than the axial velocity; (2) that axial derivatives of velocity components are much smaller than transverse derivatives of those same components; and (3) that the transverse pressure gradient is much smaller than the axial pressure gradient.

The decomposition of the flow into two distinct regions and the simplifications that we can make to the Navier-Stokes equations because the boundary layer is thin together comprise the boundary layer approximation. It is only by making this approximation that we can in fact analyze many high Reynolds number flows. Note, however, that the flows that we discuss in this chapter are laminar, so that the Reynolds numbers are not too high. A turbulent boundary layer flow, which occurs when the Reynolds number is very high, is discussed in Chapter 12 (see Section 12.2). Note also that the boundary layer approximation can be used not merely in boundary layers adjacent to walls (see Sections 11.2 and 11.3) but also in *wakes* downstream of walls and in *jets* (see Section 11.4).

11.2 FLOW PAST A FLAT PLATE

We consider the flow of an incompressible Newtonian fluid of density ρ and viscosity μ toward and past a semi-infinite flat plate of thickness H. We suppose that the flow far away from the plate (at infinity) is uniform with velocity U_∞ aligned parallel to the plate. Because of the obvious geometry, we use rectangular coordinates (x, y, z) to analyze the problem, with origin on one corner of the leading edge of the plate. The x direction is aligned with the direction of the flow at infinity, so that $\mathbf{U}_\infty = U_\infty \mathbf{i}_x$, and the y direction is aligned normal to the semi-infinite surfaces of the plate (see Fig. 11.1).

We seek to determine the flow past the plate when the Reynolds number, which we have yet to define, is large. Thus we expect that viscous effects are

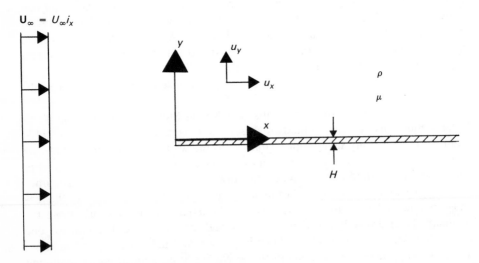

Figure 11.1. Geometry of flow past a flat plate.

negligible far from the flat plate, but that they are not negligible near it in a region called a boundary layer. Because the plate is semi-infinite, we assume that the plate is in fact of zero thickness ($H = 0$). This assumption should not lead to discrepancies at distances from the leading edge of the plate that are large compared with H. By symmetry, we henceforth confine our attention to one side ($y \geq 0$) of the plate only. We expect that the flow is steady. We also expect that it is planar, so that the only nonzero components of the velocity field **u** are the x and y components (u_x and u_y, respectively) and that they, like the pressure field p, are independent of z.

The mass conservation equation (A2.2-2) of Appendix B yields

$$\frac{\partial u_x}{\partial x} + \frac{\partial u_y}{\partial y} = 0 \tag{11.2-1}$$

The linear momentum conservation equations (A2.3-2) of Appendix B (with the transient term omitted and the gravitational term incorporated in the pressure term) yield

$$u_x \frac{\partial u_x}{\partial x} + u_y \frac{\partial u_x}{\partial y} = -\frac{1}{\rho}\frac{\partial p}{\partial x} + \nu \left(\frac{\partial^2 u_x}{\partial x^2} + \frac{\partial^2 u_x}{\partial y^2} \right) \tag{11.2-2}$$

$$u_x \frac{\partial u_y}{\partial x} + u_y \frac{\partial u_y}{\partial y} = -\frac{1}{\rho}\frac{\partial p}{\partial y} + \nu \left(\frac{\partial^2 u_y}{\partial x^2} + \frac{\partial^2 u_y}{\partial y^2} \right) \tag{11.2-3}$$

where the kinematic viscosity $\nu = \mu/\rho$. In principle, we need to specify eight boundary conditions on velocity, that is, four each on u_x and u_y. This is because Eqs. (11.2-2) and (11.2-3) are second order in u_x and u_y in both the x and y directions. Since there is no slip and no flow-through at the plate, two of these conditions are

$$u_x = 0 \qquad u_y = 0 \qquad \text{at } y = 0 \tag{11.2-4}$$

Assuming that the flow becomes uniform far from the plate, it might also be argued (not quite correctly, as we will see) that two more conditions are

$$u_x \to U_\infty \qquad u_y \to 0 \qquad \text{as } y \to \infty \tag{11.2-5}$$

Roughly speaking, this gives four boundary conditions in the y direction; we still need four in the x direction. Since the flow is a high Reynolds number flow, it might be argued that the flow upstream of the plate (i.e., for $x < 0$) is not significantly affected by the presence of the plate (this is certainly true far enough upstream), that is,

$$u_x = U_\infty \qquad u_y = 0 \qquad \text{at } x = 0 \tag{11.2-6}$$

which gives two boundary conditions in the x direction. The remaining two must presumably be imposed downstream, say as $x \to \infty$, though what these two conditions might be is not at all clear. As we will see, however, we do not in fact need to specify them.

To proceed further, and to resolve the problem of boundary conditions, we perform an *order-of-magnitude analysis*. We let δ denote the thickness of the boundary layer, that is, the dimension in the y direction of the region by the plate in which viscous and inertial effects are comparable. [Note that we are not for the moment interested in the region where inertial effects dominate viscous effects, which lies beyond the boundary layer ($y > \delta$).] Then $u_x \sim U_\infty$ and $y \sim \delta$. Also $x \sim x$ trivially, since there is no length scale along the plate other than the distance from the leading edge. Thus $\partial/\partial x \sim 1/x$ and $\partial/\partial y \sim 1/\delta$. Hence Eq. (11.2-1) yields $u_y \sim \delta U_\infty/x$. It follows from Eq. (11.2-2) that the inertial (strictly, the convection) terms $u_x \, \partial u_x/\partial x + u_y \, \partial u_x/\partial y \sim U_\infty^2/x$ and the viscous terms $\nu(\partial^2 u_x/\partial x^2 + \partial^2 u_x/\partial y^2) \sim \nu(U_\infty/x^2 + U_\infty/\delta^2) \sim \nu U_\infty/\delta^2$ since we anticipate that the boundary layer is thin (i.e., $\delta \ll x$). Because viscous and inertial effects are comparable in the boundary layer,

$$\frac{U_\infty^2}{x} \sim \frac{\nu U_\infty}{\delta^2} \tag{11.2-7}$$

Let Re_x denote the Reynolds number Re_x given by

$$\text{Re}_x = \frac{U_\infty x}{\nu} \tag{11.2-8}$$

Note that Re_x is based on the distance x from the leading edge of the plate; we have already noted that there is no other length scale along the plate. Note also that we are seeking to determine the flow past the plate when $\text{Re}_x \gg 1$. Thus we anticipate that our boundary layer analysis is invalid if $\text{Re}_x \not\gg 1$, that is, near the leading edge where $x \not\gg \nu/U_\infty$. Combining Eqs. (11.2-7) and (11.2-8), we obtain

$$\delta \sim \frac{x}{\sqrt{\text{Re}_x}} \tag{11.2-9}$$

We see that δ varies as $\sqrt{x}$; the boundary layer gets thicker further downstream along the plate. Also, because $\text{Re}_x \gg 1$, we see that $\delta \ll x$, as assumed earlier. Because $y \sim \delta$ and $u_y \sim \delta U_\infty/x$, we deduce that $y \sim x/\sqrt{\text{Re}_x}$ and $u_y \sim U_\infty/\sqrt{\text{Re}_x}$. It follows from Eq. (11.2-3) that the inertial terms $u_x \, \partial u_y/\partial x + u_y \, \partial u_y/\partial y \sim U_\infty^2 \delta/x^2$ and that the viscous terms $\nu(\partial^2 u_y/\partial x^2 + \partial^2 u_y/\partial y^2) \sim \nu(U_\infty \delta/x^3 + U_\infty/x\delta) \sim \nu U_\infty/x\delta$ since $\delta \ll x$. Because viscous and inertial effects are comparable in the boundary layer,

$$\frac{U_\infty^2 \delta}{x^2} \sim \frac{\nu U_\infty}{x\delta} \tag{11.2-10}$$

which can be rearranged to give, and is thus consistent with, Eq. (11.2-9). Finally, it follows from Eqs. (11.2-2) and (11.2-3) that $\partial p/\partial x \sim \rho U_\infty^2/x$ and $\partial p/\partial y \sim \rho U_\infty^2 \delta/x^2$. (Note, incidentally, that these order-of-magnitude estimates of $\partial p/\partial x$ and $\partial p/\partial y$ emerged from those of the other terms in the flow equations; see Section 10.5.) Hence the ratio of the transverse to the axial pressure gradients $(\partial p/\partial y)/(\partial p/\partial x) \sim \delta/x \sim 1/\sqrt{\text{Re}_x} \ll 1$.

In fact, we might have anticipated the estimate of δ in Eq. (11.2-9) from our analysis of flow past an impulsively moved flat plate in Section 10.6. The thickness of the shear layer over the impulsively moved plate is given to an order of magnitude from Eq. (10.6-34) by $\sqrt{vt}$, where t denotes the time that has elapsed since the start of the impulsive motion of the plate. We might, therefore, expect to replace t by the time taken for fluid traveling at an axial velocity U_∞ to move an axial distance x. Thus, if we replace t by x/U_∞ and loosely identify the thickness of the shear layer with that of the boundary layer, we might expect that δ $\sim \sqrt{vx/U_\infty}$, which can be rearranged to give Eq. (11.2-9).

Based on this order-of-magnitude analysis, we now make simplifications to the equations of motion (11.2-1), (11.2-2), and (11.2-3). We note first that *no* simplification can be made to the mass conservation equation (11.2-1) for, if one term is in some sense small (or large), the other term must be correspondingly small (or large), that is, both terms must be of comparable magnitude. Indeed, the only way in which this equation can be simplified is if the velocity field **u** is fully developed. This is not the case here, however: the velocity field is still developing, so no simplification can be made. On the other hand, because $(\partial p/\partial y)/(\partial p/\partial x) \sim \delta/x \ll 1$, the transverse linear momentum conservation equation (11.2-3) can be simplified to

$$\frac{\partial p}{\partial y} = 0 \qquad (11.2\text{-}11)$$

Also, because $v(\partial^2 u_x/\partial y^2)/v(\partial^2 u_x/\partial x^2) \sim x^2/\delta^2 \gg 1$, the axial linear momentum conservation equation (11.2-2) can be simplified to

$$u_x \frac{\partial u_x}{\partial x} + u_y \frac{\partial u_x}{\partial y} = -\frac{1}{\rho}\frac{\partial p}{\partial x} + v\frac{\partial^2 u_x}{\partial y^2} \qquad (11.2\text{-}12)$$

Thus the simplified equations of motion which apply inside the boundary layer are Eqs. (11.2-1), (11.2-11), and (11.2-12).

Outside the boundary layer (i.e., for $y > \delta$), in what is called the free stream, viscous effects are negligible and so the flow is effectively inviscid. Thus Bernoulli's equation holds, that is,

$$p + \tfrac{1}{2}\rho|\mathbf{u}|^2 = \zeta \qquad (11.2\text{-}13)$$

[see Eq. (9.1-13)] where ζ is a constant; recall that the gravitational term is incorporated in the pressure term. In the free stream, we assume that

$$\mathbf{u} = \mathbf{U}_\infty = U_\infty \mathbf{i}_x \qquad (11.2\text{-}14)$$

Hence, $$p + \tfrac{1}{2}\rho U_\infty^2 = \zeta \qquad (11.2\text{-}15)$$

Because U_∞ is constant, it follows immediately that the pressure p in the free stream is also constant, and hence that

$$\frac{\partial p}{\partial x} = 0 \qquad (11.2\text{-}16)$$

$$\frac{\partial p}{\partial y} = 0 \qquad \text{11.2-17)}$$

It now follows from Eq. (11.2-11), which applies inside the boundary layer, that the pressure p is constant in the boundary layer as well as in the free stream, so that p is in fact constant everywhere. Hence, in the boundary layer,

$$\frac{\partial p}{\partial x} = 0 \qquad (11.2\text{-}18)$$

In order to see this, consider two points A and B within the boundary layer with coordinates (x_1, y_1) and (x_2, y_2), respectively, and two points C and D in the free stream with coordinates (x_1, y_3) and (x_2, y_4), respectively, where $x_1 \neq x_2$ (see Fig. 11.2). It follows from Eqs. (11.2-11) and (11.2-17) that the pressure at A is the same as the pressure at C; similarly the pressure at B is the same as the pressure at D. It then follows from Eq. (11.2-16) that the pressure at C is the same as the pressure at D. Thus the pressure at A is the same as the pressure at B. Since the positions of the points A and B within the boundary layer are arbitrary, we deduce that the pressures at any two points in the boundary layer are the same and hence that the pressure is constant in the boundary layer. Thus the axial linear momentum conservation Eq. (11.2-12) simplifies to

$$u_x \frac{\partial u_x}{\partial x} + u_y \frac{\partial u_x}{\partial y} = \nu \frac{\partial^2 u_x}{\partial y^2} \qquad (11.2\text{-}19)$$

This equation and the mass conservation equation (11.2-1) comprise *Prandtl's boundary layer equations* for flow plast a flat plate. We note that we need only four boundary conditions. One is needed on u_y, since Eq. (11.2-1) is first order

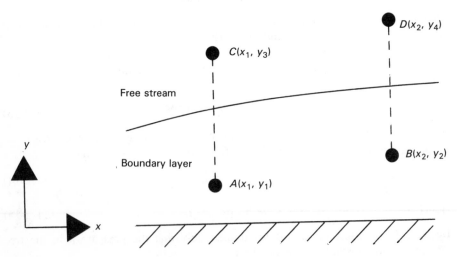

Figure 11.2. Location of points A and B within the boundary layer and C and D in the free stream: point A has the same axial coordinate as C; point B has the same axial coordinate as D.

in u_y in the y direction. Three are needed on u_x, since Eqs. (11.2-1) and (11.2-19) are both first order in u_x in the x direction and Eq. (11.2-19) is second order in u_x in the y direction. Two of these conditions, one on u_x and one on u_y, are given by Eqs. (11.2-4) (i.e., no slip and no flow-through at the plate). The other two on u_x are given by Eqs. (11.2-5) and (11.2-6), that is, by the requirement that u_x tends to U_∞ far from the plate and upstream of the plate. Thus, because Prandtl's boundary layer equations are of lower order than the Navier-Stokes equations, we need fewer boundary conditions. In particular, we do not need to specify downstream boundary conditions (as $x \to \infty$), which is fortunate since, as we noted earlier, it is not at all clear what such conditions might be.

To determine the flow within the boundary layer, we must solve the following equations:

$$\frac{\partial u_x}{\partial x} + \frac{\partial u_y}{\partial y} = 0 \qquad u_x \frac{\partial u_x}{\partial x} + u_y \frac{\partial u_x}{\partial y} = \nu \frac{\partial^2 u_x}{\partial y^2} \qquad (11.2\text{-}20)$$

subject to the following boundary conditions:

$$\begin{aligned} u_x &= 0 & u_y &= 0 & \text{at } y = 0 \\ u_x &= U_\infty & &\text{at } x = 0 & u_x &\to U_\infty & \text{as } y \to \infty \end{aligned} \qquad (11.2\text{-}21)$$

Because the flow is planar and solenoidal, there exists a stream function ψ such that

$$u_x = \frac{\partial \psi}{\partial y} \qquad u_y = -\frac{\partial \psi}{\partial x} \qquad (11.2\text{-}22)$$

(see Section 2.5). Hence the two equations (11.2-20) are replaced by the single equation:

$$\frac{\partial \psi}{\partial y} \frac{\partial^2 \psi}{\partial x \, \partial y} - \frac{\partial \psi}{\partial x} \frac{\partial^2 \psi}{\partial y^2} = \nu \frac{\partial^3 \psi}{\partial y^3} \qquad (11.2\text{-}23)$$

and boundary conditions (11.2-21) become

$$\begin{aligned} \psi &= 0 & \frac{\partial \psi}{\partial y} &= 0 & \text{at } y = 0 \\ \frac{\partial \psi}{\partial y} &= U_\infty & &\text{at } x = 0 & \frac{\partial \psi}{\partial y} &\to U_\infty & \text{as } y \to \infty \end{aligned} \qquad (11.2\text{-}24)$$

We note that, strictly, no flow-through at the plate means that ψ is constant at $y = 0$; we choose to let the constant vanish.

We now note that there is no characteristic length with which x and y can be made dimensionless. This is why we put $x \sim x$ in our order-of-magnitude analysis; we cannot use δ because it is not known in advance. Thus we seek a *similarity solution*. We define the *similarity variable* η:

$$\eta = c_1 x^{c_2} y \qquad (11.2\text{-}25)$$

and the *transformed stream function f*, which we require to be a function of η only:

$$f = c_3 x^{c_4} \psi \tag{11.2-26}$$

Here c_1, c_2, c_3, and c_4 are constants yet to be determined. In fact, c_1 and c_3 will be determined by considerations of convenience and c_2 and c_4 by the requirement that f be a function of η alone. Note that, unlike our similarity analyses in Sections 9.6 and 10.6, we not only define a similarity variable but also transform the principal flow variable. The reason for this is simply that we cannot obtain a similarity solution otherwise. (Of course, we could define a transformed stream function involving y and ψ rather than x and ψ: the choice is entirely arbitrary.) Then,

$$\frac{\partial \psi}{\partial y} = \frac{1}{c_3 x^{c_4}} \frac{df}{d\eta} \frac{\partial \eta}{\partial y} = \frac{c_1 x^{c_2}}{c_3 x^{c_4}} \frac{df}{d\eta} \tag{11.2-27}$$

Hence boundary conditions (11.2-24) yield

$$\frac{c_1}{c_3} x^{c_2 - c_4} \frac{df}{d\eta} \begin{cases} = U_\infty & \text{at } x = 0 \\ \to U_\infty & \text{as } y \to \infty \end{cases} \tag{11.2-28}$$

If f is independent of x, as is required of a similarity solution, it follows that $x^{c_2 - c_4}$ is a constant, that is,

$$c_4 = c_2 \tag{11.2-29}$$

If we also choose for convenience

$$c_3 = \frac{c_1}{U_\infty} \tag{11.2-30}$$

then boundary conditions (11.2-24) become

$$f = 0 \qquad \frac{df}{d\eta} = 0 \qquad \text{at } y = 0$$

$$\frac{df}{d\eta} = 1 \qquad \text{at } x = 0 \qquad \frac{df}{d\eta} \to 1 \qquad \text{as } y \to \infty \tag{11.2-31}$$

Because

$$\frac{\partial \psi}{\partial y} = U_\infty \frac{df}{d\eta} \qquad \frac{\partial^2 \psi}{\partial y^2} = c_1 x^{c_2} U_\infty \frac{d^2 f}{d\eta^2} \qquad \frac{\partial^3 \psi}{\partial y^3} = c_1^2 x^{2c_2} U_\infty \frac{d^3 f}{d\eta^3}$$

$$\frac{\partial \psi}{\partial x} = -\frac{c_2}{c_1} x^{-c_2 - 1} U_\infty f + c_2 x^{-1} y U_\infty \frac{df}{d\eta} \tag{11.2-32}$$

$$\frac{\partial^2 \psi}{\partial x \, \partial y} = c_1 c_2 x^{c_2 - 1} y U_\infty \frac{d^2 f}{d\eta^2}$$

Eq. (11.2-23) becomes

$$\frac{d^3f}{d\eta^3} - f\frac{d^2f}{d\eta^2}\frac{c_2 U_\infty}{vc_1^2 x^{2c_2+1}} = 0 \tag{11.2-33}$$

Because, for a similarity solution, we require that f is a function of η only, it is clearly necessary that Eq. (11.2-33) contains no terms involving x (or y; any term involving η and x can be rewritten in an obvious way as a term involving η and y). Thus it is necessary either that c_2 vanishes, which implies a trivial solution in which there is no flow, or that x^{2c_2+1} is a constant, that is, that

$$c_2 = -\tfrac{1}{2} \tag{11.2-34}$$

and hence,

$$c_4 = -\tfrac{1}{2} \tag{11.2-35}$$

It is also convenient to make the similarity variable η dimensionless. Accordingly, we choose

$$c_1 = \sqrt{\frac{U_\infty}{2v}} \tag{11.2-36}$$

(where the factor of 2 is, of course, entirely arbitrary) and hence,

$$c_3 = \frac{1}{\sqrt{2vU_\infty}} \tag{11.2-37}$$

Thus

$$\eta = y\sqrt{\frac{U_\infty}{2vx}} \tag{11.2-38}$$

and

$$f = \frac{\psi}{\sqrt{2vU_\infty x}} \tag{11.2-39}$$

We note that η and f are both dimensionless. Also, because $y \sim x/\sqrt{Re_x} = \sqrt{xv/U_\infty}$, we see that $y\sqrt{U_\infty/vx} \sim 1$. Hence $\eta \sim 1$, which shows that we have scaled η sensibly. Equation (11.2-33) becomes

$$\frac{d^3f}{d\eta^3} + f\frac{d^2f}{d\eta^2} = 0 \tag{11.2-40}$$

which is *Blasius' equation*. Boundary conditions (11.2-31) become

$$f = 0 \qquad \frac{df}{d\eta} = 0 \qquad \text{at } \eta = 0 \qquad \frac{df}{d\eta} \to 1 \qquad \text{as } \eta \to \infty \tag{11.2-41}$$

where the two boundary conditions on f as $y \to \infty$ and at $x = 0$ are combined into a single boundary condition on f as $\eta \to \infty$. Thus $y \to \infty$ is, in a sense, equivalent to $x = 0$, which is physically reasonable since conditions far from the plate are the same as those at its leading edge. Note also that boundary conditions (11.2-41), like Eq. (11.2-40), involve f (or its derivatives) and η alone, thus satisfying the essential condition for the existence of a similarity solution (see Section 10.6).

Equation (11.2-40) and boundary conditions (11.2-41) comprise a *two-point boundary value problem*, because the boundary conditions must be imposed at two different points, $\eta = 0$ and $\eta \to \infty$. Since the problem is *nonlinear* because of the term $f\, d^2f/d\eta^2$ in Eq. (11.2-40), analytical solution appears to be impossible. Numerical solution is required instead. In general, to obtain a numerical solution, we would need to

1. Guess $(d^2f/d\eta^2)|_{\eta=0}$.
2. Integrate Eq. (11.2-40) numerically, given $f|_{\eta=0}$, $(df/d\eta)|_{\eta=0}$, and $(d^2f/d\eta^2)|_{\eta=0}$.
3. Check that $(df/d\eta)|_{\eta\to\infty} \to 1$ (to within some small tolerance); if not, we would have to reguess $(d^2f/d\eta^2)|_{\eta=0}$, and so on.

In order to avoid use of such an iterative, trial-and-error scheme, we proceed as follows. Suppose that $F(\eta)$ is *any* solution of Eq. (11.2-40) [and *not* necessarily one that satisfies boundary conditions (11.2-41)]. Then it is easy to show that $f = \alpha F(\alpha\eta)$ is also a solution, where α is an arbitrary constant that can be chosen so that f satisfies boundary conditions (11.2-41). Suppose, in particular, that $F(\eta)$ satisfies the boundary conditions

$$F = 0 \qquad \frac{d}{d\eta}F(\eta) = 0 \qquad \frac{d^2}{d\eta^2}F(\eta) = 1 \qquad \text{at } \eta = 0 \quad (11.2\text{-}42)$$

Because

$$\left.\frac{df}{d\eta}\right|_{\eta\to\infty} = \alpha^2 \left.\frac{d}{d\eta}F(\alpha\eta)\right|_{\eta\to\infty} = \alpha^2 \left.\frac{d}{d\eta}F(\eta)\right|_{\eta\to\infty}$$

$$\alpha^2 = \left[\left.\frac{d}{d\eta}F(\eta)\right|_{\eta\to\infty}\right]^{-1} \qquad (11.2\text{-}43)$$

using Eqs. (11.2-41). Also

$$\left.\frac{d^2f}{d\eta^2}\right|_{\eta=0} = \alpha^3 \left.\frac{d^2}{d\eta^2}F(\alpha\eta)\right|_{\eta=0} = \alpha^3 \left.\frac{d^2}{d\eta^2}F(\eta)\right|_{\eta=0}$$

so

$$\left.\frac{d^2f}{d\eta^2}\right|_{\eta=0} = \alpha^3 \qquad (11.2\text{-}44)$$

using Eqs. (11.2-42). Thus, in order directly to determine $(d^2f/d\eta^2)|_{\eta=0}$ and avoid having iteratively to guess it, we solve the equation.

$$\frac{d^3}{d\eta^3}F(\eta) + F(\eta)\frac{d^2}{d\eta^2}F(\eta) = 0 \qquad (11.2\text{-}45)$$

subject to boundary conditions (11.2-42) by numerical integration. This comprises a one-point boundary value problem, because the boundary conditions are all imposed at one point ($\eta = 0$). It may, therefore, be solved directly (i.e., noniteratively). This enables us to determine $[d(F(\eta))/d\eta]|_{\eta\to\infty}$. In order to effect the numerical integration, any standard method may be used. We find that

$$\frac{d}{d\eta} F(\eta)\bigg|_{\eta\to\infty} \approx 1.655190 \qquad (11.2\text{-}46)$$

Hence
$$\frac{d^2 f}{d\eta^2}\bigg|_{\eta=0} \approx 0.469600 \qquad (11.2\text{-}47)$$

The solution for f (and also for $df/d\eta$ and $d^2f/d\eta^2$) may then be obtained from that for F by algebraic transformation and is given in Table 11.1 (see also Fig. 11.3). Note that, for $\eta > 5$ or so, $df/d\eta \approx 1$. Thus η is very large (i.e., effectively infinite) when $\eta > 5$ or so.

Table 11.1 Values of f, $\dfrac{df}{d\eta}$, and $\dfrac{d^2 f}{d\eta^2}$ as functions of η

η	f	$\dfrac{df}{d\eta}$	$\dfrac{d^2 f}{d\eta^2}$
0	0	0	0.4696
0.2	0.0094	0.0939	0.4693
0.4	0.0375	0.1876	0.4673
0.6	0.0844	0.2806	0.4617
0.8	0.1497	0.3720	0.4512
1.0	0.2330	0.4606	0.4344
1.2	0.3337	0.5452	0.4106
1.4	0.4507	0.6244	0.3797
1.6	0.5830	0.6967	0.3425
1.8	0.7289	0.7611	0.3004
2.0	0.8868	0.8167	0.2557
2.2	1.0549	0.8633	0.2106
2.4	1.2315	0.9011	0.1676
2.6	1.4148	0.9306	0.1286
2.8	1.6033	0.9529	0.0951
3.0	1.7956	0.9691	0.0677
3.2	1.9906	0.9804	0.0464
3.4	2.1875	0.9880	0.0305
3.6	2.3856	0.9929	0.0193
3.8	2.5845	0.9959	0.0118
4.0	2.7839	0.9978	0.0069
4.2	2.9836	0.9988	0.0039
4.4	3.1834	0.9994	0.0021
4.6	3.3833	0.9997	0.0011
4.8	3.5833	0.9999	0.0005
5.0	3.7832	0.9999	0.0003
5.2	3.9832	1.0000	0.0001
5.4	4.1832	1.0000	0.0001
5.6	4.3832	1.0000	0.0000
5.8	4.5832	1.0000	0.0000
6.0	4.7832	1.0000	0.0000
6.2	4.9832	1.0000	0.0000
6.4	5.1832	1.0000	0.0000
6.6	5.3832	1.0000	0.0000
6.8	5.5832	1.0000	0.0000
7.0	5.7832	1.0000	0.0000

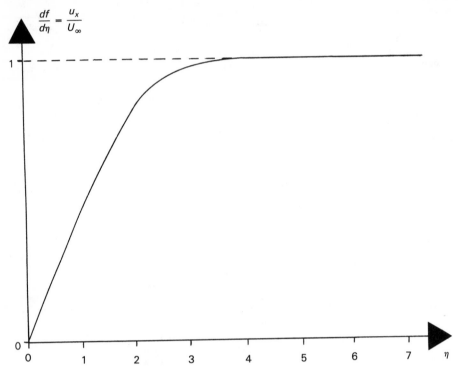

Figure 11.3. Variation with similarity variable of dimensionless axial velocity past a flat plate.

The role of the similarity variable η can be seen in Fig. 11.4, where axial velocity profiles are shown schematically at different axial positions along the plate. Along a contour of constant η, that is, of constant $y/\sqrt{x}$, $df/d\eta$ is constant, that is, u_x is constant. Thus the profiles at different positions are all transversely stretched or contracted versions of one another.

If we now substitute the solution that we have obtained back into the Navier-Stokes equations (11.2-2) and (11.2-3), we find that the terms neglected in Prandtl's boundary layer equations (11.2-20) do indeed vanish as $Re_x \rightarrow \infty$. So we may conjecture, though there is as yet no proof, that the solution is an asymptotic approximation to the exact solution for flow past a flat plate in the limit as $Re_x \rightarrow \infty$. Note, however, that the existence of this solution does not imply its stability to disturbances. Indeed, a stability analysis (corroborated by experimental evidence) suggests that the solution is unstable if Re_x exceeds some critical value $Re_{x_c} \simeq 300,000$, though the flow can be stable for higher values of Re_x if the plate, etc., are smooth enough. The resulting turbulent flow that occurs when Re_x sufficiently exceeds Re_{x_c} is discussed in Section 12.2.

It follows from the numerical solution that

$$u_y \rightarrow 0.860 \frac{U_\infty}{\sqrt{Re_x}} \qquad \text{as } y \rightarrow \infty \qquad (11.2\text{-}48)$$

Thus, while u_x tends to the free stream velocity U_∞, u_y does not vanish as $y \rightarrow \infty$. This is why, as we noted earlier, boundary condition (11.2-5) cannot (quite) be satisfied and, moreover, why our assumption of the validity of Eq. (11.2-14) is not entirely justified. The reason why the transverse velocity u_y does not tend to vanish as $y \rightarrow \infty$ can be seen using Fig. 11.5. A control surface A-C-D-B-A is identified. A and B are on the plate and C and D far from the plate (at infinity). A and C have the same axial location (the leading edge of the plate), as do B and D which are downstream of A and C, respectively.

Comparing axial velocity profiles, we see that the rate at which mass enters the control surface through surface A-C is greater than the rate at which it leaves through surface B-D. Thus, since the plate comprising surface A-B is impermeable, there must be a positive mass flow rate out of the control surface through surface C-D. Thus we may deduce not only that $u_y \neq 0$ but also that $u_y > 0$ as $y \rightarrow \infty$, in accordance with Eq. (11.2-48).

So far, we have used the imprecisely defined quantity δ to denote the boundary layer thickness. We can make the definition of thickness more precise in many different ways. One common but essentially arbitrary way is to define the thickness as the distance from the plate at which the axial velocity u_x differs from U_∞ by some specified small amount, for example, where $u_x = 0.99\, U_\infty$. A less arbitrary way is to define the displacement thickness δ^*. This is the distance that the plate would have to be moved in the y direction to maintain the same volumetric (or, more strictly, mass) flow rate past it assuming that there is slip and

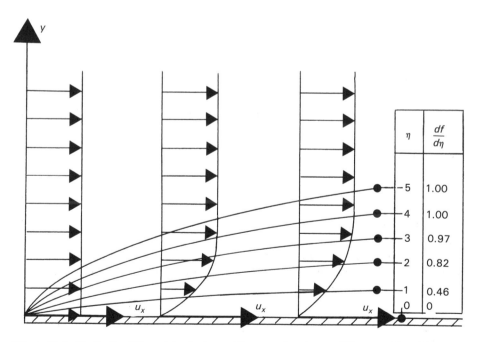

Figure 11.4. Similarity of the axial velocity profile past a flat plate at different axial positions.

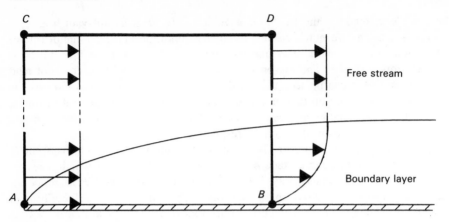

C   D

Free stream

Boundary layer

A   B

Figure 11.5. Control surface A-C-D-B-A with points A and B on the plate and C and D far from the plate: point A has the same axial position as C; point B has the same axial position as D.

hence no boundary layer (see Fig. 11.6). Thus,

$$\int_0^\infty u_x \, dy = \int_{\delta^*}^\infty U_\infty \, dy = \int_0^\infty U_\infty \, dy - U_\infty \delta^* \qquad (11.2\text{-}49)$$

so

$$\delta^* = \int_0^\infty \left(1 - \frac{u_x}{U_\infty}\right) dy = \sqrt{\frac{2\nu x}{U_\infty}} \int_0^\infty \left(1 - \frac{df}{d\eta}\right) d\eta \qquad (11.2\text{-}50)$$

or

$$\frac{\delta^*}{x} = \sqrt{\frac{2}{\text{Re}_x}} \left.(\eta - f)\right|_0^\infty \qquad (11.2\text{-}51)$$

and so (using Table 11.1),

$$\frac{\delta^*}{x} \simeq \frac{1.7208}{\sqrt{\text{Re}_x}} \qquad (11.2\text{-}52)$$

This should be compared with Eq. (11.2-9) and shows how good the order-of-magnitude estimate is.

We now estimate the drag on the plate. A convenient way of doing this is in terms of the local drag coefficient C_f. This is a dimensionless wall shear stress and is local because it varies with axial position x. It is given by

$$C_f = \frac{\left.\tau_{yx}\right|_{y=0}}{\rho U_\infty^2/2} \qquad (11.2\text{-}53)$$

The shear stress component $\tau_{yx} = \mu(\partial u_x/\partial y + \partial u_y/\partial x)$ is given by $\mu \, \partial u_x/\partial y$ at $y = 0$ because u_y vanishes at $y = 0$. Hence,

$$C_f = \sqrt{\frac{2}{\text{Re}_x}} \frac{d^2 f}{d\eta^2}\bigg|_{\eta=0} \tag{11.2-54}$$

or (using Table 11.1),

$$C_f \simeq 0.664 \sqrt{\text{Re}_x} \tag{11.2-55}$$

We also define the mean drag coefficient $\overline{C}_f$:

$$\overline{C}_f = \frac{1}{x} \int_0^x C_f \, dx \tag{11.2-56}$$

whence

$$\overline{C}_f \simeq \frac{1.328}{\sqrt{\text{Re}_x}} \tag{11.2-57}$$

Note that the mean drag coefficient up to a position x on the plate is precisely twice the local drag coefficient at that position x.

So far, we have considered uniform flow at infinity past the flat plate, for

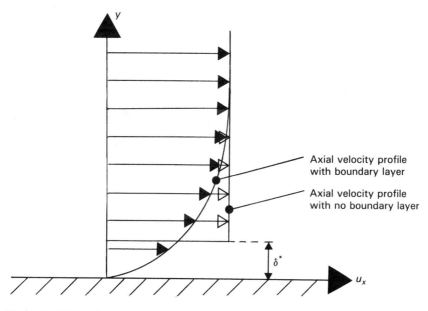

Figure 11.6. The displacement thickness is the distance that a plate has to be moved parallel to itself to maintain the same volumetric flow rate past it when there is and is not a boundary layer.

which $\partial p / \partial x = 0$ [see Eqs. (11.2-16) and (11.2-18)]. We now conclude by considering briefly cases when $\partial p / \partial x \neq 0$. At the plate, no slip and no flow-through [see boundary conditions (11.2-4)] mean that the axial linear momentum conservation equation (11.2-2) becomes

$$\frac{\partial^2 u_x}{\partial y^2} = \frac{1}{\mu} \frac{\partial p}{\partial x} \qquad \text{at } y = 0 \qquad (11.2\text{-}58)$$

Far from the plate, suppose that

$$u_x \to U_\infty(x) \qquad \text{as } y \to \infty \qquad (11.2\text{-}59)$$

Note that, since

$$\frac{\partial p}{\partial x} = -\rho U_\infty \frac{dU_\infty}{dx} \qquad \text{as } y \to \infty$$

[see Eq. (11.2-15)], $\partial p / \partial x \neq 0$ if and only if $dU_\infty / dx \neq 0$. Thus, far from the plate,

$$\frac{\partial u_x}{\partial y} \to 0 \qquad \frac{\partial^2 u_x}{\partial y^2} \to 0 \qquad \text{as } y \to \infty \qquad (11.2\text{-}60)$$

Inspecting Eq. (11.2-58), we now consider three cases:

1. If $\partial p / \partial x = 0$, then, at the plate,

$$\frac{\partial^2 u_x}{\partial y^2} = 0 \qquad \text{at } y = 0 \qquad (11.2\text{-}61)$$

so that $\partial u_x / \partial y$ is an extremum at $y = 0$.
2. If $\partial p / \partial x < 0$, then, at the plate,

$$\frac{\partial^2 u_x}{\partial y^2} < 0 \qquad \text{at } y = 0 \qquad (11.2\text{-}62)$$

so that $\partial u_x / \partial y$ decreases as y increases in the immediate vicinity of the plate.
3. If $\partial p / \partial x > 0$, then, at the plate,

$$\frac{\partial^2 u_x}{\partial y^2} > 0 \qquad \text{at } y = 0 \qquad (11.2\text{-}63)$$

so that $\partial u_x / \partial y$ increases as y increases in the immediate vicinity of the plate.

Because $\partial u_x / \partial y$ and $\partial^2 u_x / \partial y^2$ vanish far from the plate [see Eqs. (11.2-60)], the profiles of $\partial u_x / \partial y$ and hence u_x must be qualitatively as shown in Fig. 11.7. In particular, if $\partial p / \partial x$ is positive, there must be a maximum in the profile of $\partial u_x /$

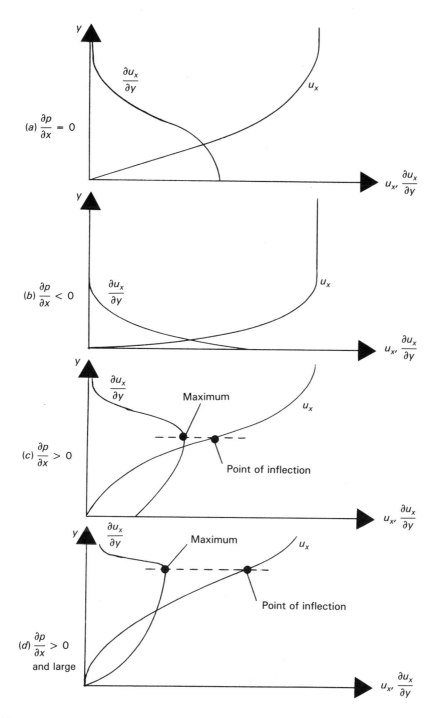

Figure 11.7. Variation of the axial velocity and transverse gradient of the axial velocity with transverse position when the axial pressure gradient is (a) zero, (b) negative, (c) positive, and (d) positive and large.

∂y and hence a point of inflection in the profile of u_x. It then follows that, provided $\partial p/\partial x$ is sufficiently large and positive, $\partial u_x/\partial y$ can vanish at the plate, that is,

$$\frac{\partial u_x}{\partial y} = 0 \qquad \text{at } y = 0 \tag{11.2-64}$$

(see Fig. 11.7). The physical significance of the vanishing of $\partial u_x/\partial y$ at the plate can be seen by considering Fig. 11.8. This shows the flow in the neighborhood of a *separation point,* which is where a *dividing streamline,* the locus of which is $x_s(y)$, meets the plate [at $x_s(0)$]. Fluid at all points (x, y) such that $x < x_s(y)$ originates far upstream $(x \to -\infty)$, whereas fluid at all points (x, y) such that $x > x_s(y)$ originates far downstream $(x \to \infty)$. This means that, at the plate,

$$\left.\begin{array}{ll} \dfrac{\partial u_x}{\partial y} > 0 & \text{for } x < x_s(0) \\[2.5em] \dfrac{\partial u_x}{\partial y} < 0 & \text{for } x > x_s(0) \end{array}\right\} \qquad \text{at } y = 0 \tag{11.2-65}$$

and hence that

$$\frac{\partial u_x}{\partial y} = 0 \qquad \text{at } x = x_s(0), \ y = 0 \tag{11.2-66}$$

Comparison of Eqs. (11.2-64) and (11.2-66) shows that flow separation occurs when the pressure increases strongly in the direction of the main flow (i.e.,

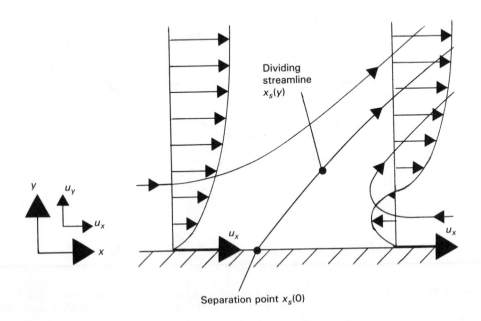

Figure 11.8. Geometry of flow in the vicinity of a separation point.

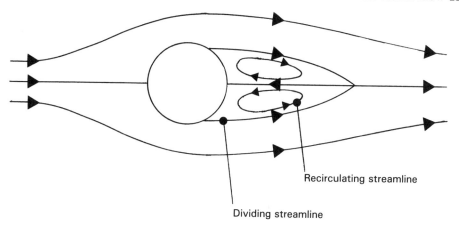

Recirculating streamline

Dividing streamline

Figure 11.9. Separation of flow past a sphere.

when there is a strong adverse pressure gradient). The physical reason for this is that an adverse pressure gradient tends to decelerate the flow: if it is strong enough, it can cause local flow in the opposite direction to the main flow and hence separation.

The importance of separation is that its presence implies significant changes in the flow about a body, for example in the drag on or heat transfer to the body. Flow around a sphere may be regarded as typical of flow about any bluff body. At a very low (strictly, zero) Reynolds number, the flow is symmetric fore and aft, and at a higher (but still low) Reynolds number, the flow is slightly asymmetric fore and aft. At a higher Reynolds number still, the flow becomes increasingly asymmetric until it eventually separates from the surface of the sphere (see Fig. 11.9) and recirculating streamlines form in a closed wake downstream of the sphere. Because the wake is closed, transported quantities such as momentum and heat can only enter and leave by a diffusion process (viscosity in the case of momentum, conduction in the case of heat) and not by convection. Accordingly, transported quantities are transferred only rather slowly between the wake and the rest of the flow. At an even higher Reynolds number, the wake becomes unstable and eventually no longer closed. At a very high Reynolds number, the flow in the vicinity of the sphere becomes turbulent and is then beyond the scope of our discussion here.

11.3 ENTRANCE LENGTH IN A PIPE

We consider the flow at constant volumetric flow rate Q of an incompressible Newtonian fluid of density ρ and viscosity μ from a very large reservoir into a very long circular pipe of radius R. Because of the obvious geometry, we use cylindrical polar coordinates (r, θ, z) to analyze the problem, with origin on the centerline of the pipe entrance and z direction aligned with the centerline (see Fig. 11.10).

Let u_z denotes the axial component of velocity. Its mean value $\bar{u}_z$ is given by

$$\bar{u}_z = \frac{1}{\pi R^2} \int_0^R u_z 2\pi r \, dr \qquad (11.3\text{-}1)$$

and is related to Q by

$$Q = \bar{u}_z \pi R^2 \qquad (11.3\text{-}2)$$

We define the Reynolds number:

$$\text{Re} = \frac{\rho \bar{u}_z 2R}{\mu} \qquad (11.3\text{-}3)$$

and assume that Re is less than a critical Reynolds number $\text{Re}_c \simeq 2000$, so that the flow in the pipe is laminar (see Section 10.2). In Section 10.2, we analyzed fully developed laminar flow in a pipe. Here we seek to obtain an estimate of the

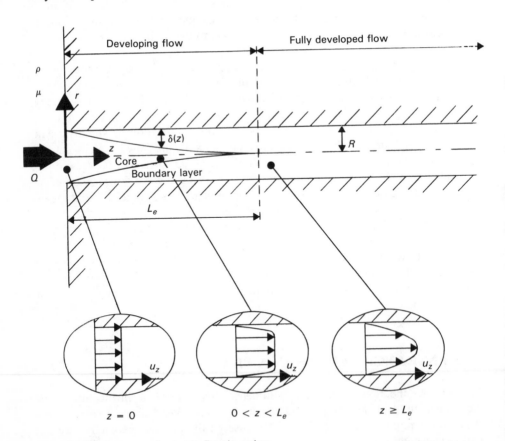

Figure 11.10. Geometry of entrance flow in a pipe.

entrance length of the flow. In other words, we seek an estimate of the distance L_e down the pipe from its entrance beyond which the flow is essentially fully developed.

We start by using dimensional analysis. The physical quantities that determine L_e are ρ, μ, R, and Q. Dimensional analysis then yields two dimensionless groups, L_e/R and Re say, and we deduce that

$$\frac{L_e}{R} = f(\text{Re}) \tag{11.3-4}$$

We see that the ratio of L_e to R is an arbitrary function f of Re alone, where Re is given by Eq. (11.3-3). We can, however, obtain no further information about the functional form of f on purely dimensional grounds.

We continue by noting that

- At the entrance to the pipe ($z = 0$), we expect that the axial velocity profile is uniform

$$u_z = \bar{u}_z \tag{11.3-5}$$

- Beyond the entrance length region of the pipe ($z \geq L_e$), the axial velocity profile is parabolic

$$u_z = 2\bar{u}_z\left(1 - \frac{r^2}{R^2}\right) \tag{11.3-6}$$

[see Eqs. (10.2-17) and (10.2-20)].

The evolution of the profile from being uniform to being parabolic (see Fig. 11.10) is intimately connected with the growth of a boundary layer on the wall on the pipe. The boundary layer forms in just the same way as it does for flow past a flat plate (see Section 11.2). Indeed, the two boundary layers would be indistinguishable were it not for the fact that the wall of the pipe is curved and that there is acceleration of the flow in the core of the pipe.

Provided the boundary layer is thin (i.e., its thickness is much less than the radius of the pipe), we would expect to be able to unroll the flow in the pipe into an equivalent flow past a flat plate and hence to be able to neglect wall curvature. The boundary layer is, of course, thin when $z \ll L_e$ but not when $z \sim L_e$ (as we will see). Although it is not, therefore, strictly justifiable to neglect wall curvature everywhere in the entrance length region, we will in fact do so in what follows.

Core acceleration occurs because the flow is bounded by the pipe walls. Since there is no slip at the walls, the fluid in the boundary layer must decelerate. Hence the fluid in the core, that is, outside the boundary layer, must accelerate in order to conserve mass. No acceleration occurs in the case of flow past a flat plate because the flow is not bounded in the same way. The result of the core acceleration is that there is a nonzero axial pressure gradient. But the centerline axial velocity at the entrance is $\bar{u}_z$ while that in the fully developed region is $2\bar{u}_z$ [see

Eqs. (11.3-5) and (11.3-6)]. Thus the centerline axial velocity only doubles between $z = 0$ and $z = L_e$. Hence we would not expect that core acceleration has a large effect on the boundary layer.

It then follows from Eqs. (11.2-9) and (11.2-52) that the thickness δ of the boundary layer is given by

$$\delta \sim \frac{z}{\sqrt{\mathrm{Re}_z}} \tag{11.3-7}$$

where the Reynolds number Re_z based on the distance z from the pipe entrance is given by

$$\mathrm{Re}_z = \frac{\rho \bar{u}_z z}{\mu} \tag{11.3-8}$$

When the boundary layer thickness δ is indistinguishable from the pipe radius R, that is, when the boundary layer has grown sufficiently to occupy the whole cross section of the pipe, the evolution of the axial velocity profile from uniform to parabolic is essentially complete. Thus we expect that $z \sim L_e$ when $\delta \to R$. If we replace z by L_e and δ by R in Eqs. (11.3-7) and (11.3-8), we obtain

$$R \sim \frac{L_e}{\sqrt{\rho \bar{u}_z L_e / \mu}} \tag{11.3-9}$$

or

$$\frac{L_e}{R} \sim \mathrm{Re} \tag{11.3-10}$$

which we may express in the alternative form

$$\frac{L_e}{R} = c\,\mathrm{Re} \tag{11.3-11}$$

Here $c \sim 1$ is a constant which cannot be determined from the dimensional and order-of-magnitude arguments that we have used here. Unless we are prepared to solve the complete flow problem, with all the difficulties which that might imply, the only way in which we can determine c is experimentally, when it is found that

$$c \simeq 0.1 \tag{11.3-12}$$

With this value of c substituted into Eq. (11.3-11), we can make an estimate of L_e. The estimate is of immediate use for flow in a pipe of length $L \gg L_e$, since the flow may then be taken to be fully developed over the whole length of the pipe, which is just what we did in Section 10.2.

11.4 FLOW OF A PLANE JET

We consider the flow of an incompressible Newtonian fluid of density ρ and viscosity μ out of a wide, shallow slit. The fluid issues steadily as a jet into a very large reservoir of the same fluid which is otherwise stationary. Since the slit is shallow, we assume that its height H is vanishingly small (i.e., $H \rightarrow 0$). In order to maintain a nonzero flow, the speed of the fluid in the slit is, therefore, infinite. Because of the obvious geometry, we use rectangular coordinates (x, y, z) to analyze the problem, with origin on the centerline of the exit of the slit. The y direction is aligned with the shallow dimension of the slit and the z direction is aligned with its wide dimension (see Fig. 11.11).

Because the slit is very wide, we expect that the jet flow is planar. Thus the z component of velocity vanishes and its x and y components u_x and u_y, respectively, are independent of z, as is the pressure p. The mass conservation equation (A2.2-2) of Appendix B yields

$$\frac{\partial u_x}{\partial x} + \frac{\partial u_y}{\partial y} = 0 \tag{11.4-1}$$

and the linear momentum conservation equations (A2.3-2) of appendix B (with the transient term omitted and the gravitational term incorporated in the pressure term) yield

$$u_x \frac{\partial u_x}{\partial x} + u_y \frac{\partial u_x}{\partial y} = -\frac{1}{\rho}\frac{\partial p}{\partial x} + \nu \left(\frac{\partial^2 u_x}{\partial x^2} + \frac{\partial^2 u_x}{\partial y^2} \right) \tag{11.4-2}$$

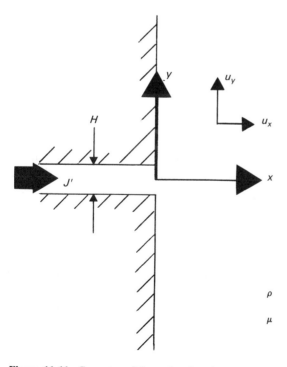

Figure 11.11. Geometry of flow of a plane jet.

$$u_x \frac{\partial u_y}{\partial x} + u_y \frac{\partial u_y}{\partial y} = -\frac{1}{\rho}\frac{\partial p}{\partial y} + \nu\left(\frac{\partial^2 u_y}{\partial x^2} + \frac{\partial^2 u_y}{\partial y^2}\right) \tag{11.4-3}$$

where the kinematic viscosity $\nu = \mu/\rho$. In order to solve these equations, we suppose that the jet spreads outward only slowly. In other words, we suppose that the jet is thin. Then the transverse velocity u_y will be very small compared with the axial velocity u_x, (i.e., $|u_y/u_x| << 1$). Hence we expect that the transverse linear momentum conservation equation (11.4-3) becomes

$$\frac{\partial p}{\partial y} = 0 \tag{11.4-4}$$

except near the slit. Far from the centerplane of the jet (i.e., as $|y| \to \infty$), we expect that the pressure p is constant. It thus follows, using Eq. (11.4-4), that p is everywhere constant (the argument is analogous to that given for constancy of p in the boundary layer and free stream in Section 11.2). The *momentum flow rate* (or *momentum flux*) J' in the x direction per unit width of slit is given by

$$J' = \rho \int_{-\infty}^{+\infty} u_x^2 \, dy \tag{11.4-5}$$

Since there are no rigid impermeable walls in the jet and p is constant, J' must be constant, that is,

$$\frac{dJ'}{dx} = 0 \tag{11.4-6}$$

It also follows that, except near the slit, the axial linear momentum conservation equation (11.4-2) simplifies to

$$u_x \frac{\partial u_x}{\partial x} + u_y \frac{\partial u_x}{\partial y} = \nu \frac{\partial^2 u_x}{\partial y^2} \tag{11.4-7}$$

Four boundary conditions are needed. Two are provided on the centerplane by symmetry (see Section 6.5); a third is provided at infinity by requiring that the kinetic energy of the flow is bounded

$$\frac{\partial u_x}{\partial y} = 0 \quad u_y = 0 \quad \text{at } y = 0 \quad u_x \to 0 \quad \text{as } |y| \to \infty \tag{11.4-8}$$

The fourth condition is provided by Eqs. (11.4-5) and (11.4-6). We recognize from Section 11.2 that Eqs. (11.4-1) and (11.4-7) are just Prandtl's boundary layer equations for flow past a flat plate: the conservation equations governing these two flows are thus identical. The flows are not the same, however, because they have different boundary conditions.

Because the flow is planar and solenoidal, there exists a stream function ψ such that

$$u_x = \frac{\partial \psi}{\partial y} \qquad u_y = -\frac{\partial \psi}{\partial x} \tag{11.4-9}$$

(see Section 2.5). Equation (11.4-1) is automatically satisfied and Eq. (11.4-7) becomes

$$\frac{\partial \psi}{\partial y} \frac{\partial^2 \psi}{\partial x \partial y} - \frac{\partial \psi}{\partial x} \frac{\partial^2 \psi}{\partial y^2} = \nu \frac{\partial^3 \psi}{\partial y^3} \tag{11.4-10}$$

Boundary conditions (11.4-8) become

$$\psi = 0 \qquad \frac{\partial^2 \psi}{dy^2} = 0 \qquad \text{at } y = 0 \qquad \frac{\partial \psi}{\partial y} \to 0 \qquad \text{as } |y| \to \infty \tag{11.4-11}$$

and Eqs. (11.4-5) and (11.4-6) become

$$J' = \rho \int_{-\infty}^{+\infty} \left(\frac{\partial \psi}{\partial y} \right)^2 dy \tag{11.4-12}$$

$$\frac{d}{dx} \left[\int_{-\infty}^{+\infty} \left(\frac{\partial \psi}{\partial y} \right)^2 dy \right] = 0 \tag{11.4-13}$$

We now note that there is no characteristic length. Thus we seek a similarity solution and define the similarity variable η:

$$\eta = c_1 x^{c_2} y \tag{11.4-14}$$

and the transformed stream function f, which we require to be a function of η only:

$$f = c_3 x^{c_4} \psi \tag{11.4-15}$$

Here c_1, c_2, c_3, and c_4 are constants yet to be determined. As we will see, they are *not* the same as the corresponding constants in Section 11.2 for, although we are seeking a similarity solution of the same equation, the boundary conditions are different. Substitution of Eqs. (11.4-14) and (11.4-15) into boundary conditions (11.4-11) yields

$$f = 0 \qquad \frac{d^2 f}{d\eta^2} = 0 \qquad \text{at } y = 0 \qquad \frac{df}{d\eta} \to 0 \qquad \text{as } |y| \to \infty \tag{11.4-16}$$

which gives no information on the constants that we wish to determine. Substitution into Eq. (11.4-13) yields

$$\frac{c_1}{c_3^2} (c_2 - 2 c_4) x^{c_2 - 2c_4 - 1} \int_{-\infty}^{+\infty} \left(\frac{df}{d\eta} \right)^2 d\eta = 0 \tag{11.4-17}$$

so that, ignoring trivial solutions,

$$c_4 = \tfrac{1}{2} c_2 \tag{11.4-18}$$

Equation (11.4-10) then becomes

$$\frac{d^3f}{d\eta^3} - \left[f\frac{d^2f}{d\eta^2} + \left(\frac{df}{d\eta}\right)^2 \right] \frac{c_2 \, x^{-(3/2)c_2-1}}{2vc_1c_3} = 0 \qquad (11.4\text{-}19)$$

Because, for a similarity solution, we require that f is a function of η only, it is clearly necessary that, ignoring trivial solutions, $x^{-(3/2)c_2-1}$ is a constant, that is,

$$c_2 = -\tfrac{2}{3} \qquad (11.4\text{-}20)$$

and hence

$$c_4 = -\tfrac{1}{3} \qquad (11.4\text{-}21)$$

We now choose to let

$$c_1 c_3 = \frac{1}{3v} \qquad (11.4\text{-}22)$$

One way of doing this is to choose

$$c_1 = \frac{1}{3v^{1/2}} \quad c_3 = \frac{1}{v^{1/2}} \qquad (11.4\text{-}23)$$

so that

$$\eta = \frac{y}{3v^{1/2}x^{2/3}} \qquad (11.4\text{-}24)$$

and

$$f = \frac{\psi}{v^{1/2}x^{1/3}} \qquad (11.4\text{-}25)$$

Note that this means that η and f are dimensional.
It thus follows that Eq. (11.4-19) becomes

$$\frac{d^3f}{d\eta^3} + f\frac{d^2f}{d\eta^2} + \left(\frac{df}{d\eta}\right)^2 = 0 \qquad (11.4\text{-}26)$$

and boundary conditions (11.4-16) become

$$f = 0 \quad \frac{d^2f}{d\eta^2} = 0 \quad \text{at } \eta = 0 \quad \frac{df}{d\eta} \to 0 \text{ as } |\eta| \to \infty \qquad (11.4\text{-}27)$$

Although the problem is nonlinear because of the terms $f\, d^2f/d\eta^2$ and $(df/d\eta)^2$ in Eq. (11.4-26), it can nevertheless be solved analytically. Thus Eq. (11.4-26) can be integrated to yield

$$\frac{d^2f}{d\eta^2} + f\frac{df}{d\eta} = c_5 \qquad (11.4\text{-}28)$$

where c_5 is a constant. Because f and $d^2f/d\eta^2$ both vanish at $\eta = 0$, it follows that

$$c_5 = 0 \qquad (11.4\text{-}29)$$

We now put

$$f(\eta) = 2\alpha F(E) \qquad \eta = \frac{E}{\alpha} \tag{11.4-30}$$

where α is an as yet unspecified constant. Then it follows that

$$\frac{d^2F}{dE^2} + 2F\frac{dF}{dE} = 0 \tag{11.4-31}$$

which can be integrated to yield

$$\frac{dF}{dE} + F^2 = c_6 \tag{11.4-32}$$

where c_6 is a constant. Because α is as yet unspecified, there is sufficient flexibility for us to put

$$c_6 = 1 \tag{11.4-33}$$

whence

$$\frac{dF}{dE} = 1 - F^2 \tag{11.4-34}$$

which can be integrated to yield

$$\tanh^{-1} F = E + c_7 \tag{11.4-35}$$

where c_7 is a constant. Because $f = 0$ at $\eta = 0$, $F = 0$ at $E = 0$ and so

$$c_7 = 0 \tag{11.4-36}$$

Therefore

$$f = 2\alpha \tanh(\alpha\eta) \tag{11.4-37}$$

We must now determine α. Since

$$u_x = \frac{\partial\psi}{\partial y} = \frac{1}{3x^{1/3}}\frac{df}{d\eta}$$

it follows that

$$u_x = \frac{2\alpha^2}{3x^{1/3}}\operatorname{sech}^2(\alpha\eta) \tag{11.4-38}$$

Equation (11.4-5) thus yields

$$J' = \rho\int_{-\infty}^{+\infty}\frac{4\alpha^4}{9x^{2/3}}\operatorname{sech}^4(\alpha\eta)\,dy \tag{11.4-39}$$

or

$$J' = \tfrac{4}{3}\alpha^3\rho\nu^{1/2}\int_{-\infty}^{+\infty}\operatorname{sech}^4 E^{\#}\,dE^{\#} \tag{11.4-40}$$

whence

$$J' = \frac{16}{9}\alpha^3\rho\nu^{1/2} \tag{11.4-41}$$

226 APPLICATIONS

Note that J' is independent of x, in accordance with Eq. (11.4-6). Thus

$$\alpha = \left(\frac{9J'}{16\rho\nu^{1/2}}\right)^{1/3} \approx 0.82548 \left(\frac{J'^2}{\rho\mu}\right)^{1/6} \qquad (11.4\text{-}42)$$

Equation (11.4-42) can be substituted into Eq. (11.4-38) to yield the axial velocity u_x. The profile of u_x is shown schematically in Fig. 11.12.

The mass flow rate M' per unit width of slit is given by

$$M' = \rho \int_{-\infty}^{+\infty} u_x \, dy \qquad (11.4\text{-}43)$$

so

$$M' = \rho \int_{-\infty}^{+\infty} \frac{2\alpha^2}{3x^{1/3}} \operatorname{sech}^2(\alpha\eta) \, dy \qquad (11.4\text{-}44)$$

or

$$M' = 2\alpha\rho\nu^{1/2} x^{1/3} \int_{-\infty}^{+\infty} \operatorname{sech}^2 E^{\#} \, dE^{\#} \qquad (11.4\text{-}45)$$

whence

$$M' = 4\alpha\rho\nu^{1/2}x^{1/3} \qquad (11.4\text{-}46)$$

or, using Eq. (11.4-42)

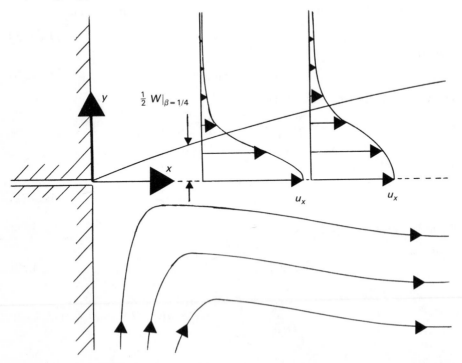

Figure 11.12. Axial velocity profile and streamlines for a plane jet.

$$M' \simeq 3.3019 \, (J'\rho\mu x)^{1/3} \qquad (11.4\text{-}47)$$

We note that M' varies as $x^{1/3}$. This may be interpreted to mean that the jet entrains ambient fluid (see the streamlines on Fig. 11.12). We note also that $M' = 0$ at $x = 0$: this is clearly unrealistic and arises because the boundary layer approximation is invalid near $x = 0$, that is, near the slit (recall that it is also invalid near the leading edge for flow past a flat plate; see Section 11.2). In fact, the boundary layer approximation holds if the local Reynolds number Re_x given by

$$\mathrm{Re}_x = \frac{M'}{\mu} \simeq 3.3019 \left(\frac{J'\rho x}{\mu^2} \right)^{1/3} \gg 1 \qquad (11.4\text{-}48)$$

that is, if

$$x \gg \frac{\mu^2}{J'\rho} \qquad (11.4\text{-}49)$$

The jet width W may be defined as twice the distance between the centerline and the position at which the axial velocity is some small fraction β of the centerline velocity:

$$W = 2 \, y\big|_{u_x(x,y)=\beta u_x(x,y=0)} \qquad 0 < \beta \ll 1 \qquad (11.4\text{-}50)$$

It follows from Eq. (11.4-38) that

$$\beta = \frac{u_x(x, y)}{u_x(x, y = 0)} = \mathrm{sech}^2 (\alpha\eta) \qquad (11.4\text{-}51)$$

so

$$W = 2 \, y\big|_{\eta=(1/\alpha)\mathrm{sech}^{-1}(\beta^{1/2})} \qquad (11.4\text{-}52)$$

or

$$W = 2 \, y\big|_{y=(3\nu^{1/2}x^{2/3}/\alpha)\mathrm{sech}^{-1}(\beta^{1/2})} \qquad (11.4\text{-}53)$$

whence, eliminating α using Eq. (11.4-42)

$$W \simeq 7.2685 \, (\mu^2 x^2/J'\rho)^{1/3} \, \mathrm{sech}^{-1} (\beta^{1/2}) \qquad (11.4\text{-}54)$$

We note that W varies as $x^{2/3}$, that is, the width of the jet increases along the jet (see Fig. 11.12). Because W varies as $x^{n<1}$, however, the width increase only slowly, in accordance with our original assumption.

We conclude by noting that experimental measurements corroborate our results, provided the jet remains laminar. This is so provided the Reynolds number for the flow in the slit given by

$$\mathrm{Re} = \frac{M' \, (x \le 0)}{\mu} \qquad (11.4\text{-}55)$$

does not exceed a critical value $\mathrm{Re}_c \simeq 30$.

TWELVE

TURBULENT FLOW

12.1 REYNOLDS' EQUATIONS

A turbulent flow is a high Reynolds number flow. The origin of the turbulence appears to be instability of the associated low Reynolds number laminar flow. Two main classes of turbulent flow must be distinguished: (1) a *wall flow,* for example flow past a flat plate or in a pipe; and (2) a *free flow,* for example flow of a jet.

We concentrate in this chapter mainly on turbulent wall flows. The details of the transition from laminar to turbulent flow are not well understood. For a wall flow, however, the main steps as the Reynolds number is increased seem to be as follows:

1. an initial, often two-dimensional, instability which leads to:
2. secondary, generally three-dimensional, motions which are themselves unstable and lead to:
3. further, almost without exception three-dimensional, motions which are themselves unstable, and so on, until:
4. intense, local, three-dimensional fluctuations called *turbulent spots* are produced which grow both in size and in number, merge and eventually:
5. the flow becomes fully turbulent.

In a turbulent flow the flow variables, such as velocity $\mathbf{u}$ and pressure p, fluctuate in space and time in an apparently random manner. We note that they do not fluctuate in a truly random manner. The fluctuations, which are often called eddies, are, as far as is known, completely predictable from the mass and linear momentum conservation equations, given appropriate boundary and initial conditions. Furthermore, it appears that many turbulent flows are dominated by large scale motions which are not random, as we will see in Section 12.2. A turbulent flow is thus *deterministic*. Recent analyses of *chaotic systems* suggest, however, that the boundary and initial conditions have to be specified to such extremely high precision that, while a turbulent flow is deterministic in principle, in practice it is not.

Because of the apparent randomness of turbulent flows, statistical flow descriptions are appropriate and we use the average value of a flow quantity f (such as $\mathbf{u}$ or p). We cannot use a volume average, such as we use for flow in porous media (see Section 8.5), because a turbulent flow comprises fluctuations over a range of length scales, as we will see. Thus no suitable averaging volume can be chosen. Similarly, we cannot use a time average because a turbulent flow can be unsteady on average. For example, a turbulent flow in a pipe to which a monotonically varying pressure drop is applied is clearly not steady on average. Instead, we use an ensemble average $\bar{f}$ which is defined by

$$\bar{f}(\mathbf{x}, t) = \lim_{N \to \infty} \left[\frac{1}{N} \sum_{i=1}^{N} f(\mathbf{x}, t)|_i \right] \tag{12.1-1}$$

where the summation is over N repetitions of an apparently identical flow, that is, one with as nearly the same boundary and initial conditions as possible (recall the earlier comment on chaotic systems). Here $\mathbf{x}$ denotes position relative to the same reference frame in each repetition and t denotes the time that has elapsed since the start of each repetition. For a stationary turbulent flow, that is, a turbulent flow that is steady on average so that

$$\frac{\partial \bar{f}}{\partial t} = 0 \tag{12.1-2}$$

for all flow variables f, the *ergodic hypothesis* is that, under certain fairly general conditions, the ensemble average of each flow variable is the same as its time average. For such a flow, we can clearly use a time average instead of an ensemble average: indeed such time averages are what are commonly used in experimental measurements.

We now decompose each flow variable f into an average, by which we will henceforth always imply ensemble average, component $\bar{f}$ defined by Eq. (12.1-1) and a fluctuating component f':

$$f = \bar{f} + f' \tag{12.1-3}$$

Since the average $\bar{\bar{f}}$ is clearly just $\bar{f}$, if we average the terms in Eq. (12.1-3) we immediately deduce that

$$\overline{f'} = 0 \tag{12.1-4}$$

that is, the fluctuating component of a flow variable has zero average. A convenient measure of a fluctuating component f' is thus given not by $\overline{f'}$ but by the root mean square $(\overline{|f'|^2})^{1/2}$. If the velocity field $\mathbf{u}$ is decomposed into two components, $\bar{\mathbf{u}}$ and $\mathbf{u}'$, the relative turbulent intensity I_r is given by

$$I_r = \frac{(\overline{|\mathbf{u}'|^2})^{1/2}}{|\bar{\mathbf{u}}|} \qquad (12.1\text{-}5)$$

In a typical turbulent flow, $I_r \sim 0.1$.

The mass and linear momentum conservation equations for an incompressible Newtonian fluid of density ρ and viscosity μ are [see Eq. (7.1-1)]

$$\nabla \cdot \mathbf{u} = 0 \qquad (12.1\text{-}6)$$

and
$$\rho \frac{\partial \mathbf{u}}{\partial t} + \rho \nabla \cdot (\mathbf{uu}) = -\nabla p + \mu \Delta \mathbf{u} \qquad (12.1\text{-}7)$$

[see Eq. (7.1-2) and note that $\mathbf{u} \cdot \nabla \mathbf{u} = \nabla \cdot (\mathbf{uu})$ since $\nabla \cdot \mathbf{u} = 0$ and that the gravitational term is incorporated in the pressure term]. Decomposition of $\mathbf{u}$ in the sense of Eq. (12.1-3) means that Eq. (12.1-6) becomes

$$\nabla \cdot (\bar{\mathbf{u}} + \mathbf{u}') = 0 \qquad (12.1\text{-}8)$$

or
$$\nabla \cdot \bar{\mathbf{u}} + \nabla \cdot \mathbf{u}' = 0 \qquad (12.1\text{-}9)$$

If we now average the terms in this equation, we obtain

$$\overline{\nabla \cdot \bar{\mathbf{u}}} + \overline{\nabla \cdot \mathbf{u}'} = 0 \qquad (12.1\text{-}10)$$

Because the averaging operator defined by Eq. (12.1-1) commutes with the divergence operator (i.e., the order in which they operate is irrelevant), it now follows that

$$\nabla \cdot \bar{\bar{\mathbf{u}}} + \nabla \cdot \overline{\mathbf{u}'} = 0 \qquad (12.1\text{-}11)$$

But $\bar{\bar{\mathbf{u}}} = \bar{\mathbf{u}}$ and $\overline{\mathbf{u}'} = \mathbf{0}$ so

$$\nabla \cdot \bar{\mathbf{u}} = 0 \qquad (12.1\text{-}12)$$

Similarly, decomposition of $\mathbf{u}$ and p in the sense of Eq. (12.1-3) means that Equation (12.1-7) becomes

$$\rho \frac{\partial}{\partial t} (\bar{\mathbf{u}} + \mathbf{u}') + \rho \nabla \cdot (\overline{\bar{\mathbf{u}}\bar{\mathbf{u}} + \bar{\mathbf{u}}\mathbf{u}' + \mathbf{u}'\bar{\mathbf{u}} + \mathbf{u}'\mathbf{u}'})$$
$$= -\nabla(\bar{p} + p') + \mu \Delta(\bar{\mathbf{u}} + \mathbf{u}') \qquad (12.1\text{-}13)$$

whence, averaging the terms in this equation, we obtain

$$\rho \frac{\partial \bar{\mathbf{u}}}{\partial t} + \nabla \cdot (\rho \overline{\mathbf{uu}}) = -\nabla \bar{p} + \mu \Delta \bar{\mathbf{u}} + \nabla \cdot (-\rho \overline{\mathbf{u}'\mathbf{u}'})$$

$$\qquad\qquad \uparrow \qquad\quad \uparrow \qquad\quad \uparrow \qquad \uparrow \qquad\qquad \uparrow \qquad (12.1\text{-}14)$$
$$\qquad\qquad (a) \qquad\quad (b) \qquad\; (c) \qquad (d) \qquad\qquad (e)$$

Equations (12.1-12) and (12.1-14) for the average flow are called *Reynolds' equations*. They are given in component form in rectangular, cylindrical polar, and spherical polar coordinates in Section A2.6 of Appendix B. We note that Eq. (12.1-12) for $\bar{\mathbf{u}}$ is the same as Eq. (12.1-6) for $\mathbf{u}$. Equation (12.1-14) for $\bar{\mathbf{u}}$ and $\bar{p}$ differs, however, from Eq. (12.1-7) for $\mathbf{u}$ and p: there is an extra term involving the tensor $-\rho\overline{\mathbf{u}'\mathbf{u}'}$ in the former. Thus, in addition to the transient term (a), the convection term (b), the pressure term (c) and the viscous term (d), there is a new term (e). The viscous term $\mu\Delta\bar{\mathbf{u}}$ can be written $\nabla\cdot\bar{\mathbf{\tau}}$, where $\bar{\mathbf{\tau}}$ denotes the average component of the stress $\mathbf{\tau}$. By analogy, $-\rho\overline{\mathbf{u}'\mathbf{u}'}$ is called the Reynolds stress. Its presence in term (e) of Eq. (12.1-14) illustrates the closure problem in turbulence: there are more unknowns than equations. Thus there are four equations (12.1-12) and (12.1-14) (one mass and three components of linear momentum). But there are ten variables $\bar{p}$, $\bar{\mathbf{u}}$ (three components) and $-\rho\overline{\mathbf{u}'\mathbf{u}'}$ (nine components, of which only six are independent since the Reynolds stress is symmetric). In fact,

A closure problem always arises when a nonlinear system is averaged since averaging loses information.

Thus, when the nonlinear inertial term $\nabla\cdot(\mathbf{u}\mathbf{u})$ in Eq. (12.1-7) is averaged, *two* terms result: $\nabla\cdot(\overline{\mathbf{u}\mathbf{u}})$ and $\nabla\cdot(\overline{\mathbf{u}'\mathbf{u}'})$. In order to see why $\overline{\mathbf{u}'\mathbf{u}'}\neq\mathbf{0}$, suppose that $\mathbf{u}$ can have with equal probability one of only two values, $-\mathbf{c}$ and $+\mathbf{c}$ (where $\mathbf{c}\neq\mathbf{0}$ is a constant). Then it is clear that $\bar{\mathbf{u}}=\mathbf{0}$. It is also clear that $\mathbf{u}'=\pm\mathbf{c}$ so that $\mathbf{u}'\mathbf{u}'=\mathbf{c}\mathbf{c}$. Hence $\overline{\mathbf{u}'}=\mathbf{0}$ but $\overline{\mathbf{u}'\mathbf{u}'}=\mathbf{c}\mathbf{c}\neq\mathbf{0}$. Thus the average value of a product of fluctuating components is *not* zero, though the product of average values of fluctuating components is zero. Except in isolated cases,

The average of a product is not the same as the product of averages.

In order to eliminate the closure problem in turbulence, we need an independent equation for the Reynolds stress. Thus we need what we might call a turbulence model (i.e., a constitutive equation; see Chapter 5) for the Reynolds stress. Such a model cannot, however, be obtained from the flow equations. By forming the dyadic product [see Eq. (A1.2-21) of Appendix A] of $\mathbf{u}'$ with each of the terms in Eq. (12.1-13) and then averaging, we could obtain an equation for $\partial(-\rho\overline{\mathbf{u}'\mathbf{u}'}/\partial t$. This equation would, however, comprise a term involving $\rho\overline{\mathbf{u}'\mathbf{u}'\mathbf{u}'}$; similarly, an equation for $\partial(\rho\overline{\mathbf{u}'\mathbf{u}'\mathbf{u}'})/\partial t$ would comprise a term involving $\rho\overline{\mathbf{u}'\mathbf{u}'\mathbf{u}'\mathbf{u}'}$, and so on. We see then that to obtain an equation for the Reynolds stress, we would need to solve an infinite set of equations. This would require infinite labor and moreover give us no more information than the original unaveraged Eqs. (12.1-7), thus defeating our object of considering only average flow variables. An equation for the Reynolds stress must be obtained using other than conservation arguments. We discuss how this can be done in Section 12.2.

In order to gain insight into the mechanics of a turbulent flow, it is useful to obtain equations for the average kinetic energies per unit mass of the average flow $\frac{1}{2}\bar{\mathbf{u}}\cdot\bar{\mathbf{u}}$ and of the fluctuating flow $\frac{1}{2}\overline{\mathbf{u}'\cdot\mathbf{u}'}$.

- For the average flow, we form the dot product of $\bar{\mathbf{u}}$ with each of the terms in Eq. (12.1-14) and obtain

$$\rho \frac{\partial}{\partial t} \left(\frac{1}{2} \bar{\mathbf{u}} \cdot \bar{\mathbf{u}} \right) + \rho \bar{\mathbf{u}} \cdot \nabla \left(\frac{1}{2} \bar{\mathbf{u}} \cdot \bar{\mathbf{u}} \right)$$

$$\underset{(a)}{\uparrow} \qquad\qquad \underset{(b)}{\uparrow}$$

$$= -\nabla \cdot (\bar{p}\bar{\mathbf{u}}) + \mu \nabla \cdot (\bar{\mathbf{e}} \cdot \bar{\mathbf{u}}) - \frac{1}{2} \mu \bar{\mathbf{e}} : \bar{\mathbf{e}} + \frac{1}{2} \rho \overline{\mathbf{u}'\mathbf{u}'} : \bar{\mathbf{e}} - \nabla \cdot (\bar{\mathbf{u}} \cdot \rho \overline{\mathbf{u}'\mathbf{u}'}) \quad (12.1\text{-}15)$$

$$\underset{(c)}{\uparrow} \qquad \underset{(d)}{\uparrow} \qquad \underset{(e)}{\uparrow} \qquad \underset{(f)}{\uparrow} \qquad \underset{(g)}{\uparrow}$$

where $\bar{\mathbf{e}}$ denotes the average component of the rate of strain $\mathbf{e}$.

- For the fluctuating flow, we form the dot product of $\mathbf{u}'$ with each of the terms in Eq. (12.1-13) and then average to obtain

$$\rho \frac{\partial}{\partial t} \left(\frac{1}{2} \overline{\mathbf{u}' \cdot \mathbf{u}'} \right) + \rho \bar{\mathbf{u}} \cdot \nabla \left(\frac{1}{2} \overline{\mathbf{u}' \cdot \mathbf{u}'} \right)$$

$$\cdot \underset{(a)}{\uparrow} \qquad\qquad \underset{(b)}{\uparrow}$$

$$= -\nabla \cdot (\overline{p'\mathbf{u}'}) + \mu \nabla \cdot (\overline{\mathbf{e}' \cdot \mathbf{u}'}) - \frac{1}{2} \mu \overline{\mathbf{e}' : \mathbf{e}'} - \frac{1}{2} \rho \overline{\mathbf{u}'\mathbf{u}'} : \bar{\mathbf{e}} - \nabla \cdot \left(\frac{1}{2} \rho \overline{\mathbf{u}' \cdot \mathbf{u}'\mathbf{u}'} \right)$$

$$\underset{(c)}{\uparrow} \qquad \underset{(d)}{\uparrow} \qquad \underset{(e)}{\uparrow} \qquad \underset{(f)}{\uparrow} \qquad\qquad \underset{(g)}{\uparrow}$$

$$(12.1\text{-}16)$$

where $\mathbf{e}'$ denotes the fluctuating component of $\mathbf{e}$.

The terms in these equations may be interpreted as follows. Terms (a) and (b) are the transient and convection terms, respectively. Terms (c) and (d) represent transport by pressure and viscous stress, respectively. Terms (e) are the viscous dissipation terms. Terms (f) represent the change of kinetic energy by the working of the Reynolds stress against the mean rate of strain. Terms (g) represent transport by velocity fluctuations: note that in Eq. (12.1-16) this term involves $\overline{\mathbf{u}' \cdot \mathbf{u}'\mathbf{u}'}$, which illustrates the closure problem that we discussed earlier.

We now perform an order-of-magnitude analysis of Eqs. (12.1-15) and (12.1-16). Let k denote the average kinetic energy per unit mass of the fluctuating flow

$$k = \tfrac{1}{2} \overline{\mathbf{u}' \cdot \mathbf{u}'} \qquad (12.1\text{-}17)$$

We then identify $\sqrt{k}$ as a characteristic velocity u_c and assume not only that $|\mathbf{u}'| \sim \sqrt{k}$, which is obvious, but also that $|\bar{\mathbf{u}}| \sim \sqrt{k}$ which is strictly the case only if no other velocity characterizes the turbulence. We also identify as a characteristic length l_c the length scale Λ of the largest fluctuations, which is determined by the overall flow geometry (so that, for flow in a pipe of radius R, $\Lambda \sim R$). Then the Reynolds number is given by

$$\mathrm{Re} = \frac{\rho \sqrt{k} \, \Lambda}{\mu} \qquad (12.1\text{-}18)$$

Examining Eq. (12.1-15), it follows that

$$\rho \frac{\partial}{\partial t} \left(\frac{1}{2} \bar{\mathbf{u}} \cdot \bar{\mathbf{u}} \right) \sim \rho \bar{\mathbf{u}} \cdot \nabla \left(\frac{1}{2} \bar{\mathbf{u}} \cdot \bar{\mathbf{u}} \right) \sim \frac{1}{2} \rho \overline{\mathbf{u}' \mathbf{u}'} : \bar{\mathbf{e}} \sim \nabla \cdot (\bar{\mathbf{u}} \cdot \rho \overline{\mathbf{u}' \mathbf{u}'}) \sim \frac{\rho k^{3/2}}{\Lambda} \quad (12.1\text{-}19)$$

and

$$\mu \nabla \cdot (\bar{\mathbf{e}} \cdot \bar{\mathbf{u}}) \sim \frac{1}{2} \mu \bar{\mathbf{e}} : \bar{\mathbf{e}} \sim \frac{\mu k}{\Lambda^2} = \frac{1}{\text{Re}} \frac{\rho k^{3/2}}{\Lambda} \quad (12.1\text{-}20)$$

Thus, for a turbulent flow in which Re $\gg$ 1, the viscous terms (d) and (e) in Eq. (12.1-15) are negligible, just as they are for an inviscid flow. (Just as for an inviscid flow, however, viscous effects can be important: near a wall, for example, the relevant length scale is given not by the overall flow geometry but by the distance from the wall, so that Re $\ggg$ 1.) We may conclude, therefore, that the gross structure of a turbulent flow is virtually independent of viscosity.

Examining Eq. (12.1-16), it follows that

$$\rho \frac{\partial}{\partial t} \left(\frac{1}{2} \overline{\mathbf{u}' \cdot \mathbf{u}'} \right) \sim \rho \bar{\mathbf{u}} \cdot \nabla \left(\frac{1}{2} \overline{\mathbf{u}' \cdot \mathbf{u}'} \right) \sim \frac{1}{2} \rho \overline{\mathbf{u}' \mathbf{u}'} : \bar{\mathbf{e}}$$

$$\sim \nabla \cdot \left(\frac{1}{2} \rho \overline{\mathbf{u}' \cdot \mathbf{u}' \mathbf{u}'} \right) \sim \frac{\rho k^{3/2}}{\Lambda} \quad (12.1\text{-}21)$$

We cannot, however, obtain order-of-magnitude estimates of $\mu \nabla \cdot (\overline{\mathbf{e}' \cdot \mathbf{u}'})$ and $\frac{1}{2} \mu \overline{\mathbf{e}' : \mathbf{e}'}$ immediately, because the fluctuating component of the velocity gradient and hence of the rate of strain varies as the inverse of the unknown length scale of the fluctuations. What is clear, however, is that the viscous terms (d) and (e) in Eq. (12.1-16) are not negligible, even if Re $\gg$ 1. This is because, if they were, the continuous transfer of kinetic energy from the average flow to the fluctuating flow would cause the fluctuations to become larger and larger and eventually to become unbounded. The transfer is effected by terms (f) in Eqs. (12.1-15) and (12.1-16), which are the same except for their signs. They thus serve to exchange kinetic energy between the average and the fluctuating flows, generally from the former to the latter. The mean length scale of the fluctuations, which is called the Taylor microscale λ_T, is such that

$$\mu \nabla \cdot (\overline{\mathbf{e}' \cdot \mathbf{u}'}) \sim \frac{1}{2} \mu \overline{\mathbf{e}' : \mathbf{e}'} \sim \frac{\mu k}{\lambda_T^2} \quad (12.1\text{-}22)$$

But, in order that the fluctuations do not become unbounded:

$$\mu \nabla \cdot (\overline{\mathbf{e}' \cdot \mathbf{u}'}) \sim \frac{1}{2} \mu \overline{\mathbf{e}' : \mathbf{e}'} \sim \frac{\rho k^{3/2}}{\Lambda} \quad (12.1\text{-}23)$$

so

$$\lambda_T \sim \frac{\Lambda}{\text{Re}^{1/2}} \quad (12.1\text{-}24)$$

Physically, the viscous terms (d) and (e) in Eq. (12.1-16), in particular the latter which is the product of the density ρ and the dissipation ϵ given by

$$\epsilon = \frac{1}{2} \frac{\mu}{\rho} \overline{\mathbf{e}' : \mathbf{e}'} \quad (12.1\text{-}25)$$

provide a mechanism by which kinetic energy is converted into internal energy (i.e., into heat).

Turbulent flows are essentially dissipative and need a continuous supply of energy if they are not to decay.

For flow in a pipe, such a source of energy is provided by the pressure drop in the pipe.

   A turbulent flow comprises fluctuations of many different length scales. This is why, as we noted earlier, we cannot use a volume average and must instead use an ensemble average. The largest scales are determined, as we have already noted, by the overall flow geometry: the smallest are determined by the viscosity of the fluid, all other variables being held constant. The large scales interact with the average flow most effectively and abstract kinetic energy from it; this energy is fed into the fluctuating motion and passed from larger to smaller scales in a process called the *energy cascade*. One of the essential features of this process appears to be amplification of vorticity $\boldsymbol{\omega}$ by $\mathbf{e}$. For an incompressible Newtonian fluid, the vorticity transport equation (4.3-16) is

$$\rho \frac{\partial \boldsymbol{\omega}}{\partial t} + \rho \mathbf{u} \cdot \nabla \boldsymbol{\omega} - \rho \boldsymbol{\omega} \cdot \nabla \mathbf{u} = \mu \Delta \boldsymbol{\omega} \qquad (12.1\text{-}26)$$

But

$$\boldsymbol{\omega} \cdot \nabla \mathbf{u} = \tfrac{1}{2} \boldsymbol{\omega} \cdot (\mathbf{e} + \mathbf{w}) = \tfrac{1}{2} \boldsymbol{\omega} \cdot \mathbf{e} - \tfrac{1}{2} \boldsymbol{\omega} . \boldsymbol{\omega} = \tfrac{1}{2} \boldsymbol{\omega} \cdot \mathbf{e} \qquad (12.1\text{-}27)$$

where $\mathbf{w}$ denotes the vorticity tensor (which is to be distinguished from $\boldsymbol{\omega}$, the vorticity vector: see Section 2.4). Thus,

$$\rho \frac{\partial \boldsymbol{\omega}}{\partial t} + \rho \mathbf{u} \cdot \nabla \boldsymbol{\omega} - \frac{1}{2} \rho \boldsymbol{\omega} \cdot \mathbf{e} = \mu \Delta \boldsymbol{\omega} \qquad (12.1\text{–}28)$$

The term $\tfrac{1}{2} \rho \boldsymbol{\omega} \cdot \mathbf{e}$ is responsible for vorticity amplification. In order to see this, we introduce the concept of a *vortex line*. This is a line that is instantaneously at every point aligned with the local axis of rotation of the fluid, that is, with $\boldsymbol{\omega}$ (see Section 2.4; recall from Section 2.5 that a streamline is similarly aligned with $\mathbf{u}$). We now consider the local straining in the flow, which is given by the local $\mathbf{e}$. If $\boldsymbol{\omega} \cdot \mathbf{e} = 0$, then there is no straining of vortex lines. If, however, $\boldsymbol{\omega} \cdot \mathbf{e} \neq 0$, then there is straining of vortex lines. Depending on the value of $\boldsymbol{\omega} \cdot \mathbf{e}$, the vortex lines are either extended or contracted. Conservation of mass, which implies conservation of volume for an incompressible fluid, then means that extended vortex lines move closer together and contracted ones move further apart (see Fig. 12.1). Viscous effects may be neglected since the gross structure of a turbulent flow is virtually independent of viscosity, as we have already seen. Then conservation of angular momentum means that, if the vortex lines move closer together, the vorticity increases and vice versa. This is because the magnitude of the angular momentum varies as Ωr^2, where Ω denotes the magnitude of the angular velocity and r denotes the distance from the axis of rotation. If r decreases then constancy of Ωr^2 means that Ω increases; if r increases, Ω decreases. Hence, if the vortex lines are extended, the vorticity increases: this is vorticity amplifi-

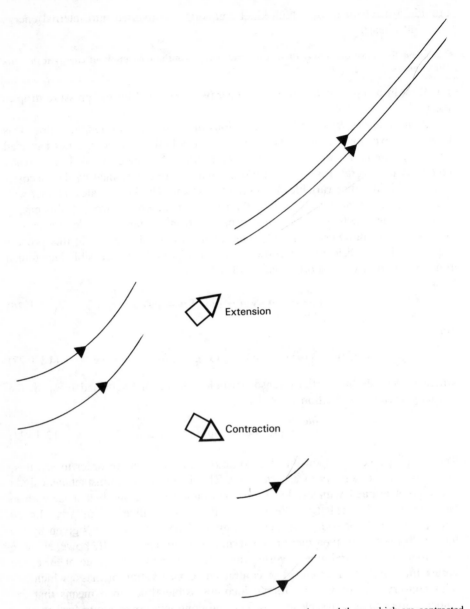

Figure 12.1. Vortex lines which are extended move closer together and those which are contracted move further apart.

cation. The increase of vorticity as vortex lines move closer together is, therefore, precisely analogous to the increase of velocity as streamlines move closer together (see Section 2.5). For any two-dimensional flow, however, $\boldsymbol{\omega} \cdot \mathbf{e} = 0$ and no vorticity amplification can occur; hence we deduce that turbulence is essentially three dimensional.

The kinetic energy is transferred down the energy cascade to smaller a. smaller length scales, so that velocity gradients get larger and larger. Eventually, the length scale is sufficiently small for the energy to be dissipated by viscous action: the lower the viscosity, the smaller the small length scales. The length scale λ of the smallest fluctuations, which is called the Kolmogorov microscale (or henceforth just the microscale), is related to the length scale Λ of the largest by

$$\lambda \sim \frac{\Lambda}{Re^{3/4}} \tag{12.1-29}$$

(as we will show in Section 12.4) so that λ varies as $\mu^{3/4}$. Comparison of Eqs. (12.1-24) and (12.1-29) reveals that

$$\frac{\lambda}{\lambda_T} \sim \frac{1}{Re^{1/4}} \tag{12.1-30}$$

that is, the mean length scale of the fluctuations is of order $Re^{1/4}$ times the length scale of the smallest. There are no fluctuations with length scale smaller than λ, since fluctuations of such a small size would be dissipated quickly by viscous action. Note that the rate at which energy is dissipated depends not on the magnitude of the viscosity but on the rate at which energy is transferred down the energy cascade. The main role of viscosity is to determine the smallest length scale λ. Also, because the cascade process is essentially random and three dimensional, directions associated with large-scale motions (such as the axial direction for turbulent flow in a pipe) tend not to be transmitted to small-scale motions, which tend, therefore, to be isotropic (i.e., nondirectional).

12.2 FLOW PAST A FLAT PLATE

We consider again the problem that we considered in Section 11.2 (see also Fig. 12.2). The uniform flow far from the flat plate is at velocity $U_\infty \mathbf{i}_x$ and is of an incompressible Newtonian fluid of density ρ and viscosity μ. Re_x based on the distance x from the leading edge of the plate is given by

$$Re_x = \frac{U_\infty x}{\nu} \tag{12.2-1}$$

where the kinematic viscosity $\nu = \mu/\rho$. A boundary layer forms by the plate.

- If Re_x is less than some critical value $Re_{x_c} \simeq 300,000$, the flow in the boundary layer is laminar, as we noted in Section 11.2.
- If Re_x exceeds Re_{x_c} but is less than some second critical value $Re_{x_C} \simeq 3,000,000$, the flow in the boundary layer is transitional, with the flow at each point in the boundary layer intermittently laminar and turbulent and consequently very difficult to analyze.
- If Re_x exceeds Re_{x_C}, the flow in the boundary layer is turbulent.

$$\mathbf{U}_\infty = U_\infty \mathbf{i}_x$$

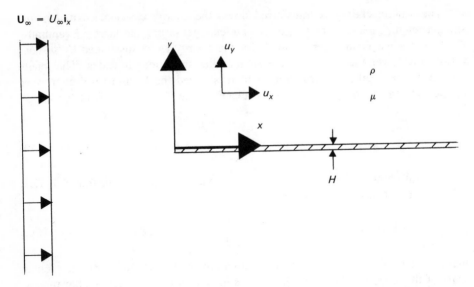

Figure 12.2. Geometry of flow past a flat plate.

These boundary layers are shown schematically in Fig. 12.3. We seek to determine the flow within the turbulent boundary layer.

We noted in Section 12.1 that a turbulent flow comprises fluctuations of many different length scales. We suppose that the flow far upstream of the plate is quite uniform so that it is free of fluctuations in velocity $\mathbf{u}$ and pressure p. As a result we may assume that, in the free stream (i.e., the region outside the boundary layer), fluctuations in $\mathbf{u}$ and p are negligible. Hence not only are viscous stresses negligible in the free stream, so that the flow there is effectively inviscid, but so also are the Reynolds stresses. All dimensions of the fluctuations in a turbulent flow are comparable, for otherwise the fluctuations would effectively not be three dimensional and the flow would not be truly turbulent (see Section 12.1).

It then follows that the length scale Λ of the largest fluctuations in the bound-

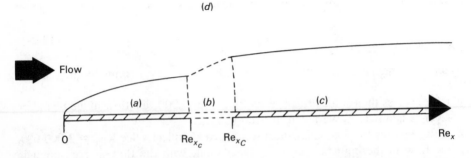

Figure 12.3. Structure of flow past a flat plate: (*a*) is the laminar boundary layer, (*b*) the transitional boundary layer, (*c*) the turbulent boundary layer, and (*d*) the free stream.

ary layer, which is determined by the overall flow geometry, is of the same order of magnitude as the thickness δ of the boundary layer. Experimental observation reveals that $\delta \ll x$: thus Λ is not of the same order of magnitude as x. It is, therefore, appropriate to define a second Reynolds number based on δ given by

$$\mathrm{Re}_\delta = \frac{U_\infty \delta}{\nu} = \frac{\mathrm{Re}_x \delta}{x} \qquad (12.2\text{-}2)$$

It then follows from Eq. (12.1-29) that the length scale λ of the smallest fluctuations is given by

$$\lambda \sim \frac{\delta}{\mathrm{Re}_\delta^{3/4}} \qquad (12.2\text{-}3)$$

In order to simplify our analysis, we now assume that the plate is smooth, that is, we assume that the length scale of any roughness associated with the surface of the plate is very much less than that of any other relevant length scale and hence is negligible.

We expect that the average flow is stationary [see Eq. (12.1-2)] and planar. Thus, decomposing the flow variables in the sense of Eq. (12.1-3), the mass conservation equation (A2.6-3) of Appendix B yields

$$\frac{\partial \overline{u}_x}{\partial x} + \frac{\partial \overline{u}_y}{\partial y} = 0 \qquad (12.2\text{-}4)$$

and the linear momentum conservation equations (A2.6-4) of Appendix B (with the gravitational term incorporated in the pressure term) yield

$$\overline{u}_x \frac{\partial \overline{u}_x}{\partial x} + \overline{u}_y \frac{\partial \overline{u}_x}{\partial y} = -\frac{1}{\rho}\frac{\partial \overline{p}}{\partial x} + \nu \left(\frac{\partial^2 \overline{u}_x}{\partial x^2} + \frac{\partial^2 \overline{u}_x}{\partial y^2} \right) - \left(\frac{\partial \overline{u'_x u'_x}}{\partial x} + \frac{\partial \overline{u'_y u'_x}}{\partial y} \right) \qquad (12.2\text{-}5)$$

$$\overline{u}_x \frac{\partial \overline{u}_y}{\partial x} + \overline{u}_y \frac{\partial \overline{u}_y}{\partial y} = -\frac{1}{\rho}\frac{\partial \overline{p}}{\partial y} + \nu \left(\frac{\partial^2 \overline{u}_y}{\partial x^2} + \frac{\partial^2 \overline{u}_y}{\partial y^2} \right) - \left(\frac{\partial \overline{u'_x u'_y}}{\partial x} + \frac{\partial \overline{u'_y u'_y}}{\partial y} \right) \qquad (12.2\text{-}6)$$

Note that the velocity field **u** has components given by

$$\mathbf{u} = (\overline{u}_x + u'_x)\mathbf{i}_x + (\overline{u}_y + u'_y)\mathbf{i}_y + (\overline{u}_z + u'_z)\mathbf{i}_z \qquad (12.2\text{-}7)$$

Because the average flow is planar, $\overline{u}_z$, $\overline{u'_x u'_z} = \overline{u'_z u'_x}$, and $\overline{u'_y u'_z} = \overline{u'_z u'_y}$ all vanish, and the z derivatives of $\overline{u}_x$, $\overline{u}_y$, $\overline{u'_x u'_x}$, $\overline{u'_x u'_y} = \overline{u'_y u'_x}$, and $\overline{u'_y u'_y}$ all vanish. Also the z derivative of $\overline{u'_z u'_z}$ vanishes but $\overline{u'_z u'_z}$ does not itself vanish: turbulence is, as we noted in Section 12.1 essentially three dimensional.

The boundary conditions for the flow are no slip and no flow-through at the plate, that is,

$$\overline{u}_x = 0 \qquad \overline{u}_y = 0 \qquad \overline{u'_x u'_x} = 0 \qquad \overline{u'_x u'_y} = \overline{u'_y u'_x} = 0$$
$$\overline{u'_y u'_y} = 0 \qquad \text{at } y = 0 \qquad (12.2\text{-}8)$$

and uniform flow far from the plate (at infinity), that is,

$$\bar{u}_x \to U_\infty \qquad \overline{u'_x u'_x} \to 0 \qquad \overline{u'_x u'_y} = \overline{u'_y u'_x} \to 0$$

$$\overline{u'_y u'_y} \to 0 \qquad \text{as } y \to \infty \qquad\qquad\qquad (12.2\text{-}9)$$

(Note that we do not seek to impose a condition on $\bar{u}_y$ as $y \to \infty$. Such a condition is unnecessary and, as we recall from Section 11.2, it is certainly incorrect to require that $\bar{u}_y \to 0$ as $y \to \infty$: mass conservation means that $\bar{u}_y > 0$ as $y \to \infty$.) There should also, presumably, be appropriate upstream matching conditions with the flow in the transitional boundary layer, and some downstream conditions, though what these might be is not at all clear. Fortunately, we will not in fact need such conditions.

In order to proceed further, we perform an order-of-magnitude analysis. In the boundary layer, $\bar{u}_x \sim U_\infty$, $x \sim x$ (trivially), and $y \sim \delta$. Hence Eq. (12.2-4) yields $\bar{u}_y \sim \delta U_\infty / x$. We now note that $u'_x \sim u'_y \sim \sqrt{k}$, where k denotes the average kinetic energy per unit mass of the fluctuating flow

$$k = \tfrac{1}{2} \overline{\mathbf{u}' \cdot \mathbf{u}'} \qquad\qquad\qquad (12.2\text{-}10)$$

Thus it follows from Eq. (12.2-5) that

- The inertial terms $\bar{u}_x \, \partial \bar{u}_x / \partial x + \bar{u}_y \, \partial \bar{u}_x / \partial y \sim U_\infty^2 / x$.
- The viscous terms $v \, (\partial^2 \bar{u}_x / \partial x^2 + \partial^2 \bar{u}_x / \partial y^2) \sim v \, (U_\infty / x^2 + U_\infty / \delta^2) \sim v U_\infty / \delta^2$, since we anticipate that the boundary layer is thin (i.e., that $\delta \ll x$).
- The Reynolds stress terms $- (\partial \overline{u'_x u'_x} / \partial x + \partial \overline{u'_y u'_x} / \partial y) \sim (k/x + k/\delta) \sim k/\delta$, since $\delta \ll x$.

It thus follows that Eq. (12.2-5) can be simplified to

$$\bar{u}_x \frac{\partial \bar{u}_x}{\partial x} + \bar{u}_y \frac{\partial \bar{u}_x}{\partial y} = -\frac{1}{\rho} \frac{\partial \bar{p}}{\partial x} + v \frac{\partial^2 \bar{u}_x}{\partial y^2} - \frac{\partial \overline{u'_y u'_x}}{\partial y} \qquad (12.2\text{-}11)$$

It follows from Eq. (12.2-6) that

- The inertial terms $\bar{u}_x \, \partial \bar{u}_y / \partial x + \bar{u}_y \, \partial \bar{u}_y / \partial y \sim U_\infty^2 \, \delta / x^2$.
- The viscous terms $v \, (\partial^2 \bar{u}_y / \partial x^2 + \partial^2 \bar{u}_y / \partial y^2) \sim v \, (U_\infty \delta / x^3 + U_\infty / (\delta x) \sim v U_\infty / \delta x$, since $\delta \ll x$.
- The Reynolds stress terms $- (\partial \overline{u'_x u'_y} / \partial x + \partial \overline{u'_y u'_y} / \partial y) \sim (k/x + k/\delta) \sim k/\delta$ since $\delta \ll x$.

It thus follows that Eq. (12.2-6) can be simplified to

$$0 = -\frac{1}{\rho} \frac{\partial \bar{p}}{\partial y} - \frac{\partial \overline{u'_y u'_y}}{\partial y} \qquad\qquad\qquad (12.2\text{-}12)$$

which, since $\overline{u'_y u'_y}$ vanishes at the plate [see boundary conditions (12.2-8)], can be integrated to yield

$$\bar{p} = \bar{p}_0 - \rho \overline{u'_y u'_y} \qquad\qquad\qquad (12.2\text{-}13)$$

Here $\bar{p}_0$ denotes the average pressure at the plate; it is at most a function of axial position x.

Outside the boundary layer, in the free stream, the flow is assumed to be effectively inviscid, with uniform velocity $U_\infty \mathbf{i}_x$. Thus Bernoulli's equation holds, that is,

$$p + \tfrac{1}{2}\rho U_\infty^2 = \zeta \qquad (12.2\text{-}14)$$

[see Eq. (9.1-13)] where ζ is a constant; recall that the gravitational term is incorporated in the pressure term. Since fluctuations in the free stream are assumed to be negligible, it follows that $\bar{p} = p$ there. Hence, matching Eqs. (12.2-13) and (12.2-14) and using boundary conditions (12.2-9)

$$\bar{p}_0 + \tfrac{1}{2}\rho U_\infty^2 = \zeta \qquad (12.2\text{-}15)$$

so that $\bar{p}_0$ is, in fact, independent of x. Thus, in the boundary layer:

$$\bar{p} + \tfrac{1}{2}\rho U_\infty^2 = \zeta - \rho \overline{u_y' u_y'} \qquad (12.2\text{-}16)$$

whence

$$\frac{1}{\rho}\frac{\partial \bar{p}}{\partial x} = -\frac{\partial \overline{u_y' u_y'}}{\partial x} \qquad (12.2\text{-}17)$$

Thus Eq. (12.2-11) becomes

$$\bar{u}_x \frac{\partial \bar{u}_x}{\partial x} + \bar{u}_y \frac{\partial \bar{u}_x}{\partial y} = \nu \frac{\partial^2 \bar{u}_x}{\partial y^2} - \frac{\partial \overline{u_y' u_x'}}{\partial y} + \frac{\partial \overline{u_y' u_y'}}{\partial x} \qquad (12.2\text{-}18)$$

Because $-\partial \overline{u_y' u_x'}/\partial y \sim k/\delta$, while $\partial \overline{u_y' u_y'}/\partial x \sim k/x$, Eq. (12.2-18) may be simplified finally to yield

$$\bar{u}_x \frac{\partial \bar{u}_x}{\partial x} + \bar{u}_y \frac{\partial \bar{u}_x}{\partial y} = \nu \frac{\partial^2 \bar{u}_x}{\partial y^2} - \frac{\partial \overline{u_y' u_x'}}{\partial y} \qquad (12.2\text{-}19)$$

Note that, in the turbulent boundary layer, the average axial pressure gradient is negligible but the average transverse one is not. In the laminar boundary layer, in contrast, both gradients are negligible (see Section 11.2).

The flow within the boundary layer is, therefore, determined by the following equations:

$$\frac{\partial \bar{u}_x}{\partial x} + \frac{\partial \bar{u}_y}{\partial y} = 0 \qquad \bar{u}_x \frac{\partial \bar{u}_x}{\partial x} + \bar{u}_y \frac{\partial \bar{u}_x}{\partial y} = \nu \frac{\partial^2 \bar{u}_x}{\partial y^2} - \frac{\partial \overline{u_y' u_x'}}{\partial y} \qquad (12.2\text{-}20)$$

The boundary conditions are

$$\bar{u}_x = 0 \qquad \bar{u}_y = 0 \qquad \overline{u_y' u_x'} = 0 \qquad \text{at } y = 0$$
$$\bar{u}_x \to U_\infty \text{ as } y \to \infty \qquad (12.2\text{-}21)$$

together with a suitable upstream matching condition on $\bar{u}_x$.

We now seek to solve Eqs. (12.2-20) subject to boundary conditions (12.2-21) and an upstream condition on $\bar{u}_x$, whatever that might be. Clearly, no solution can be obtained without an equation for $\overline{u_y' u_x'}$, i.e. without a turbulence model.

One way to obtain such an equation is to define an eddy viscosity v_t, by analogy with the kinematic viscosity v

$$- \overline{u_y' u_x'} = v_t \frac{\partial \overline{u_x}}{\partial y} \tag{12.2-22}$$

where the subscript t indicates a turbulent quantity. An example of the use of an eddy viscosity is mentioned in Section 10.7. We note tht v_t is a flow, and not a fluid, property: it can and does vary with position. We note also that Eq. (12.2-22) is not a physical statement. Thus, for example, it implies that the Reynolds stress vanishes when the average velocity gradient vanishes (unless, of course, v_t becomes unbounded which seems physically unreasonable). This means that there can be no Reynolds stress at symmetry lines and planes (see Section 6.5). All experimental evidence contradicts this. Indeed, Eq. (12.2-22) is not so much a physical statement as a definition of v_t: we have replaced ignorance of $\overline{u_y' u_x'}$ by ignorance of v_t. Another way to obtain an equation for $\overline{u_y' u_x'}$ is to assume that u_x' and u_y' are each proportional to $\partial \overline{u_x}/\partial y$, whence

$$- \overline{u_y' u_x'} = l^2 \left(\frac{\partial \overline{u_x'}}{\partial y} \right)^2 \tag{12.2-23}$$

where l is called the mixing length. The difficulty is that we have again replaced ignorance of $\overline{u_y' u_x'}$, this time by ignorance of l. We are, moreover, assuming that a single length scale characterizes the turbulence, which is the case only exceptionally.

As a result of these difficulties, we will adopt a different approach. We will use physical, dimensional analysis and order-of-magnitude arguments to obtain most, but not all, of the solution that we seek. To obtain all of the solution, we would have to use a specific turbulence model. An example is the k-ϵ model, in which pseudoconservation equations are solved for the average kinetic energy per unit mass of the fluctuating flow k and the dissipation ϵ [see Eqs. (12.2-10) and (12.1.25)]. These models are, however, of such complexity that numerical methods must be used to solve the flow equations; accordingly, we do not use them here.

In order to proceed, we decompose the turbulent boundary layer into (1) a thin *inner layer* by the plate in which we expect that viscous effects are significant; (2) a thicker *outer layer* in which we expect that viscous effects are insignificant; the inner and outer layers coincide in (3) an *overlap layer*. These layers are shown schematically in Fig. 12.4. We assume that the boundary layer thickness δ varies only slowly with axial position x, that is,

$$\left| \frac{d\delta}{dx} \right| << 1 \tag{12.2-24}$$

The length scale of the largest fluctuations is of the same order of magnitude as δ. As a result, the length scale over which δ varies is much larger than all other relevant length scales and hence is irrelevant. We now analyze each layer in turn.

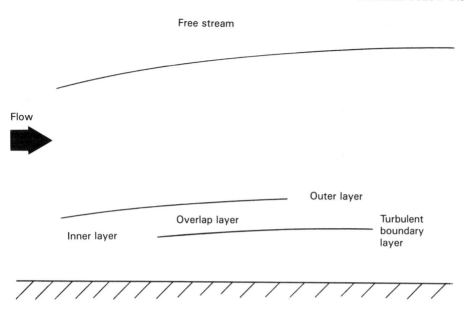

Figure 12.4. Decomposition of turbulent boundary layer by a flat plate.

In the inner layer, we expect physically that $\overline{u}_x$ and $\overline{u'_y u'_x}$ depend on local flow quantities only. Thus $\overline{u}_x$ and $\overline{u'_y u'_x}$ should depend only on the following quantities, which are relevant to the inner layer:

- the distance y from the plate: because we expect that inner layer is much thinner than the boundary layer (i.e., that $y \ll \delta$), δ is not a quantity that is relevant in the inner layer
- the kinematic viscosity ν: recall that what distinguishes the inner from the outer layer is that viscous effects are important only in the former
- the shear stress $\tau = \tau_{yx}|_{y=0}$ exerted by the fluid on the plate: if there were slip at the plate and hence no such stress, there would be no boundary layer
- the density ρ.

Thus we expect that $\overline{u}_x$ and $\overline{u'_y u'_x}$ are functions of y, ν, τ and ρ. Dimensional analysis then yields

$$\frac{\overline{u}_x}{u_\tau} = f\left(\frac{y u_\tau}{\nu}\right) \qquad \frac{\overline{u'_y u'_x}}{u_\tau^2} = F\left(\frac{y u_\tau}{\nu}\right) \tag{12.2-25}$$

where u_τ denotes the friction velocity given by

$$u_\tau = \sqrt{\frac{\tau}{\rho}} \tag{12.2-26}$$

It is conventional now to define

$$u^+ = \frac{\overline{u}_x}{u_\tau} \qquad y^+ = \frac{y u_\tau}{\nu} \qquad (12.2\text{-}27)$$

Thus u^+ is a dimensionless axial velocity and y^+ is the ratio of the geometric length scale y to the dynamic length scale ν/u_τ and can be thought of as a Reynolds number based on distance from the plate. Then Eqs. (12.2-25) can be written in the form

$$u^+ = f(y^+) \qquad \frac{\overline{u'_y u'_x}}{u_\tau^2} = F(y^+) \qquad (12.2\text{-}28)$$

Equations (12.2-25) and (12.2-28) are expressions of what is called the *law of the wall*. So far, we have used physical and dimensional analysis arguments. We now use order-of-magnitude arguments and note that it follows from Eqs. (12.2-25) that the inertial terms

$$\overline{u}_x \frac{\partial \overline{u}_x}{\partial x} + \overline{u}_y \frac{\partial \overline{u}_x}{\partial y} \sim \frac{u_\tau^2}{x} = \frac{u_\tau^3}{\nu} \frac{\nu}{u_\tau x}$$

the viscous term

$$\nu \frac{\partial^2 \overline{u}_x}{\partial y^2} \sim \frac{\nu u_\tau}{y^2} = \frac{u_\tau^3}{\nu} \left(\frac{\nu}{u_\tau y} \right)^2$$

and the Reynolds stress term

$$-\frac{\partial \overline{u'_y u'_x}}{\partial y} \sim \frac{u_\tau^2}{y} = \frac{u_\tau^3}{\nu} \frac{\nu}{u_\tau y}$$

Since we assume that $y \ll \delta$, and since y and ν/u_τ must be comparable in the inner layer (recall that viscous effects are significant only in the inner layer) it follows that $\nu/u_\tau \ll \delta$. But $\delta \ll x$, so $\nu/u_\tau \ll x$. Thus inertial effects are insignificant in the inner layer and the second of Eqs. (12.2-20) can be simplified to

$$\nu \frac{\partial^2 \overline{u}_x}{\partial y^2} - \frac{\partial \overline{u'_y u'_x}}{\partial y} = 0 \qquad (12.2\text{-}29)$$

which can be integrated to yield

$$\nu \frac{\partial \overline{u}_x}{\partial y} - \overline{u'_y u'_x} = c_1 \qquad (12.2\text{-}30)$$

where c_1 is a constant. We immediately deduce that the overall shear stress, that is, the sum of the viscous stress $\mu \, \partial \overline{u}_x/\partial y$ and the Reynolds stress $-\rho \overline{u'_y u'_x}$, is constant in the inner layer. Boundary conditions (12.2-21) mean that the Reynolds stress vanishes at the plate. Hence the shear stress on the plate is purely viscous and it follows that $c_1 = \tau/\rho$. Thus, in a sublayer very close to the plate, Eq. (12.2-30) becomes

$$\mu \frac{\partial \overline{u}_x}{\partial y} = \tau \tag{12.2-31}$$

which may be integrated to yield

$$\overline{u}_x = \frac{\tau y}{\mu} \tag{12.2-32}$$

or

$$u^+ = y^+ \tag{12.2-33}$$

Because of this linear relationship between u^+ and y^+, the sublayer very close to the plate is called the linear sublayer.

In the outer layer, we expect physically that $(U_\infty - \overline{u}_x)$ and $\overline{u'_y u'_x}$ depend on local flow quantities only, which are (1) the thickness δ of the boundary layer; (2) the distance y from the plate (we expect that $y \sim \delta$); (3) the shear stress τ exerted by the fluid on the plate; and (4) the density ρ. Viscous effects are important only in the inner layer, so the kinematic viscosity ν is *not* a flow quantity that is relevant in the outer layer. Note that we are considering the velocity *defect* $(U_\infty - \overline{u}_x)$, and not $\overline{u}_x$. This is because the effect of the plate on the outer layer is manifested solely in the shear stress τ: the fact that there is no slip at the plate (i.e., that $\overline{u}_x = 0$ at $y = 0$) is relevant to the inner layer but not to the outer layer. In the outer layer, it is, therefore, deviations from the free-stream velocity U_∞ that are of concern. Thus we expect that $(U_\infty - \overline{u}_x)$ and $\overline{u'_y u'_x}$ are functions only of δ, y, τ, and ρ. Dimensional analysis then yields

$$\frac{U_\infty - \overline{u}_x}{u_\tau} = g\left(\frac{y}{\delta}\right) \qquad \frac{\overline{u'_y u'_x}}{u_\tau^2} = G\left(\frac{y}{\delta}\right) \tag{12.2-34}$$

This equation is an expression of what is called the *defect law*.

In the overlap layer (also called the matching layer, the intermediate layer, or the inertial sublayer), we expect that Eqs. (12.2-25) and (12.2-34) hold simultaneously since the overlap layer is part of both the inner and outer layers. Hence,

$$f\left(\frac{y u_\tau}{\nu}\right) = \frac{U_\infty}{u_\tau} - g\left(\frac{y}{\delta}\right) \tag{12.2-35}$$

or

$$f\left(\frac{y}{\delta} \frac{\delta u_\tau}{\nu}\right) = \frac{U_\infty}{u_\tau} - g\left(\frac{y}{\delta}\right) \tag{12.2-36}$$

For this equation to hold everywhere in the overlap layer, the functions f and g must be logarithmic. This is because the logarithm is the only function for which multiplying the argument by a constant is equivalent to adding a constant to the function. Thus, in terms of the inner layer variables,

$$\frac{\overline{u}_x}{u_\tau} = A \ln \left(\frac{yu_\tau}{\nu}\right) + B \tag{12.2-37}$$

or
$$u^+ = A \ln (y^+) + B \tag{12.2-38}$$

and, in terms of the outer layer variables,

$$\frac{U_\infty - \overline{u}_x}{u_\tau} = C - A \ln \left(\frac{y}{\delta}\right) \tag{12.2-39}$$

where A, B, and C are constants which, without further information, we cannot determine; this is the penalty that we have incurred by failing to obtain a turbulence model. Empirically, however, it is found that

$$A \simeq 2.4\text{--}2.5 \qquad B \simeq 5.0\text{--}5.5 \qquad C \simeq 2.3\text{--}2.8 \tag{12.2-40}$$

We note, incidentally, that it follows from Eq. (12.2-37) that

$$\frac{\partial \overline{u}_x}{\partial y} = \frac{Au_\tau}{y} \tag{12.2-41}$$

Because $c_1 = \tau/\rho = u_\tau^2$, it thus follows from Eq. (12.2-30) that

$$-\overline{u'_y u'_x} \simeq u_\tau^2 \tag{12.2-42}$$

in the overlap layer, since $y >> \nu/u_\tau$ there. It then follows from Eq. (12.2-22) that

$$\nu_\tau \simeq \frac{yu_\tau}{A} \tag{12.2-43}$$

showing that ν_t varies with position, as we noted earlier.

In fact, we can show that the axial velocity profile is logarithmic in the overlap layer without reference to the law of the wall or the defect law. The overlap layer is too far from the plate for ν to be a relevant quantity and too close to the plate for δ to be a relevant quantity. Thus we expect that $\partial \overline{u}_x/\partial y$ [*not* $\overline{u}_x$ or ($U_\infty - \overline{u}_x$)] depends only on y, τ, and ρ. Dimensional analysis then yields

$$\frac{\partial \overline{u}_x}{\partial y} = c_2 \frac{u_\tau}{y} \tag{12.2-44}$$

where c_2 is a constant. This equation may be integrated to yield

$$\frac{\overline{u}_x}{u_\tau} = c_2 \ln y + c_3 \tag{12.2-45}$$

where c_3 is another constant. Clearly, Eq. (12.2-45) is of the same logarithmic form as Eqs. (12.2-37) and (12.2-39).

The structure of the various layers and sublayers comprising the boundary layer and the free stream may be summarized as follows (see also Fig. 12.5; the

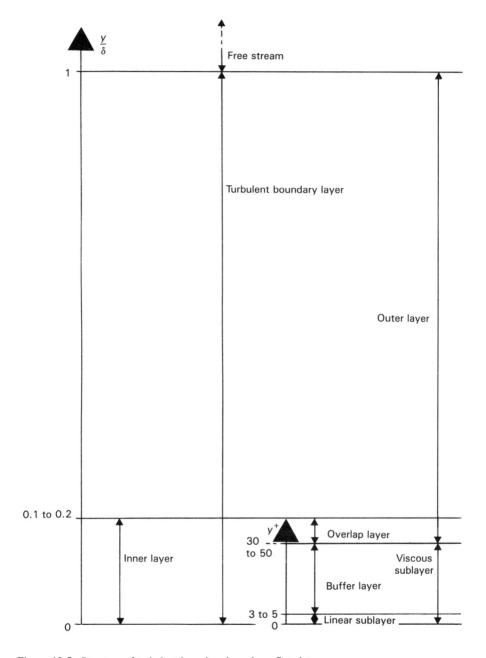

Figure 12.5. Structure of turbulent boundary layer by a flat plate.

values of y/δ and y^+ at the edges of the various layers and sublayers are empirical in origin).

- In the *inner layer,* inertial effects are insignificant.
- In the *viscous sublayer,* which is that part of the inner layer *not* in the overlap layer, viscous effects are significant (note that this sublayer is sometimes called the laminar sublayer which is completely misleading since the flow in it is in no sense laminar).
- In the *linear sublayer,* Reynolds stress effects are insignificant.
- In the *buffer layer,* which is that part of the viscous sublayer not in the linear sublayer, viscous and Reynolds stress effects are comparable.
- In the *overlap layer,* inertial and viscous effects are both insignificant.
- In the *outer layer,* viscous effects are insignificant.
- In the *free stream,* viscous and Reynolds stress effects are insignificant.

The profile of the average axial velocity $\overline{u}_x$ is shown schematically in Fig. 12.6. The dominant component of the average rate of production $-\frac{1}{2}\rho\overline{\mathbf{u}'\mathbf{u}'}{:}\overline{\mathbf{e}}$ of the kinetic energy of the fluctuating flow [see term (f) in Eq. (12.1-16)] is given

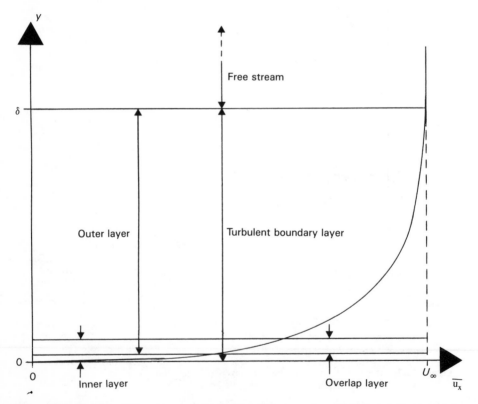

Figure 12.6. Variation of the average axial velocity with transverse position in a turbulent boundary layer by a flat plate.

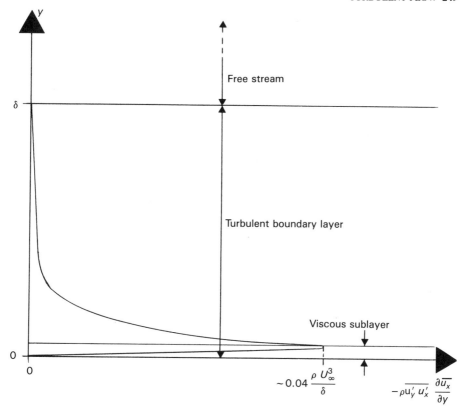

Figure 12.7. Variation of the dominant component of the average rate of production of kinetic energy of the fluctuating flow with transverse position in a turbulent boundary layer by a flat plate.

by $-\rho\overline{u_y'u_x'}\,\partial\overline{u}_x/\partial y$. Its profile is shown schematically in Fig. 12.7. Note that $-\rho\overline{u_y'u_x'}\,\partial\overline{u}_x/\partial y > 0$, so that the kinetic energy is transferred *from* the average *to* the fluctuating flow (see Section 12.1). Note also the very sharp maximum, the location of which appears to coincide with the outer edge of the viscous sublayer.

The drag on the plate is given in dimensionless form by the local drag coefficient C_f:

$$C_f = \frac{\tau}{\rho U_\infty^2/2} = \frac{2u_\tau^2}{U_\infty^2} \tag{12.2-46}$$

If we combine Eqs. (12.2-37) and (12.2-39), we obtain

$$\frac{U_\infty}{u_\tau} = A \ln\left(\frac{\delta u_\tau}{\nu}\right) + B + C \tag{12.2-47}$$

whence

$$\sqrt{\frac{2}{C_f}} = A \ln\left(\mathrm{Re}_\delta \sqrt{\frac{C_f}{2}}\right) + B + C \tag{12.2-48}$$

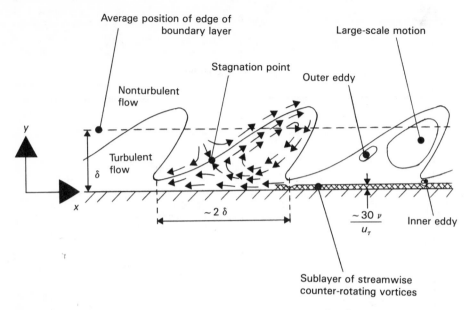

Figure 12.8. Organized motions in a turbulent boundary layer by a flat plate.

where Re_δ is the Reynolds number based on δ given by Eq. (12.2-2). Note that we did not need a turbulence model in order to obtain this (implicit) equation for C_f; empirical data [see Eqs. (12.2-40)] provide estimates for A, B, and C.

We conclude by noting, as mentioned in Section 12.1, that the unaveraged large-scale motions in the turbulent boundary layer appear not to be random but instead to be highly organized or structured. Recognition of such organization is important because it enables us to improve our physical understanding of turbulent flows. This in turn is important because it may enable us to make progress in the formulation of turbulence models. Relative to a reference frame that moves with respect to the plate at a velocity of approximately $0.8U_\infty \mathbf{i}_x$, there is considerable experimental evidence to support the structure shown in Fig. 12.8. There is a sharp demarcation between turbulent and nonturbulent parts of the flow. The main structural features within the turbulent part of the flow can be summarized as follows:

1. Immediately beside the plate, there is a sublayer which is approximately coincident with the viscous sublayer. It comprises a fluctuating array of *streamwise counter-rotating vortices*. These are long, narrow regions of rotating fluid, the long dimensions and axes of rotation of which are approximately aligned with the x direction, and the senses of rotation in which alternate. The vortices are of order $1000\ v/u_\tau$ long, $100\ v/u_\tau$ wide, and $30\ v/u_\tau$ high.

2. Slightly above the vortices, but still rather close to the plate, there is a layer in which there are fairly regular bursts of very intense fluctuations called *inner eddies*. These are of order $30\ v/u_\tau$ long and $15\ v/u_\tau$ high. Their axial velocity

relative to the plate is of order 0.6 U_∞ and their persistence length is of order δ, where the persistence length of a structure is roughly speaking the axial distance moved through by that structure before it loses its identity. The time between bursts is of order 6 δ/U_∞.

3. Far from the plate, but still within the turbulent part of the flow, there are very intense fluctuations called *outer eddies*. They are mainly located near the stagnation point, where the fluid is approximately stationary relative to the moving reference frame, on the upstream portions of the undulatory interface between the turbulent and nonturbulent parts of the flow. They are of order 200 v/u_τ long and 100 v/u_τ high. Their axial velocity relative to the plate is of order 0.9 U_∞ and their persistence length is of order 1000 v/u_τ.

4. Far from the plate, but again still within the turbulent part of the flow, there are *large-scale motions* of order δ long, δ wide, and δ high. Their axial velocity relative to the plate is of order 0.8 U_∞ and their persistence length and hence axial spacing is of order 2 δ.

One of the most interesting features of these structures within the turbulent part of the flow is that some of the parameters that characterize them are determined by the outer layer length scale δ, while others are determined by the inner layer length scale v/u_τ. It is particularly interesting that the time between bursts of the inner eddies is determined by the outer layer length scale, since the inner eddies are located rather close to the plate. This suggests that the bursts affect nearly all of the turbulent part of the flow.

12.3 FLOW IN A PIPE

We consider again the problem that we considered in Section 10.2 (see also Fig. 12.9). A volumetric flow rate Q is induced in the pipe of radius R and length L ($>>R$) by a constant pressure drop ($p_0 - p_L$). The flow is of an incompressible Newtonian fluid of density ρ and viscosity μ. The Reynolds number is given by

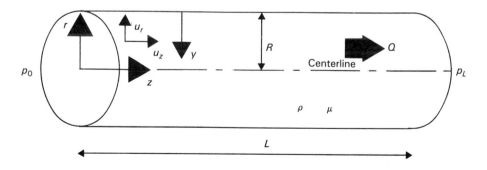

Figure 12.9. Geometry of flow in a pipe.

$$\text{Re} = \frac{2Q}{\pi \nu R} \tag{12.3-1}$$

where the kinematic viscosity $\nu = \mu/\rho$. If Re is less than some critical value $\text{Re}_c \simeq 2000$, the flow in the pipe is laminar (see Section 10.2). If Re exceeds Re_c but is less than some second critical $\text{Re}_C \simeq 4000$, the flow in the pipe is transitional. If Re exceeds Re_C, the flow in the pipe is turbulent, and it is this flow that we seek to determine.

The length scale Λ of the largest fluctuations in the turbulent flow in the pipe is of the same order of magnitude as the pipe radius R. The length scale λ of the smallest fluctuations is given from Eq. (12.1-29) by

$$\lambda \sim \frac{R}{\text{Re}^{3/4}} \tag{12.3-2}$$

In order to simplify our analysis, we now assume that the wall of the pipe is smooth, that is, that the length scale of any roughness is very much less than that of any other relevant length scale. Later we consider the case when the wall of the pipe is rough.

Because the pressure drop is constant, we expect that the turbulent flow is stationary, that is, that it is steady on average (see Eq. (12.1-2). Because of the obvious symmetry of the pipe, we expect that the average flow is axisymmetric and swirl-free. Because the pipe is long, we expect that the average flow is fully developed everywhere except near the entrance ($z = 0$) and exit ($z = L$) of the pipe. Thus, in particular, we expect that

$$\frac{\partial \overline{u}_z}{\partial z} = 0 \tag{12.3-3}$$

The mass conservation equation (A2.6-5) of Appendix B yields

$$\frac{1}{r} \frac{\partial}{\partial r} (r \overline{u}_r) + \frac{\partial \overline{u}_z}{\partial z} = 0 \tag{12.3-4}$$

whence

$$\frac{1}{r} \frac{\partial}{\partial r} (r \overline{u}_r) = 0 \tag{12.3-5}$$

which can be integrated to yield

$$r \overline{u}_r = 0 \tag{12.3-6}$$

since no flow-through at the pipe wall ($r = R$) means that $\overline{u}_r$ vanishes there. Hence,

$$\overline{u}_r = 0 \tag{12.3-7}$$

The linear momentum conservation equations (A2.6-6) of Appendix B (with the gravitational term incorporated in the pressure term) yield

$$0 = -\frac{\partial \bar{p}}{\partial r} - \rho \frac{1}{r} \frac{d}{dr} (r \overline{u_r' u_r'}) + \rho \frac{1}{r} \overline{u_\theta' u_\theta'} \tag{12.3-8}$$

$$0 = -\frac{\partial \bar{p}}{\partial z} + \mu \frac{1}{r} \frac{d}{dr} \left(r \frac{d\bar{u}_z}{dr} \right) - \rho \frac{1}{r} \frac{d}{dr} (r \overline{u_r' u_z'}) \tag{12.3-9}$$

Note that, although the average flow is axisymmetric and swirl-free, $\overline{u_\theta' u_\theta'} \neq 0$: turbulence is, as we noted in Section 12.1, essentially three dimensional. The boundary conditions for the flow are no slip and no flow-through at the wall of the pipe

$$\bar{u}_z = 0 \qquad \overline{u_r' u_z'} = 0 \qquad \overline{u_r' u_r'} = 0 \qquad \overline{u_\theta' u_\theta'} = 0 \qquad \text{at } r = R \tag{12.3-10}$$

and symmetry at the centerline of the pipe

$$\frac{d\bar{u}_z}{dr} = 0 \qquad \overline{u_r' u_z'} = 0 \qquad \text{at } r = 0 \tag{12.3-11}$$

We can integrate Eq. (12.3-8) to yield, on imposition of boundary conditions (12.3-10),

$$\bar{p} + \rho \overline{u_r' u_r'} + \rho \int_R^r \frac{1}{r} (\overline{u_r' u_r'} - \overline{u_\theta' u_\theta'}) \, dr = \bar{p}_w \tag{12.3-12}$$

where $\bar{p}_w$ denotes the average pressure at the wall, which is a function of axial position z. Note that $\bar{p}$ varies radially as well as axially. There is, accordingly, a possible ambiguity in the definition of the pressure drop $(p_0 - p_L)$. Provided p_0 and p_L refer to the *same* radial position, however, there is no ambiguity. Accordingly, we choose to define p_0 and p_L to be the average wall pressures at the entrance and exit of the pipe, respectively:

$$p_0 = \bar{p}_w|_{z=0} \qquad p_L = \bar{p}_w|_{z=L} \tag{12.3-13}$$

It follows from Eq. (12.3-12), therefore, that

$$\frac{\partial \bar{p}}{\partial z} = \frac{d\bar{p}_w}{dz} = -\frac{p_0 - p_L}{L} \tag{12.3-14}$$

and so Eq. (12.3-9) becomes

$$0 = \frac{p_0 - p_L}{L} + \mu \frac{1}{r} \frac{d}{dr} \left(r \frac{d\bar{u}_z}{dr} \right) - \rho \frac{1}{r} \frac{d}{dr} (r \overline{u_r' u_z'}) \tag{12.3-15}$$

which may be integrated to yield, on imposition of boundary conditions (12-3-11):

$$0 = r \frac{p_0 - p_L}{2L} + \mu \frac{d\bar{u}_z}{dr} - \rho \overline{u_r' u_z'} \tag{12.3-16}$$

For convenience, we prefer now to use not the distance r from the axis of the pipe but the distance y from its wall

$$y = R - r \tag{12.3-17}$$

As a result, it is convenient to consider the velocity component in the y, as opposed to the r, direction; in particular,

$$u_y' = -u_r' \tag{12.3-18}$$

Then Eq. (12.3-16) becomes

$$0 = (R - y)\frac{p_0 - p_L}{2L} - \mu\frac{d\bar{u}_z}{dy} + \rho\overline{u_y'u_z'} \tag{12.3-19}$$

or

$$\nu\frac{d\bar{u}_z}{dy} - \overline{u_y'u_z'} = (R - y)\frac{p_0 - p_L}{2\rho L} \tag{12.3-20}$$

An overall force balance on the fluid in the pipe yields

$$\pi R^2(p_0 - p_L) = 2\pi RL\tau \tag{12.3-21}$$

where $\tau = \tau_{rz}|_{r=R}$ denotes the wall shear stress and is independent of axial position z because the flow is fully developed. Just as in Section 12.2, we define the friction velocity u_τ

$$u_\tau = \sqrt{\frac{\tau}{\rho}} \tag{12.3-22}$$

so Eq. (12.3-21) becomes

$$\frac{p_0 - p_L}{2\rho L} = \frac{u_\tau^2}{R} \tag{12.3-23}$$

Thus Eq. (12.3-20) becomes finally

$$\nu\frac{d\bar{u}_z}{dy} - \overline{u_y'u_z'} = u_\tau^2\left(1 - \frac{y}{R}\right) \tag{12.3-24}$$

and boundary conditions (12.3-10) become

$$\bar{u}_z = 0 \qquad \overline{u_y'u_z'} = 0 \qquad \text{at } y = 0 \tag{12.3-25}$$

We now seek to solve Eq. (12.3-24) subject to boundary conditions (12.3-25). Clearly, we cannot obtain a complete solution without an equation for $\overline{u_y'u_z'}$. We can, however, obtain a partial solution by decomposing the flow into layers, just as we did for the turbulent boundary layer in Section 12.2. These layers comprise a thin surface layer by the wall (in which viscous effects are significant) and a thicker core layer about the centerline (in which viscous effects are insignificant); they coincide in an overlap layer (see Fig. 12.10). Note that the surface and core layers are the analogues of the inner and outer layers, respectively, which we discussed in Section 12.2. We analyze each layer in turn, making extensive use of the corresponding analyses in Section 12.2.

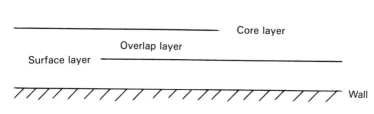

Figure 12.10. Decomposition of turbulent flow in a pipe.

In the surface layer, we expect that $\overline{u}_z$ and $\overline{u_y'u_z'}$ depend only on y, ν, τ, and ρ. Dimensional analysis then yields

$$\frac{\overline{u}_z}{u_\tau} = f\left(\frac{yu_\tau}{\nu}\right) \qquad \frac{\overline{u_y'u_z'}}{u_\tau^2} = F\left(\frac{yu_\tau}{\nu}\right) \tag{12.3-26}$$

and, since $y \ll R$, Eq. (12.3-24) becomes

$$\nu\frac{d\overline{u}_z}{dy} - \overline{u_y'u_z'} = u_\tau^2 = \frac{\tau}{\rho} \tag{12.3-27}$$

At the wall, boundary conditions (12.3-25) mean that the Reynolds stress vanishes. Thus the shear stress on the pipe wall is purely viscous. We recall that the shear stress on the plate discussed in Section 12.2 is also purely viscous. In general, for a Newtonian fluid,

The shear stress on any rigid, impermeable wall is purely viscous, whether the flow past the wall is laminar or turbulent.

In the core layer, we expect that $(\overline{u}_z|_{y=R} - \overline{u}_z)$ and $\overline{u_y'u_z'}$ depend only on R, y, τ, and ρ. Here $\overline{u}_z|_{y=R}$ denotes the average centerline axial velocity: $y = R$ is equivalent to $r = 0$. Dimensional analysis then yields

$$\frac{\overline{u}_z|_{y=R} - \overline{u}_z}{u_\tau} = g\left(\frac{y}{R}\right) \qquad \frac{\overline{u_y'u_z'}}{u_\tau^2} = G\left(\frac{y}{R}\right) \tag{12.3-28}$$

In the overlap layer, we deduce using either of the arguments given in Section 12.2 that, in terms of surface layer variables,

$$\frac{\overline{u}_z}{u_\tau} = A \ln\left(\frac{yu_\tau}{\nu}\right) + B \tag{12.3-29}$$

and, in terms of core layer variables

$$\frac{\overline{u}_z|_{y=R} - \overline{u}_z}{u_\tau} = C - A \ln\frac{y}{R} \tag{12.3-30}$$

Here A, B, and C are constants which, just as in Section 12.2, we cannot determine without further information. Empirically, however, it is found that

$$A \simeq 2.4\text{--}2.5 \qquad B \simeq 5.0\text{--}5.5 \qquad C \simeq 0.6\text{--}1.0 \qquad (12.3\text{-}31)$$

We note immediately, by comparing Eqs. (12.2-40) and (12.3-31), that the values of A and B are roughly the same for flow in a turbulent boundary layer on a flat plate as for turbulent flow in a pipe, whereas the values of C are certainly different. In general,

In the vicinity of a rigid, impermeable wall, the nature of a turbulent shear flow is virtually independent of the overall flow geometry.

Here, by a turbulent shear flow, we mean a turbulent flow for which (1) the principal average velocity component is in a direction approximately parallel to the wall; and (2) the gradient of the principal average velocity component is greatest in a direction approximately normal to the wall. Thus the flow in the inner layer of a turbulent boundary layer on a flat plate is virtually identical to that in the surface layer of a turbulent flow in a pipe. In contrast, the flow in the outer layer of a turbulent boundary layer on a flat plate is different from that in the core layer of a turbulent flow in a pipe: turbulent shear flows are not all the same.

We now relate the pressure drop $(p_0 - p_L)$ to the volumetric flow rate Q in the pipe. A convenient way of doing this is by use of the friction factor f [not to be confused with the function f in Eq. (12.3-26)], which may be interpreted as a dimensionless wall shear stress. It is given by

$$f = \frac{\tau}{\frac{1}{2}\rho Q^2/\pi^2 R^4} = \frac{2u_\tau^2}{Q^2/\pi^2 R^4} \qquad (12.3\text{-}32)$$

If we combine Eqs. (12.3-29) and (12.3-30), we obtain

$$\frac{\bar{u}_z|_{y=R}}{u_\tau} = A \ln\left(\frac{Ru_\tau}{\nu}\right) + B + C \qquad (12.3\text{-}33)$$

By definition

$$Q = \int_0^R 2\pi r \bar{u}_z \, dr = \int_0^R 2\pi (R - y)\bar{u}_z \, dy \qquad (12.3\text{-}34)$$

If we assume that Eq. (12.3-30) holds not just in the overlap layer but everywhere (i.e., for $0 \le y \le R$; there is experimental evidence which suggests that this is a reasonable assumption)

$$Q = \int_0^R 2\pi R\left(1 - \frac{y}{R}\right)\left[\bar{u}_z|_{y=R} - u_\tau\left(C - A \ln\frac{y}{R}\right)\right] dy \qquad (12.3\text{-}35)$$

Hence,

$$Q = \pi R^2 \bar{u}_z|_{y=R} - (C + \tfrac{3}{2}A)\pi R^2 u_\tau \qquad (12.3\text{-}36)$$

or, using Eq. (12.3-33),

$$Q = \pi R^2 u_\tau \left(A \ln \left(\frac{R u_\tau}{\nu} \right) + B - \frac{3}{2} A \right) \tag{12.3-37}$$

Thus, using Eq. (12.3-32),

$$\sqrt{\frac{2}{f}} = A \left[\ln \left(\frac{1}{2} \operatorname{Re} \sqrt{\frac{f}{2}} \right) - \frac{3}{2} \right] + B \tag{12.3-38}$$

If we substitute for A and B from Eqs. (12.3-31), we obtain

$$\frac{1}{\sqrt{f}} \approx 1.7 \ln (\operatorname{Re} \sqrt{f}) - 0.7 \tag{12.3-39}$$

which is called the von Karman-Nikuradse equation. Experimentally, it is found to be corroborated, perhaps with slight adjustment of the values of the numerical constants, provided the Reynolds number is large enough. Note that, although Eq. (13.3-39) is implicit in f, $\operatorname{Re}\sqrt{f}$ is independent of Q so that, given the pressure drop $(p_0 - p_L)$, the volumetric flow rate Q may be determined directly. For laminar flow in a pipe, it follows from Eq. (10.2-22) that

$$f = \frac{16}{\operatorname{Re}} \tag{12.3-40}$$

A graph of friction factor f versus Re is given in Fig. 12.11.

We conclude by modifying our analysis for the case when the wall of the pipe is rough, not smooth. A convenient measure of the length scale k of the roughness (see Fig. 12.12) is the root mean square roughness height [not to be confused with the average kinetic energy per unit mass of the fluctuating flow given by Eq. (12.1-17)]. It is usually the case in practice that k/R is sufficiently small that the roughness does not affect the flow in the core layer, that is, we may assume that the velocity defect law

$$\frac{\bar{u}_z|_{y=R} - \bar{u}_z}{u_\tau} = g\left(\frac{y}{R}\right) \tag{12.3-41}$$

[see Eqs. (12.3-28)] still holds. In the surface layer, on the other hand, the additional length scale k must be added to the variables $(y, \nu, \tau, \text{and } \rho)$ upon which $\bar{u}_z$ depends. Dimensional analysis then yields

$$\frac{\bar{u}_z}{u_\tau} = f_1\left(\frac{y u_\tau}{\nu}, \operatorname{Re}_k\right) = f_2\left(\frac{y}{k}, \operatorname{Re}_k\right) \tag{12.3-42}$$

where the roughness Reynolds number is given by

$$\operatorname{Re}_k = \frac{k u_\tau}{\nu} \tag{12.3-43}$$

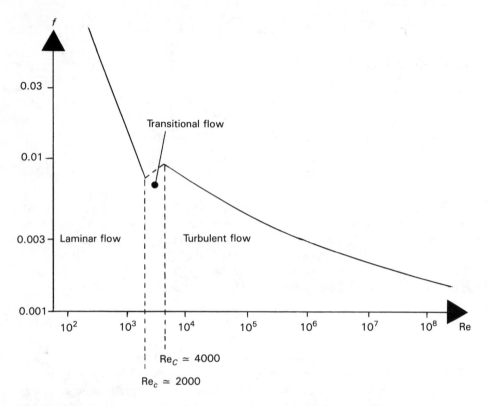

Figure 12.11. Variation of friction factor with Reynolds number for smooth pipes.

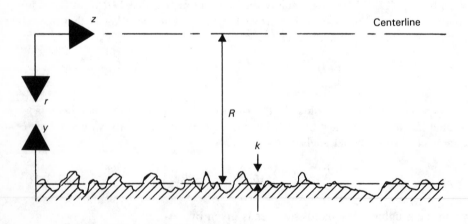

Figure 12.12. Roughness on a pipe wall.

In the overlap layer, we expect that Eqs. (12.3-41) and (12.3-42) hold simultaneously. We thus deduce that the functions f_1, f_2, and g must be logarithmic in y and that the functions f_1 and f_2 must be additive in Re_k. Thus, in terms of surface layer variables,

$$\bar{u}_z/u_\tau = A \ln \left(\frac{y u_\tau}{\nu} \right) + B_1(\text{Re}_k) = A \ln \left(\frac{y}{k} \right) + B_2(\text{Re}_k) \qquad (12.3\text{-}44)$$

and, in terms of core layer variables,

$$\frac{\bar{u}_z|_{y=R} - \bar{u}_z}{u_\tau} = C - A \ln \frac{y}{R} \qquad (12.3\text{-}45)$$

Note that Eqs. (12.3-30) and (12.3-45) are identical; the constants A and C are, moreover, given by Eqs. (12.3-31) in both cases. The first of Eqs. (12.3-44) differs, however, from Eq. (12.3-29) because B_1 is a function of Re_k and not a constant. Clearly, if $\text{Re}_k \to 0$, so that the wall of the pipe is smooth, we would expect that $B_1 \to B$ where B is given by Eqs. (12.3-31). In practice, roughness appears to have no effect, and the pipe wall is said to be hydraulically smooth, if Re_k is less than a lower critical value $\text{Re}_{k_c} \simeq 4$. Thus if the roughness elements on the wall do not protrude much beyond the linear sublayer (see Fig. 12.5), the effects of roughness are effectively subsumed in the viscous sublayer and may be neglected. On the other hand, if $\text{Re}_k \to \infty$ (but $k \ll R$, so that a distinct surface layer still exists), we would expect that B_2 becomes independent of Re_k (i.e., that it tends to a constant). In practice, roughness ceases to have an apparent effect on B_2 when Re_k exceeds an upper critical value $\text{Re}_{k_c} \simeq 40$. Thus, if the roughness elements on the wall protrude much beyond the viscous sublayer (see Fig. 12.5), the effects of viscosity are effectively obscured by those of roughness and the flow in the surface layer is essentially independent of viscosity. When $\text{Re}_k > \text{Re}_{k_c}$, the pipe wall is, therefore, said to be fully rough and it is found empirically that

$$B_2 \simeq 6.0\text{--}11.0 \qquad (12.3\text{-}46)$$

The relatively large uncertainty in the value of B_2 arises partly as a result of experimental uncertainty [recall that there is uncertainty in the values of A, B, and C in Eqs. (12.3-31)]. It arises mainly because B_2 is not a unique function of Re_k but varies slightly with the type of roughness. An empirical equation for B_1 which is approximately valid for all values of Re_k is

$$B_1 \simeq B - A \ln (1.0 + 0.3\, \text{Re}_k) \qquad (12.3\text{-}47)$$

We can now generalize the von Karman-Nikuradse equation (12.3-39) to pipes with rough walls by noting that, if we replace B in Eqs. (12.3-37) and (12.3-38) by B_1 given by Eq. (12.3-47), we obtain

$$\frac{1}{\sqrt{f}} \simeq 1.7 \ln \left[\frac{\text{Re}\sqrt{f}}{1.0 + 0.1(k/R)\text{Re}\sqrt{f}} \right] - 0.7 \qquad (12.3\text{-}48)$$

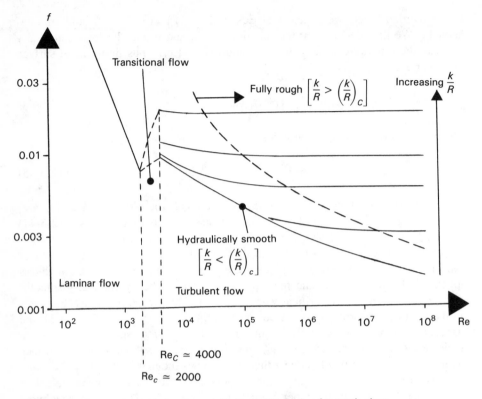

Figure 12.13. Variation of friction factor with Reynolds number for rough pipes.

This equation is found to be corroborated experimentally, perhaps with slight adjustment of the numerical constants, provided the Reynolds number is large enough. In a turbulent flow, if the relative roughness (k/R) is less than some lower critical value $(k/R)_c \simeq 20/\text{Re}$, f is essentially independent of k, while, if it exceeds some upper critical value $(k/R)_c \simeq 2000/\text{Re}$, f is essentially independent of Re. In a laminar flow, roughness has no effect (provided $k/R \ll 1$) because the viscous region occupies the whole cross section of the pipe. This can be seen in Fig. 12.13, which is a graph of f versus Re known as a Moody chart. Such a chart is based on a single type of roughness, typically that of pipes lined with sand grains. Because both the magnitude and type of the roughness are important, it is common in practice to use not the actual roughness k but the equivalent roughness k_e.

12.4 BREAKUP OF A SPHERICAL DROP

We consider a spherical drop of diameter D of an incompressible, Newtonian liquid of density $\rho\dagger$ and viscosity $\mu\dagger$. The drop is in a stream of an immiscible, incompressible, Newtonian liquid of density ρ and viscosity μ flowing at characteristic velocity u_c in a flow geometry of characteristic length scale $l_c >> D$. Thus, if the flow is in a pipe, u_c might be the average velocity of the liquid stream and l_c might be the radius of the pipe. The interfacial tension of the two liquids is σ. We assume that $\rho \simeq \rho\dagger$, so that buoyancy or gravitational effects are insignificant.

We suppose that

$$\text{Re} = \frac{\rho u_c l_c}{\mu} >> 1 \qquad (12.4\text{-}1)$$

so that the flow of the liquid stream is turbulent. In certain circumstances, the drop may break up. This is because the velocity in the stream fluctuates, apparently randomly, with position. Provided $D \lll \lambda$, where λ denotes the length scale of the smallest fluctuations, different pressures are, as a result, exerted at different positions on the surface of the drop. If the effect of the interfacial tension is not sufficiently large, the drop breaks up. We seek to determine when such breakup occurs. Because we have no turbulence model (i.e., no equation for the Reynolds stress), we do not seek an exact condition for breakup. Instead, in the spirit of our analyses in Sections 12.2 and 12.3, we use physical, dimensional analysis and order-of-magnitude arguments to obtain a condition which indicates approximately when breakup occurs.

We start by noting that, provided $D >> \lambda$, there are three distinct length scales characterizing the flow. The largest is the macroscale l_c which is, to an order of magnitude, the length scale Λ of the largest fluctuations. The physical quantities associated with the macroscale are the characteristic stream velocity u_c, the stream density ρ (but not its viscosity μ: recall that the gross structure of a turbulent flow is virtually independent of viscosity), and the dissipation ϵ.

The intermediate length scale is the drop scale D, the physical quantities associated with which are the drop density $\rho\dagger$ and viscosity $\mu\dagger$, the stream density ρ (but not its viscosity μ, for the same reason as before), the interfacial tension σ and the dissipation ϵ. The smallest is the microscale λ, the physical quantities associated with which are the stream density ρ and viscosity μ (viscous effects are significant on this length scale) and the dissipation ϵ.

With the trivial exception of the density ρ of the liquid stream, the only quantity that is significant on all three length scales is the dissipation given by

$$\epsilon = \frac{1}{2}\frac{\mu}{\rho}\overline{\mathbf{e}':\mathbf{e}'} \qquad (12.4\text{-}2)$$

where $\mathbf{e}'$ denotes the fluctuating component of the rate of strain $\mathbf{e}$. Because it is impossible for physical processes occurring on two disparate length scales to interact directly, there must be at least one physical quantity that is common to both length scales if the processes are to interact at all. Thus it is the transfer of kinetic energy down to the energy cascade (see Section 12.1) that is the key to relating the physical processes occurring on the three different length scales l_c (or Λ), D, and λ. Kinetic energy is abstracted from the average flow on the macroscale l_c

and converted into kinetic energy in the fluctuating flow again on the macroscale. The kinetic energy of the fluctuating flow is then passed from larger to smaller length scales by the cascade process at a rate essentially given by ϵ. This must be so for, if it is passed down at a rate different from that at which it is dissipated, the kinetic energy would not be dissipated by viscous action on the microscale λ. Hence the smallest fluctuations would have a length scale different from λ, which is a contradiction. The kinetic energy passes down, through the drop scale D, where the fluctuations may be sufficiently intense for the drop to break up, and hence to the microscale λ, where it is dissipated by viscous action.

Because of its central physical role, we continue, therefore, by obtaining two order-of-magnitude estimates of ϵ, one on the macroscale and the other on the microscale. The kinetic energy of the fluctuating flow is abstracted from that of the average flow on the macroscale l_c. The quantities relevant on this length scale are ϵ, u_c, l_c, and ρ. Dimensional analysis then yields the single dimensionless group $\epsilon l_c / u_c^3$. Note that this group is independent of ρ: no dimensionless group incorporating ρ can be devised on the macroscale. Hence the density of the liquid stream is *not* in fact a relevant quantity on the macroscale. Because only one dimensionless group is relevant on the macroscale, the group must be a constant. Moreover that constant must be of unit order of magnitude (see Section 7.2), that is,

$$\frac{\epsilon l_c}{u_c^3} \sim 1 \tag{12.4-3}$$

In order to see why, we note that any dimensionless group can be interpreted as a ratio of two quantities. For example, the Reynolds number can be interpreted as the ratio of inertial forces to viscous forces (see Section 7.2). The group $\epsilon l_c / u_c^3$ can be interpreted as the ratio of the rate at which kinetic energy is passed down the energy cascade to the rate at which energy is abstracted from the average flow. If these two rates do not match, there would be an imbalance and hence a net surplus or deficit of kinetic energy of the fluctuating flow on the macroscale. Such an imbalance is clearly impossible, from which we deduce that $\epsilon l_c / u_c^3$ must be of unit order of magnitude. The kinetic energy of the fluctuating flow, which is generated on the macroscale l_c, is dissipated on the microscale λ. The quantities relevant on the microscale are ϵ, ρ, μ and λ. Dimensional analysis then yields the single dimensionless group $\lambda \epsilon^{1/4} \rho^{3/4} / \mu^{3/4}$. Thus we deduce that

$$\frac{\lambda \epsilon^{1/4} \rho^{3/4}}{\mu^{3/4}} \sim 1 \tag{12.4-4}$$

It then follows, by eliminating ϵ between Eqs. (12.4-3) and (12.4-4) that

$$\lambda \sim \frac{l_c}{\mathrm{Re}^{3/4}} \tag{12.4-5}$$

or, since $\Lambda \sim l_c$,

$$\lambda \sim \frac{\Lambda}{\mathrm{Re}^{3/4}} \qquad (12.4\text{-}6)$$

which is Eq. (12.1-29). Note that, because we are assuming the existence of three disparate length scales l_c, D, and λ, it follows from Eq. (12.4-5) that we require that

$$l_c \gg D \gg \frac{l_c}{\mathrm{Re}^{3/4}} \qquad (12.4\text{-}7)$$

We now consider the physical processes occurring on the drop scale D. We recall that the physical quantities relevant on this length scale are ϵ, σ, $\rho\dagger$, $\mu\dagger$, ρ, and D. The interfacial or surface force F_s tending to maintain the sphericity, and hence integrity of the drop is given by

$$F_s \sim \sigma D \qquad (12.4\text{-}8)$$

This is because the only physical quantities relevant to the interfacial force are σ, D, and F_s. Dimensionless analysis then yields the single dimensionless group $F_s/\sigma D$, from which Eq. (12.4-8) immediately follows. The order of magnitude of the pressure force F_p tending to distort and hence break up the drop can be estimated as follows. The origin of the pressure force F_p is the velocity fluctuations on the drop scale which cause pressure fluctuations that act on the surface of the drop to produce the pressure force. Let $[u^2]$ denote a characteristic difference of speed squared at different positions on the surface of the drop (i.e., at positions separated by distances of the order of magnitude of the drop diameter D). Because viscous effects are significant only on the microscale in the liquid stream, the flow on the drop scale is effectively inviscid there. Accordingly, the difference in pressures $[p]$ at different positions on the outside surface of the drop is given approximately from Bernoulli's equation (9.1-13) by

$$[p] \sim \rho[u^2] \qquad (12.4\text{-}9)$$

It then follows immediately that

$$F_p \sim [p]D^2 \sim \rho[u^2]D^2 \qquad (12.4\text{-}10)$$

We now note that the reason why there are velocity fluctuations on the drop scale D is that kinetic energy is abstracted from the average flow on the macroscale l_c and passed via the drop scale to the microscale λ where it is dissipated. Thus we seek to relate $[u^2]$, which is relevant on the drop scale, to ϵ. Dimensional analysis then yields the single dimensionless group $[u^2]/D^{2/3}\epsilon^{2/3}$ from which we deduce that

$$\frac{[u^2]}{D^{2/3}\epsilon^{2/3}} \sim 1 \qquad (12.4\text{-}11)$$

If we now eliminate $[u^2]$ between Eqs. (12.4-10) and (12.4-11) and then eliminate ϵ using Eq. (12.4-3), we obtain

$$F_p \sim \rho u_c^2 D^2 \left(\frac{D}{l_c}\right)^{2/3} \tag{12.4-12}$$

We now note that F_s varies as D while F_p varies as $D^{8/3}$. Thus, as D increases, F_s increases, but F_p increases more quickly. There is, therefore, a critical value D_c of D at which F_p equals F_s and which is given from Eqs. (12.4-8) and (12.4-12) by

$$D_c \sim \frac{l_c}{\text{We}^{3/5}} \tag{12.4-13}$$

where the Weber number is given by

$$\text{We} = \frac{\rho u_c^2 l_c}{\sigma} \tag{12.4-14}$$

[see Eq. (7.2-13)]. For $D > D_c$, $F_p > F_s$ and the force tending to break up the drop exceeds that tending to maintain its integrity. Thus a drop of diameter $D < D_c$ is expected to be stable to turbulent fluctuations while one of diameter $D > D_c$ is expected to be unstable and to break up. Note, incidentally, that D_c is independent of the viscosity μ of the liquid stream. This is to have been expected since viscous effects are relevant only on the microscale, that is, on a length scale much smaller than that of the drop [see Eq. (12.4-7)]. Note also that this analysis could be applied to the breakup of a nonspherical drop, provided all dimensions of the drop are comparable (as is usually the case in practice), since it is only necessary that D is a characteristic dimension of the drop, for example the diameter $(6V/\pi)^{1/3}$ of the sphere of the same volume V as the drop. Similarly, it could be applied to the breakup of a dilute suspension of drops; provided the volume fraction of drops is very small, each drop is essentially independent of the others, which can, therfore, be regarded as being at infinity (see Section 6.4). Note finally that we have only considered one possible mode of drop breakup. Other modes exist, both laminar (as a result of shear or extension; see Section 2.5) and turbulent.

THIRTEEN

POSTLUDE

13.1 SOLUTION METHODOLOGY

Most of the problems solved in Chapters 8 through 12 illustrate variants of a solution method that is applicable to other fluid mechanics problems and indeed to other physical problems. Although the method is not always the same (partly because the problems themselves differ and partly because the levels to which the problems can or need be solved differ), it is, nevertheless, useful to summarize its usual main steps.

1. The geometry of the problem is used to select an appropriate reference frame and coordinate system.
2. The symmetry of the problem is used to suggest symmetry in the solution.
3. Extreme values of the dimensionless groups formed from the parameters of the problem are used to simplify the governing equations by the method given in Section 7.3 (note that different simplifications can be appropriate in different regions of flows in which there are multiple length scales).
4. The boundary and initial conditions are used to suggest an appropriate form for the solution.
5. The solution is obtained by an appropriate combination of inspection, use of trial solutions, physical arguments, appeal to experimental observation, and experience (note that, by an appropriate combination, we often mean any combination that enables us to obtain a solution: the more difficult the problem, the less fastidious we must usually be about how we solve it).

Selected problems are presented in Appendix C which give practice in the use of this solution method. Others are presented that supplement the material in Chapters 1 through 7.

13.2 CONCLUSION

The level of the discussion of the fundamentals of fluid mechanics in Chapters 1 through 7 is adequate for most engineering and scientific purposes, though it is not the deepest one attainable. Many applications are examined in Chapters 8 through 12, yet they comprise no more than a very small (and rather special) fraction of those that are of interest. Selected references are, however, presented in Appendix D to texts that discuss fundamentals at a deeper level and examine many other applications. These references can conveniently form the basis for a more advanced study of the subject.

ALGEBRA AND CALCULUS OF SCALAR, VECTOR, AND TENSOR FIELDS

A1.1 FIELDS

We define here typical scalar, vector, and tensor fields. We let s denote a scalar field; $\mathbf{v}$, $\mathbf{v}_1$, and $\mathbf{v}_2$ denote vector fields; and $\mathbf{T}$, $\mathbf{T}_1$, and $\mathbf{T}_2$ denote tensor fields. We let (ξ_1, ξ_2, ξ_3) denote a three-dimensional, orthogonal, right-handed coordinate system. Although any cyclical permutation of ξ_1, ξ_2, and ξ_3 (see Section 1.2) is permitted in what follows in Sections A1.2 and A1.3 (where we summarize the algebra and calculus, respectively, of scalar, vector, and tensor fields), we specifically identify ξ_1, ξ_2, and ξ_3 in rectangular coordinates thus:

$$\xi_1 = x \qquad \xi_2 = y \qquad \xi_3 = z \qquad \text{(A1.1-1)}$$

in cylindrical polar coordinates thus:

$$\xi_1 = r \qquad \xi_2 = \theta \qquad \xi_3 = z \qquad \text{(A1.1-2)}$$

and in spherical polar coordinates thus:

$$\xi_1 = r \qquad \xi_2 = \theta \qquad \xi_3 = \alpha \qquad \text{(A1.1-3)}$$

In component form, the vector field $\mathbf{v}$ is given by

$$\mathbf{v} = v_{\xi_1}\mathbf{i}_{\xi_1} + v_{\xi_2}\mathbf{i}_{\xi_2} + v_{\xi_3}\mathbf{i}_{\xi_3} \qquad \text{(A1.1-4)}$$

(note that $\mathbf{v}$ has three components, one associated with each coordinate direction), with obvious analogous expressions for $\mathbf{v}_1$ and $\mathbf{v}_2$, and the tensor field $\mathbf{T}$ is given by

267

$$\mathbf{T} = T_{\xi_1\xi_1}\mathbf{i}_{\xi_1}\mathbf{i}_{\xi_1} + T_{\xi_1\xi_2}\mathbf{i}_{\xi_1}\mathbf{i}_{\xi_2} + T_{\xi_1\xi_3}\mathbf{i}_{\xi_1}\mathbf{i}_{\xi_3} + T_{\xi_2\xi_1}\mathbf{i}_{\xi_2}\mathbf{i}_{\xi_1}$$
$$+ T_{\xi_2\xi_2}\mathbf{i}_{\xi_2}\mathbf{i}_{\xi_2} + T_{\xi_2\xi_3}\mathbf{i}_{\xi_2}\mathbf{i}_{\xi_3} + T_{\xi_3\xi_1}\mathbf{i}_{\xi_3}\mathbf{i}_{\xi_1} + T_{\xi_3\xi_2}\mathbf{i}_{\xi_3}\mathbf{i}_{\xi_2} + T_{\xi_3\xi_3}\mathbf{i}_{\xi_3}\mathbf{i}_{\xi_3} \quad \text{(A1.1-5)}$$

(note that $\mathbf{T}$ has nine components, one associated with each pair of coordinate directions), with obvious analogous expressions for $\mathbf{T}_1$ and $\mathbf{T}_2$.

A1.2 ALGEBRA

We give here certain algebraic properties of the scalar, vector, and tensor fields defined in Section A1.1. We define first two quantities related to a tensor field $\mathbf{T}$. They are the transpose $\mathbf{T}^T$ of $\mathbf{T}$, which is defined by

$$\mathbf{T}^T = T_{\xi_1\xi_1}\mathbf{i}_{\xi_1}\mathbf{i}_{\xi_1} + T_{\xi_2\xi_1}\mathbf{i}_{\xi_1}\mathbf{i}_{\xi_2} + T_{\xi_3\xi_1}\mathbf{i}_{\xi_1}\mathbf{i}_{\xi_3} + T_{\xi_1\xi_2}\mathbf{i}_{\xi_2}\mathbf{i}_{\xi_1}$$
$$+ T_{\xi_2\xi_2}\mathbf{i}_{\xi_2}\mathbf{i}_{\xi_2} + T_{\xi_3\xi_2}\mathbf{i}_{\xi_2}\mathbf{i}_{\xi_3} + T_{\xi_1\xi_3}\mathbf{i}_{\xi_3}\mathbf{i}_{\xi_1} + T_{\xi_2\xi_3}\mathbf{i}_{\xi_3}\mathbf{i}_{\xi_2} + T_{\xi_3\xi_3}\mathbf{i}_{\xi_3}\mathbf{i}_{\xi_3} \quad \text{(A1.2-1)}$$

and the determinant $\det(\mathbf{T})$ of $\mathbf{T}$, which is defined by

$$\det(\mathbf{T}) = \det(\mathbf{T}^T) = \begin{vmatrix} T_{\xi_1\xi_1} & T_{\xi_1\xi_2} & T_{\xi_1\xi_3} \\ T_{\xi_2\xi_1} & T_{\xi_2\xi_2} & T_{\xi_2\xi_3} \\ T_{\xi_3\xi_1} & T_{\xi_3\xi_2} & T_{\xi_3\xi_3} \end{vmatrix} \quad \text{(A1.2-2)}$$

that is,
$$\det(\mathbf{T}) = \det(\mathbf{T}^T) = T_{\xi_1\xi_1}(T_{\xi_2\xi_2}T_{\xi_3\xi_3} - T_{\xi_2\xi_3}T_{\xi_3\xi_2})$$
$$+ T_{\xi_1\xi_2}(T_{\xi_2\xi_3}T_{\xi_3\xi_1} - T_{\xi_2\xi_1}T_{\xi_3\xi_3}) + T_{\xi_1\xi_3}(T_{\xi_2\xi_1}T_{\xi_3\xi_2} - T_{\xi_2\xi_2}T_{\xi_3\xi_1}) \quad \text{(A1.2-3)}$$

We next define multiplication of vector and tensor fields by a scalar field

$$s\mathbf{v} = \mathbf{v}s = sv_{\xi_1}\mathbf{i}_{\xi_1} + sv_{\xi_2}\mathbf{i}_{\xi_2} + sv_{\xi_3}\mathbf{i}_{\xi_3} \quad \text{(A1.2-4)}$$

$$s\mathbf{T} = \mathbf{T}s = sT_{\xi_1\xi_1}\mathbf{i}_{\xi_1}\mathbf{i}_{\xi_1} + sT_{\xi_1\xi_2}\mathbf{i}_{\xi_1}\mathbf{i}_{\xi_2} + sT_{\xi_1\xi_3}\mathbf{i}_{\xi_1}\mathbf{i}_{\xi_3} + sT_{\xi_2\xi_1}\mathbf{i}_{\xi_2}\mathbf{i}_{\xi_1} + sT_{\xi_2\xi_2}\mathbf{i}_{\xi_2}\mathbf{i}_{\xi_2}$$
$$+ sT_{\xi_2\xi_3}\mathbf{i}_{\xi_2}\mathbf{i}_{\xi_3} + sT_{\xi_3\xi_1}\mathbf{i}_{\xi_3}\mathbf{i}_{\xi_1} + sT_{\xi_3\xi_2}\mathbf{i}_{\xi_3}\mathbf{i}_{\xi_2} + sT_{\xi_3\xi_3}\mathbf{i}_{\xi_3}\mathbf{i}_{\xi_3} \quad \text{(A1.2-5)}$$

and addition of two vector fields and of two tensor fields

$$\mathbf{v}_1 + \mathbf{v}_2 = \mathbf{v}_2 + \mathbf{v}_1 = (v_{1\xi_1} + v_{2\xi_1})\mathbf{i}_{\xi_1} + (v_{1\xi_2} + v_{2\xi_2})\mathbf{i}_{\xi_2} + (v_{1\xi_3} + v_{2\xi_3})\mathbf{i}_{\xi_3} \quad \text{(A1.2-6)}$$

$$\mathbf{T}_1 + \mathbf{T}_2 = \mathbf{T}_2 + \mathbf{T}_1 = (T_{1\xi_1\xi_1} + T_{2\xi_1\xi_1})\mathbf{i}_{\xi_1}\mathbf{i}_{\xi_1} + (T_{1\xi_1\xi_2} + T_{2\xi_1\xi_2})\mathbf{i}_{\xi_1}\mathbf{i}_{\xi_2}$$
$$+ (T_{1\xi_1\xi_3} + T_{2\xi_1\xi_3})\mathbf{i}_{\xi_1}\mathbf{i}_{\xi_3} + (T_{1\xi_2\xi_1} + T_{2\xi_2\xi_1})\mathbf{i}_{\xi_2}\mathbf{i}_{\xi_1}$$
$$+ (T_{1\xi_2\xi_2} + T_{2\xi_2\xi_2})\mathbf{i}_{\xi_2}\mathbf{i}_{\xi_2} + (T_{1\xi_2\xi_3} + T_{2\xi_2\xi_3})\mathbf{i}_{\xi_2}\mathbf{i}_{\xi_3}$$
$$+ (T_{1\xi_3\xi_1} + T_{2\xi_3\xi_1})\mathbf{i}_{\xi_3}\mathbf{i}_{\xi_1} + (T_{1\xi_3\xi_2} + T_{2\xi_3\xi_2})\mathbf{i}_{\xi_3}\mathbf{i}_{\xi_2}$$
$$+ (T_{1\xi_3\xi_3} + T_{2\xi_3\xi_3})\mathbf{i}_{\xi_3}\mathbf{i}_{\xi_3} \quad \text{(A1.2-7)}$$

We now go on to define some products of vector and tensor fields. The dot product is defined by

$$\mathbf{v}_1 \cdot \mathbf{v}_2 = \mathbf{v}_2 \cdot \mathbf{v}_1 = v_{1\xi_1}v_{2\xi_1} + v_{1\xi_2}v_{2\xi_2} + v_{1\xi_3}v_{2\xi_3} \quad \text{(A1.2-8)}$$

$$\mathbf{v} \cdot \mathbf{T} = \mathbf{T}^T \cdot \mathbf{v} = (v_{\xi_1}T_{\xi_1\xi_1} + v_{\xi_2}T_{\xi_2\xi_1} + v_{\xi_3}T_{\xi_3\xi_1})\mathbf{i}_{\xi_1}$$
$$+ (v_{\xi_1}T_{\xi_1\xi_2} + v_{\xi_2}T_{\xi_2\xi_2} + v_{\xi_3}T_{\xi_3\xi_2})\mathbf{i}_{\xi_2}$$
$$+ (v_{\xi_1}T_{\xi_1\xi_3} + v_{\xi_2}T_{\xi_2\xi_3} + v_{\xi_3}T_{\xi_3\xi_3})\mathbf{i}_{\xi_3} \quad \text{(A1.2-9)}$$

$$\mathbf{T}_1 \cdot \mathbf{T}_2 = \mathbf{T}_2^T \cdot \mathbf{T}_1^T = (T_{1\xi_1\xi_1}T_{2\xi_1\xi_1} + T_{1\xi_1\xi_2}T_{2\xi_2\xi_1} + T_{1\xi_1\xi_3}T_{2\xi_3\xi_1})\mathbf{i}_{\xi_1}\mathbf{i}_{\xi_1}$$
$$+ (T_{1\xi_2\xi_1}T_{2\xi_1\xi_1} + T_{1\xi_2\xi_2}T_{2\xi_2\xi_1} + T_{1\xi_2\xi_3}T_{2\xi_3\xi_1})\mathbf{i}_{\xi_2}\mathbf{i}_{\xi_1}$$
$$+ (T_{1\xi_3\xi_1}T_{2\xi_1\xi_1} + T_{1\xi_3\xi_2}T_{2\xi_2\xi_1} + T_{1\xi_3\xi_3}T_{2\xi_3\xi_1})\mathbf{i}_{\xi_3}\mathbf{i}_{\xi_1}$$
$$+ (T_{1\xi_1\xi_1}T_{2\xi_1\xi_2} + T_{1\xi_1\xi_2}T_{2\xi_2\xi_2} + T_{1\xi_1\xi_3}T_{2\xi_3\xi_2})\mathbf{i}_{\xi_1}\mathbf{i}_{\xi_2}$$
$$+ (T_{1\xi_2\xi_1}T_{2\xi_1\xi_2} + T_{1\xi_2\xi_2}T_{2\xi_2\xi_2} + T_{1\xi_2\xi_3}T_{2\xi_3\xi_2})\mathbf{i}_{\xi_2}\mathbf{i}_{\xi_2}$$
$$+ (T_{1\xi_3\xi_1}T_{2\xi_1\xi_2} + T_{1\xi_3\xi_2}T_{2\xi_2\xi_2} + T_{1\xi_3\xi_3}T_{2\xi_3\xi_2})\mathbf{i}_{\xi_3}\mathbf{i}_{\xi_2}$$
$$+ (T_{1\xi_1\xi_1}T_{2\xi_1\xi_3} + T_{1\xi_1\xi_2}T_{2\xi_2\xi_3} + T_{1\xi_1\xi_3}T_{2\xi_3\xi_3})\mathbf{i}_{\xi_1}\mathbf{i}_{\xi_3}$$
$$+ (T_{1\xi_2\xi_1}T_{2\xi_1\xi_3} + T_{1\xi_2\xi_2}T_{2\xi_2\xi_3} + T_{1\xi_2\xi_3}T_{2\xi_3\xi_3})\mathbf{i}_{\xi_2}\mathbf{i}_{\xi_3}$$
$$+ (T_{1\xi_3\xi_1}T_{2\xi_1\xi_3} + T_{1\xi_3\xi_2}T_{2\xi_2\xi_3} + T_{1\xi_3\xi_3}T_{2\xi_3\xi_3})\mathbf{i}_{\xi_3}\mathbf{i}_{\xi_3} \qquad \text{(A1.2-10)}$$

Note that the dot product of two vector fields is a scalar field (so that it is then also called the scalar product), that of a vector field and a tensor field is a vector field and that of two tensor fields is a tensor field.

The double dot product is defined by

$$\mathbf{T}_1 : \mathbf{T}_2 = \mathbf{T}_2 : \mathbf{T}_1 = T_{1\xi_1\xi_1}T_{2\xi_1\xi_1} + T_{1\xi_1\xi_2}T_{2\xi_2\xi_1} + T_{1\xi_1\xi_3}T_{2\xi_3\xi_1} + T_{1\xi_2\xi_1}T_{2\xi_1\xi_2}$$
$$+ T_{1\xi_2\xi_2}T_{2\xi_2\xi_2} + T_{1\xi_2\xi_3}T_{2\xi_3\xi_2} + T_{1\xi_3\xi_1}T_{2\xi_1\xi_3}$$
$$+ T_{1\xi_3\xi_2}T_{2\xi_2\xi_3} + T_{1\xi_3\xi_3}T_{2\xi_3\xi_3} \qquad \text{(A1.2-11)}$$

Note that the double dot product of two tensor fields is a scalar field. In fact the double dot product is also given by

$$\mathbf{T}_1 : \mathbf{T}_2 = \text{trace}(\mathbf{T}_1 \cdot \mathbf{T}_2) \qquad \text{(A1.2-12)}$$

where the trace of $\mathbf{T}$, $\text{trace}(\mathbf{T})$, is given by

$$\text{trace}(\mathbf{T}) = T_{\xi_1\xi_1} + T_{\xi_2\xi_2} + T_{\xi_3\xi_3} \qquad \text{(A1.2-13)}$$

The magnitude $|\mathbf{v}|$ of $\mathbf{v}$ is defined by

$$|\mathbf{v}| = \sqrt{\mathbf{v} \cdot \mathbf{v}} \qquad \text{(A1.2-14)}$$

and the magnitude $|\mathbf{T}|$ of $\mathbf{T}$ is defined by

$$|\mathbf{T}| = \sqrt{\tfrac{1}{2}\mathbf{T} : \mathbf{T}^T} \qquad \text{(A1.2-15)}$$

The inverse $\mathbf{T}^{-1}$ of $\mathbf{T}$ is defined (assuming that it exists) by

$$\mathbf{T}^{-1} \cdot \mathbf{T} = \mathbf{I} = \mathbf{T} \cdot \mathbf{T}^{-1} \qquad \text{(A1.2-16)}$$

where $\mathbf{I}$ denotes the unit tensor, which is given by

$$\mathbf{I} = \mathbf{i}_{\xi_1}\mathbf{i}_{\xi_1} + \mathbf{i}_{\xi_2}\mathbf{i}_{\xi_2} + \mathbf{i}_{\xi_3}\mathbf{i}_{\xi_3} \qquad \text{(A1.2-17)}$$

and is such that

$$\mathbf{I} \cdot \mathbf{v} = \mathbf{v} = \mathbf{v} \cdot \mathbf{I} \qquad \mathbf{I} \cdot \mathbf{T} = \mathbf{T} = \mathbf{T} \cdot \mathbf{I} \qquad \text{(A1.2-18)}$$

for all vector fields $\mathbf{v}$ and tensor fields $\mathbf{T}$. The cross product is defined by

$$\mathbf{v}_1 \wedge \mathbf{v}_2 = -\mathbf{v}_2 \wedge \mathbf{v}_1 = (v_{1\xi_2}v_{2\xi_3} - v_{1\xi_3}v_{2\xi_2})\mathbf{i}_{\xi_1} + (v_{1\xi_3}v_{2\xi_1} - v_{1\xi_1}v_{2\xi_3})\mathbf{i}_{\xi_2}$$
$$+ (v_{1\xi_1}v_{2\xi_2} - v_{1\xi_2}v_{2\xi_1})\mathbf{i}_{\xi_3} \qquad \text{(A1.2-19)}$$

Note that the cross product (so called because $\mathbf{v}_1 {\scriptstyle\wedge} \mathbf{v}_2$ is sometimes written $\mathbf{v}_1 \times \mathbf{v}_2$) of two vector fields is a vector field (so that it is then also called the vector product). More strictly, it is a pseudovector (or axial vector) field, that is, it exists only in three-dimensional space and is defined only when the handedness of the coordinate system is specified: in a right-handed system, which we will always use here (as we noted in Section 1.2), $\mathbf{v}_1 {\scriptstyle\wedge} \mathbf{v}_2$ is aligned in a direction normal to both $\mathbf{v}_1$ and $\mathbf{v}_2$ in such a way that $\mathbf{v}_1$, $\mathbf{v}_2$ and $\mathbf{v}_1 {\scriptstyle\wedge} \mathbf{v}_2$ form a right-handed set of vectors. Note that

$$\mathbf{i}_{\xi_1} {\scriptstyle\wedge} \mathbf{i}_{\xi_2} = \mathbf{i}_{\xi_3} \qquad \mathbf{i}_{\xi_2} {\scriptstyle\wedge} \mathbf{i}_{\xi_3} = \mathbf{i}_{\xi_1} \qquad \mathbf{i}_{\xi_3} {\scriptstyle\wedge} \mathbf{i}_{\xi_1} = \mathbf{i}_{\xi_2} \qquad \text{(A1.2-20)}$$

The dyadic product is defined by

$$\begin{aligned}
\mathbf{v}_1 \mathbf{v}_2 = (\mathbf{v}_2 \mathbf{v}_1)^T &= v_{1\xi_1} v_{2\xi_1} \mathbf{i}_{\xi_1} \mathbf{i}_{\xi_1} + v_{1\xi_2} v_{2\xi_1} \mathbf{i}_{\xi_2} \mathbf{i}_{\xi_1} \\
&+ v_{1\xi_3} v_{2\xi_1} \mathbf{i}_{\xi_3} \mathbf{i}_{\xi_1} + v_{1\xi_1} v_{2\xi_2} \mathbf{i}_{\xi_1} \mathbf{i}_{\xi_2} + v_{1\xi_2} v_{2\xi_2} \mathbf{i}_{\xi_2} \mathbf{i}_{\xi_2} + v_{1\xi_3} v_{2\xi_2} \mathbf{i}_{\xi_3} \mathbf{i}_{\xi_2} \\
&+ v_{1\xi_1} v_{2\xi_3} \mathbf{i}_{\xi_1} \mathbf{i}_{\xi_3} + v_{1\xi_2} v_{2\xi_3} \mathbf{i}_{\xi_2} \mathbf{i}_{\xi_3} + v_{1\xi_3} v_{2\xi_3} \mathbf{i}_{\xi_3} \mathbf{i}_{\xi_3}
\end{aligned} \qquad \text{(A1.2-21)}$$

Note that the dyadic product of two vector fields is a tensor field (so that it is then also called the tensor product). It is defined in general by

$$\mathbf{v} \cdot (\mathbf{v}_1 \mathbf{v}_2) = (\mathbf{v} \cdot \mathbf{v}_1)\mathbf{v}_2 \qquad \text{(A1.2-22)}$$

or equivalently by

$$(\mathbf{v}_1 \mathbf{v}_2)^T \cdot \mathbf{v} = \mathbf{v}_2(\mathbf{v}_1 \cdot \mathbf{v}) \qquad \text{(A1.2-23)}$$

for all vector fields $\mathbf{v}$. Special examples of dyadic products are $\mathbf{i}_{\xi_1} \mathbf{i}_{\xi_1}$, $\mathbf{i}_{\xi_1} \mathbf{i}_{\xi_2}$, $\mathbf{i}_{\xi_1} \mathbf{i}_{\xi_3}$, $\mathbf{i}_{\xi_2} \mathbf{i}_{\xi_1}$, $\mathbf{i}_{\xi_2} \mathbf{i}_{\xi_2}$, $\mathbf{i}_{\xi_2} \mathbf{i}_{\xi_3}$, $\mathbf{i}_{\xi_3} \mathbf{i}_{\xi_1}$, $\mathbf{i}_{\xi_3} \mathbf{i}_{\xi_2}$, and $\mathbf{i}_{\xi_3} \mathbf{i}_{\xi_3}$, in terms of which a tensor $\mathbf{T}$ is defined [see Eq. (A1.1-5)]. They are analogous to $\mathbf{i}_{\xi_1}$, $\mathbf{i}_{\xi_2}$, and $\mathbf{i}_{\xi_3}$, in terms of which a vector $\mathbf{v}$ is defined [see Eq. (A1.1-4)]. Note that the dyadic product of two vector fields can be regarded as an ordered pair of those vector fields since, except in degenerate cases, $\mathbf{v}_1 \mathbf{v}_2$ is *not* the same as $\mathbf{v}_2 \mathbf{v}_1$.

Scalar, vector, and tensor fields may be transformed from a rectangular coordinate system to a cylindrical or a spherical polar coordinate system by making use of the transformation properties of these fields. Indeed, such properties in fact define these fields. We assume that the cylindrical polar coordinates (r, θ, z) are related to the rectangular coordinates (x, y, z) by Eqs. (1.2-1) and (1.2-2) and that the spherical polar coordinates (r, θ, α) are related to the rectangular coordinates (x, y, z) by Eqs. (1.2-3) and (1.2-4) (see also Fig. 1.2). The single component of a scalar field s is, of course, unaltered by change of coordinate system. It follows from elementary trigonometric considerations that the components of a vector field $\mathbf{v}$ and a tensor field $\mathbf{T}$ in cylindrical polar coordinates (r, θ, z) are related to those in rectangular coordinates (x, y, z) as follows

$$\mathbf{v}|_{(r,\theta,z)} = \mathbf{v}|_{(x,y,z)} \cdot \mathbf{R}_c = \mathbf{R}_c^T \cdot \mathbf{v}|_{(x,y,z)} \qquad \text{(A1.2-24)}$$

$$\mathbf{v}|_{(x,y,z)} = \mathbf{v}|_{(r,\theta,z)} \cdot \mathbf{R}_c^T = \mathbf{R}_c \cdot \mathbf{v}|_{(r,\theta,z)} \qquad \text{(A1.2-25)}$$

$$\mathbf{T}|_{(r,\theta,z)} = \mathbf{R}_c^T \cdot \mathbf{T}|_{(x,y,z)} \cdot \mathbf{R}_c \qquad \text{(A1.2-26)}$$

$$\mathbf{T}|_{(x,y,z)} = \mathbf{R}_c \cdot \mathbf{T}|_{(r,\theta,z)} \cdot \mathbf{R}_c^T \qquad \text{(A1.2-27)}$$

where

$$\mathbf{v}|_{(x,y,z)} = v_x \mathbf{i}_x + v_y \mathbf{i}_y + v_z \mathbf{i}_z \qquad \text{(A1.2-28)}$$

$$\mathbf{v}|_{(r,\theta,z)} = v_r \mathbf{i}_r + v_\theta \mathbf{i}_\theta + v_z \mathbf{i}_z \qquad \text{(A1.2-29)}$$

$$\mathbf{T}|_{(x,y,z)} = T_{xx} \mathbf{i}_x \mathbf{i}_x + T_{xy} \mathbf{i}_x \mathbf{i}_y + T_{xz} \mathbf{i}_x \mathbf{i}_z$$
$$+ T_{yx} \mathbf{i}_y \mathbf{i}_x + T_{yy} \mathbf{i}_y \mathbf{i}_y + T_{yz} \mathbf{i}_y \mathbf{i}_z + T_{zx} \mathbf{i}_z \mathbf{i}_x + T_{zy} \mathbf{i}_z \mathbf{i}_y + T_{zz} \mathbf{i}_z \mathbf{i}_z \quad \text{(A1.2-30)}$$

$$\mathbf{T}|_{(r,\theta,z)} = T_{rr} \mathbf{i}_r \mathbf{i}_r + T_{r\theta} \mathbf{i}_r \mathbf{i}_\theta + T_{rz} \mathbf{i}_r \mathbf{i}_z$$
$$+ T_{\theta r} \mathbf{i}_\theta \mathbf{i}_r + T_{\theta\theta} \mathbf{i}_\theta \mathbf{i}_\theta + T_{\theta z} \mathbf{i}_\theta \mathbf{i}_z + T_{zr} \mathbf{i}_z \mathbf{i}_r + T_{z\theta} \mathbf{i}_z \mathbf{i}_\theta + T_{zz} \mathbf{i}_z \mathbf{i}_z \quad \text{(A1.2-31)}$$

The tensor $\mathbf{R}_c$ is given by

$$\mathbf{R}_c = \cos\theta \mathbf{i}_x \mathbf{i}_r + \sin\theta \mathbf{i}_y \mathbf{i}_r - \sin\theta \mathbf{i}_x \mathbf{i}_\theta + \cos\theta \mathbf{i}_y \mathbf{i}_\theta + \mathbf{i}_z \mathbf{i}_z \qquad \text{(A1.2-32)}$$

Similarly, the components of a vector field $\mathbf{v}$ and a tensor field $\mathbf{T}$ in spherical polar coordinates (r, θ, α) are related to those in rectangular coordinates (x, y, z) as follows

$$\mathbf{v}|_{(r,\theta,\alpha)} = \mathbf{v}|_{(x,y,z)} \cdot \mathbf{R}_s = \mathbf{R}_s^T \cdot \mathbf{v}|_{(x,y,z)} \qquad \text{(A1.2-33)}$$

$$\mathbf{v}|_{(x,y,z)} = \mathbf{v}|_{(r,\theta,\alpha)} \cdot \mathbf{R}_s^T = \mathbf{R}_s \cdot \mathbf{v}|_{(r,\theta,\alpha)} \qquad \text{(A1.2-34)}$$

$$\mathbf{T}|_{(r,\theta,\alpha)} = \mathbf{R}_s^T \cdot \mathbf{T}|_{(x,y,z)} \cdot \mathbf{R}_s \qquad \text{(A1.2-35)}$$

$$\mathbf{T}|_{(x,y,z)} = \mathbf{R}_s \cdot \mathbf{T}|_{(r,\theta,\alpha)} \cdot \mathbf{R}_s^T \qquad \text{(A1.2-36)}$$

where $\mathbf{v}|_{(x,y,z)}$ and $\mathbf{T}|_{(x,y,z)}$ are given by Eqs. (A1.2-28) and (A1.2-30), respectively, and

$$\mathbf{v}|_{(r,\theta,\alpha)} = v_r \mathbf{i}_r + v_\theta \mathbf{i}_\theta + v_\alpha \mathbf{i}_\alpha \qquad \text{(A1.2-37)}$$

$$\mathbf{T}|_{(r,\theta,\alpha)} = T_{rr} \mathbf{i}_r \mathbf{i}_r + T_{r\theta} \mathbf{i}_r \mathbf{i}_\theta + T_{r\alpha} \mathbf{i}_r \mathbf{i}_\alpha$$
$$+ T_{\theta r} \mathbf{i}_\theta \mathbf{i}_r + T_{\theta\theta} \mathbf{i}_\theta \mathbf{i}_\theta + T_{\theta\alpha} \mathbf{i}_\theta \mathbf{i}_\alpha + T_{\alpha r} \mathbf{i}_\alpha \mathbf{i}_r + T_{\alpha\theta} \mathbf{i}_\alpha \mathbf{i}_\theta + T_{\alpha\alpha} \mathbf{i}_\alpha \mathbf{i}_\alpha \quad \text{(A1.2-38)}$$

The tensor $\mathbf{R}_s$ is given by

$$\mathbf{R}_s = \sin\theta \cos\alpha \mathbf{i}_x \mathbf{i}_r + \sin\theta \sin\alpha \mathbf{i}_y \mathbf{i}_r + \cos\theta \mathbf{i}_z \mathbf{i}_r$$
$$+ \cos\theta \cos\alpha \mathbf{i}_x \mathbf{i}_\theta + \cos\theta \sin\alpha \mathbf{i}_y \mathbf{i}_\theta - \sin\theta \mathbf{i}_z \mathbf{i}_\theta$$
$$- \sin\alpha \mathbf{i}_x \mathbf{i}_\alpha + \cos\alpha \mathbf{i}_y \mathbf{i}_\alpha \qquad \text{(A1.2-39)}$$

A1.3 CALCULUS

We give here certain differential and integral properties of the scalar, vector, and tensor fields defined in Section A1.1. We assume that all the fields are sufficiently smooth to be differentiable twice with respect to any space variable. In particular, we consider a scalar field s that has a single component that is, of course, independent of the coordinate system; we also consider a vector field $\mathbf{v}$ and a tensor field $\mathbf{T}$ given in rectangular coordinates by Eqs. (A1.1-4) and (A1.1-

5), respectively, with ξ_1, ξ_2, and ξ_3 given by Eq. (A1.1-1); cylindrical polar coordinates by Eqs. (A1.1-4) and (A1.1-5), respectively, with ξ_1, ξ_2, and ξ_3 given by Eq. (A1.1-2); and spherical polar coordinates by Eqs. (A1.1-4) and (A1.1-5), respectively, with ξ_1, ξ_2, and ξ_3 given by Eq. (A1.1-3).

We now define some differential operators. Note that the forms of these operators differ in rectangular, cylindrical polar, and spherical polar coordinates. The reason for this is associated with the changes with spatial position in the directions of $\mathbf{i}_r$ and $\mathbf{i}_\theta$ in cylindrical polar coordinates and of $\mathbf{i}_r$, $\mathbf{i}_\theta$, and $\mathbf{i}_\alpha$ in spherical polar coordinates (see Section 1.2).

The gradient is defined in rectangular coordinates by

$$\nabla s = \frac{\partial s}{\partial x} \mathbf{i}_x + \frac{\partial s}{\partial y} \mathbf{i}_y + \frac{\partial s}{\partial z} \mathbf{i}_z \qquad (A1.3-1)$$

$$\begin{aligned}
\nabla \mathbf{v} = &\frac{\partial v_x}{\partial x} \mathbf{i}_x \mathbf{i}_x + \frac{\partial v_x}{\partial y} \mathbf{i}_y \mathbf{i}_x + \frac{\partial v_x}{\partial z} \mathbf{i}_z \mathbf{i}_x \\
&+ \frac{\partial v_y}{\partial x} \mathbf{i}_x \mathbf{i}_y + \frac{\partial v_y}{\partial y} \mathbf{i}_y \mathbf{i}_y + \frac{\partial v_y}{\partial z} \mathbf{i}_z \mathbf{i}_y + \frac{\partial v_z}{\partial x} \mathbf{i}_x \mathbf{i}_z + \frac{\partial v_z}{\partial y} \mathbf{i}_y \mathbf{i}_z + \frac{\partial v_z}{\partial z} \mathbf{i}_z \mathbf{i}_z \quad (A1.3-2)
\end{aligned}$$

in cylindrical polar coordinates by

$$\nabla s = \frac{\partial s}{\partial r} \mathbf{i}_r + \frac{1}{r} \frac{\partial s}{\partial \theta} \mathbf{i}_\theta + \frac{\partial s}{\partial z} \mathbf{i}_z \qquad (A1.3-3)$$

$$\begin{aligned}
\nabla \mathbf{v} = &\frac{\partial v_r}{\partial r} \mathbf{i}_r \mathbf{i}_r + \left(\frac{1}{r} \frac{\partial v_r}{\partial \theta} - \frac{1}{r} v_\theta \right) \mathbf{i}_\theta \mathbf{i}_r + \frac{\partial v_r}{\partial z} \mathbf{i}_z \mathbf{i}_r \\
&+ \frac{\partial v_\theta}{\partial r} \mathbf{i}_r \mathbf{i}_\theta + \left(\frac{1}{r} \frac{\partial v_\theta}{\partial \theta} + \frac{1}{r} v_r \right) \mathbf{i}_\theta \mathbf{i}_\theta + \frac{\partial v_\theta}{\partial z} \mathbf{i}_z \mathbf{i}_\theta \\
&+ \frac{\partial v_z}{\partial r} \mathbf{i}_r \mathbf{i}_z + \frac{1}{r} \frac{\partial v_z}{\partial \theta} \mathbf{i}_\theta \mathbf{i}_z + \frac{\partial v_z}{\partial z} \mathbf{i}_z \mathbf{i}_z \qquad (A1.3-4)
\end{aligned}$$

and in spherical polar coordinates by

$$\nabla s = \frac{\partial s}{\partial r} \mathbf{i}_r + \frac{1}{r} \frac{\partial s}{\partial \theta} \mathbf{i}_\theta + \frac{1}{r \sin \theta} \frac{\partial s}{\partial \alpha} \mathbf{i}_\alpha \qquad (A1.3-5)$$

$$\begin{aligned}
\nabla \mathbf{v} = &\frac{\partial v_r}{\partial r} \mathbf{i}_r \mathbf{i}_r + \left(\frac{1}{r} \frac{\partial v_r}{\partial \theta} - \frac{1}{r} v_\theta \right) \mathbf{i}_\theta \mathbf{i}_r + \left(\frac{1}{r \sin \theta} \frac{\partial v_r}{\partial \alpha} - \frac{1}{r} v_\alpha \right) \mathbf{i}_\alpha \mathbf{i}_r \\
&+ \frac{\partial v_\theta}{\partial r} \mathbf{i}_r \mathbf{i}_\theta + \left(\frac{1}{r} \frac{\partial v_\theta}{\partial \theta} + \frac{1}{r} v_r \right) \mathbf{i}_\theta \mathbf{i}_\theta + \left(\frac{1}{r \sin \theta} \frac{\partial v_\theta}{\partial \alpha} - \frac{\cot \theta}{r} v_\alpha \right) \mathbf{i}_\alpha \mathbf{i}_\theta \\
&+ \frac{\partial v_\alpha}{\partial r} \mathbf{i}_r \mathbf{i}_\alpha + \frac{1}{r} \frac{\partial v_\alpha}{\partial \theta} \mathbf{i}_\theta \mathbf{i}_\alpha + \left(\frac{1}{r \sin \theta} \frac{\partial v_\alpha}{\partial \alpha} + \frac{1}{r} v_r + \frac{\cot \theta}{r} v_\theta \right) \mathbf{i}_\alpha \mathbf{i}_\alpha \quad (A1.3-6)
\end{aligned}$$

Note that the gradient of a scalar field is a vector field and that of a vector field is a tensor field.

The divergence is defined in rectangular coordinates by

$$\nabla \cdot \mathbf{v} = \frac{\partial v_x}{\partial x} + \frac{\partial v_y}{\partial y} + \frac{\partial v_z}{\partial z} \tag{A1.3-7}$$

$$\nabla \cdot \mathbf{T} = \left(\frac{\partial T_{xx}}{\partial x} + \frac{\partial T_{yx}}{\partial y} + \frac{\partial T_{zx}}{\partial z} \right) \mathbf{i}_x + \left(\frac{\partial T_{xy}}{\partial x} + \frac{\partial T_{yy}}{\partial y} + \frac{\partial T_{zy}}{\partial z} \right) \mathbf{i}_y$$
$$+ \left(\frac{\partial T_{xz}}{\partial x} + \frac{\partial T_{yz}}{\partial y} + \frac{\partial T_{zz}}{\partial z} \right) \mathbf{i}_z \tag{A1.3-8}$$

in cylindrical polar coordinates by

$$\nabla \cdot \mathbf{v} = \frac{1}{r} \frac{\partial}{\partial r} (r v_r) + \frac{1}{r} \frac{\partial v_\theta}{\partial \theta} + \frac{\partial v_z}{\partial z} \tag{A1.3-9}$$

$$\nabla \cdot \mathbf{T} = \left(\frac{1}{r} \frac{\partial}{\partial r} (r T_{rr}) + \frac{1}{r} \frac{\partial T_{\theta r}}{\partial \theta} + \frac{\partial T_{zr}}{\partial z} - \frac{1}{r} T_{\theta \theta} \right) \mathbf{i}_r$$
$$+ \left(\frac{1}{r^2} \frac{\partial}{\partial r} (r^2 T_{r\theta}) + \frac{1}{r} \frac{\partial T_{\theta \theta}}{\partial \theta} + \frac{\partial T_{z\theta}}{\partial z} + \frac{1}{r} (T_{\theta r} - T_{r\theta}) \right) \mathbf{i}_\theta$$
$$+ \left(\frac{1}{r} \frac{\partial}{\partial r} (r T_{rz}) + \frac{1}{r} \frac{\partial T_{\theta z}}{\partial \theta} + \frac{\partial T_{zz}}{\partial z} \right) \mathbf{i}_z \tag{A1.3-10}$$

and in spherical polar coordinates by

$$\nabla \cdot \mathbf{v} = \frac{1}{r^2} \frac{\partial}{\partial r} (r^2 v_r) + \frac{1}{r \sin \theta} \frac{\partial}{\partial \theta} (\sin \theta v_\theta) + \frac{1}{r \sin \theta} \frac{\partial v_\alpha}{\partial \alpha} \tag{A1.3-11}$$

$$\nabla \cdot \mathbf{T} = \left(\frac{1}{r^2} \frac{\partial}{\partial r} (r^2 T_{rr}) + \frac{1}{r \sin \theta} \frac{\partial}{\partial \theta} (\sin \theta T_{\theta r}) + \frac{1}{r \sin \theta} \cdot \frac{\partial T_{\alpha r}}{\partial \alpha} \right.$$
$$\left. - \frac{1}{r} (T_{\theta \theta} + T_{\alpha \alpha}) \right) \mathbf{i}_r + \left(\frac{1}{r^3} \frac{\partial}{\partial r} \cdot (r^3 T_{r\theta}) \right.$$
$$+ \frac{1}{r \sin \theta} \frac{\partial}{\partial \theta} (\sin \theta T_{\theta \theta}) + \frac{1}{r \sin \theta} \cdot \frac{\partial T_{\alpha \theta}}{\partial \alpha}$$
$$\left. - \frac{1}{r} (\cot \theta T_{\alpha \alpha} + T_{r\theta} - T_{\theta r}) \right) \mathbf{i}_\theta$$
$$+ \left(\frac{1}{r^3} \frac{\partial}{\partial r} (r^3 T_{r\alpha}) + \frac{1}{r \sin \theta} \frac{\partial}{\partial \theta} \cdot (\sin \theta T_{\theta \alpha}) \right.$$
$$\left. + \frac{1}{r \sin \theta} \frac{\partial T_{\alpha \alpha}}{\partial \alpha} + \frac{1}{r} (\cot \theta T_{\alpha \theta} + T_{\alpha r} - T_{r\alpha}) \right) \mathbf{i}_\alpha \tag{A1.3-12}$$

Note that the divergence of a vector field is a scalar field and that of a tensor field is a vector field. The curl is defined in rectangular coordinates by

$$\nabla_\lambda \mathbf{v} = \left(\frac{\partial v_z}{\partial y} - \frac{\partial v_y}{\partial z} \right) \mathbf{i}_x + \left(\frac{\partial v_x}{\partial z} - \frac{\partial v_z}{\partial x} \right) \mathbf{i}_y + \left(\frac{\partial v_y}{\partial x} - \frac{\partial v_x}{\partial y} \right) \mathbf{i}_z \tag{A1.3-13}$$

in cylindrical coordinates by

$$\nabla_{\wedge}\mathbf{v} = \left(\frac{1}{r}\frac{\partial v_z}{\partial \theta} - \frac{\partial v_\theta}{\partial z}\right)\mathbf{i}_r + \left(\frac{\partial v_r}{\partial z} - \frac{\partial v_z}{\partial r}\right)\mathbf{i}_\theta + \left(\frac{1}{r}\frac{\partial}{\partial r}(rv_\theta) - \frac{1}{r}\frac{\partial v_r}{\partial \theta}\right)\mathbf{i}_z \quad \text{(A1.3-14)}$$

and in spherical polar coordinates by

$$\nabla_{\wedge}\mathbf{v} = \left(\frac{1}{r \sin \theta}\frac{\partial}{\partial \theta}(\sin \theta v_\alpha) - \frac{1}{r \sin \theta}\frac{\partial v_\theta}{\partial \alpha}\right)\mathbf{i}_r$$

$$+ \left(\frac{1}{r \sin \theta}\frac{\partial v_r}{\partial \alpha} - \frac{1}{r}\frac{\partial}{\partial r}(rv_\alpha)\right)\mathbf{i}_\theta + \left(\frac{1}{r}\frac{\partial}{\partial r}(rv_\theta) - \frac{1}{r}\frac{\partial v_r}{\partial \theta}\right)\mathbf{i}_\alpha \quad \text{(A1.3-15)}$$

Note that the curl of a vector field is a vector field or, more strictly, a pseudo-vector field, that is, it is, like the cross product, defined only when the handedness of the coordinate system is specified. The scalar Laplacian is defined in rectangular coordinates by

$$\nabla^2 s = \frac{\partial^2 s}{\partial x^2} + \frac{\partial^2 s}{\partial y^2} + \frac{\partial^2 s}{\partial z^2} \quad \text{(A1.3-16)}$$

in cylindrical polar coordinates by

$$\nabla^2 s = \frac{1}{r}\frac{\partial}{\partial r}\left(r\frac{\partial s}{\partial r}\right) + \frac{1}{r^2}\frac{\partial^2 s}{\partial \theta^2} + \frac{\partial^2 s}{\partial z^2} \quad \text{(A1.3-17)}$$

and in spherical polar coordinates by

$$\nabla^2 s = \frac{1}{r^2}\frac{\partial}{\partial r}\left(r^2\frac{\partial s}{\partial r}\right) + \frac{1}{r^2 \sin \theta}\frac{\partial}{\partial \theta}\left(\sin \theta \frac{\partial s}{\partial \theta}\right) + \frac{1}{r^2 \sin^2 \theta}\frac{\partial^2 s}{\partial \alpha^2} \quad \text{(A1.3-18)}$$

Note that the scalar Laplacian of a scalar field is a scalar field. The vector Laplacian is defined in rectangular coordinates by

$$\Delta\mathbf{v} = \left(\frac{\partial^2 v_x}{\partial x^2} + \frac{\partial^2 v_x}{\partial y^2} + \frac{\partial^2 v_x}{\partial z^2}\right)\mathbf{i}_x$$

$$+ \left(\frac{\partial^2 v_y}{\partial x^2} + \frac{\partial^2 v_y}{\partial y^2} + \frac{\partial^2 v_y}{\partial z^2}\right)\mathbf{i}_y + \left(\frac{\partial^2 v_z}{\partial x^2} + \frac{\partial^2 v_z}{\partial y^2} + \frac{\partial^2 v_z}{\partial z^2}\right)\mathbf{i}_z \quad \text{(A1.3-19)}$$

in cylindrical polar coordinates by

$$\Delta\mathbf{v} = \left\{\frac{\partial}{\partial r}\left[\frac{1}{r}\frac{\partial}{\partial r}(rv_r)\right] + \frac{1}{r^2}\frac{\partial^2 v_r}{\partial \theta^2} - \frac{2}{r^2}\frac{\partial v_\theta}{\partial \theta} + \frac{\partial^2 v_r}{\partial z^2}\right\}\mathbf{i}_r$$

$$+ \left\{\frac{\partial}{\partial r}\left[\frac{1}{r}\frac{\partial}{\partial r}(rv_\theta)\right] + \frac{1}{r^2}\frac{\partial^2 v_\theta}{\partial \theta^2} + \frac{2}{r^2}\frac{\partial v_r}{\partial \theta} + \frac{\partial^2 v_\theta}{\partial z^2}\right\}\mathbf{i}_\theta$$

$$+ \left\{\frac{1}{r}\frac{\partial}{\partial r}\left(r\frac{\partial v_z}{\partial r}\right) + \frac{1}{r^2}\frac{\partial^2 v_z}{\partial \theta^2} + \frac{\partial^2 v_z}{\partial z^2}\right\}\mathbf{i}_z \quad \text{(A1.3-20)}$$

and in spherical polar coordinates by

$$\mathbf{\Delta v} = \left\{ \frac{1}{r^2}\frac{\partial}{\partial r}\left(r^2\frac{\partial v_r}{\partial r}\right) + \frac{1}{r^2 \sin\theta}\frac{\partial}{\partial\theta}\left(\sin\theta\frac{\partial v_r}{\partial\theta}\right) + \frac{1}{r^2\sin^2\theta}\frac{\partial^2 v_r}{\partial\alpha^2} \right.$$
$$\left. - \frac{2}{r^2}v_r - \frac{2}{r^2}\frac{\partial v_\theta}{\partial\theta} - \frac{2\cot\theta}{r^2}v_\theta - \frac{2}{r^2\sin\theta}\frac{\partial v_\alpha}{\partial\alpha} \right\}\mathbf{i}_r$$
$$+ \left\{ \frac{1}{r^2}\frac{\partial}{\partial r}\left(r^2\frac{\partial v_\theta}{\partial r}\right) + \frac{1}{r^2\sin\theta}\frac{\partial}{\partial\theta}\left(\sin\theta\frac{\partial v_\theta}{\partial\theta}\right) + \frac{1}{r^2\sin^2\theta}\frac{\partial^2 v_\theta}{\partial\alpha^2} \right.$$
$$\left. + \frac{2}{r^2}\frac{\partial v_r}{\partial\theta} - \frac{1}{r^2\sin^2\theta}v_\theta - \frac{2\cos\theta}{r^2\sin^2\theta}\frac{\partial v_\alpha}{\partial\alpha} \right\}\mathbf{i}_\theta$$
$$+ \left\{ \frac{1}{r^2}\frac{\partial}{\partial r}\left(r^2\frac{\partial v_\alpha}{\partial r}\right) + \frac{1}{r^2\sin\theta}\frac{\partial}{\partial\theta}\left(\sin\theta\frac{\partial v_\alpha}{\partial\theta}\right) + \frac{1}{r^2\sin^2\theta}\frac{\partial^2 v_\alpha}{\partial\alpha^2} \right.$$
$$\left. - \frac{1}{r^2\sin^2\theta}v_\alpha + \frac{2}{r^2\sin\theta}\frac{\partial v_r}{\partial\alpha} + \frac{2\cos\theta}{r^2\sin^2\theta}\frac{\partial v_\theta}{\partial\alpha} \right\}\mathbf{i}_\alpha \qquad \text{(A1.3-21)}$$

Note that the vector Laplacian of a vector field is a vector field. Note also that, in rectangular coordinates, and only in rectangular coordinates, each component of the vector Laplacian of a vector field is just the same as the scalar Laplacian of that component of the vector field. In arbitrary coordinates, the vector Laplacian is defined by

$$\mathbf{\Delta v} = \mathbf{\nabla}(\mathbf{\nabla}\cdot\mathbf{v}) - \mathbf{\nabla}_{\!\wedge}(\mathbf{\nabla}_{\!\wedge}\mathbf{v}) \qquad \text{(A1.3-22)}$$

It is for this reason that we distinguish between the symbol ∇^2 for the scalar Laplacian and the symbol $\mathbf{\Delta}$ for the vector Laplacian.

Irrespective of the coordinate system, it follows from these definitions that

$$\mathbf{\nabla}_{\!\wedge}\mathbf{\nabla}s = 0 \qquad \text{(A1.3-23)}$$
$$\mathbf{\nabla}\cdot\mathbf{\nabla}_{\!\wedge}\mathbf{v} = 0 \qquad \text{(A1.3-24)}$$

and also that

$$\mathbf{\nabla}(\mathbf{v}_1\cdot\mathbf{v}_2) = (\mathbf{v}_2\cdot\mathbf{\nabla})\mathbf{v}_1 + (\mathbf{v}_1\cdot\mathbf{\nabla})\mathbf{v}_2 + \mathbf{v}_2{}_{\wedge}(\mathbf{\nabla}_{\!\wedge}\mathbf{v}_1) + \mathbf{v}_1{}_{\wedge}(\mathbf{\nabla}_{\!\wedge}\mathbf{v}_2) \quad \text{(A1.3-25)}$$

$$\mathbf{\nabla}\cdot(\mathbf{v}_1\mathbf{v}_2) = \mathbf{v}_1\cdot\mathbf{\nabla}\mathbf{v}_2 + \mathbf{v}_2(\mathbf{\nabla}\cdot\mathbf{v}_1) \qquad \text{(A1.3-26)}$$

$$s\mathbf{I}:\mathbf{\nabla}\mathbf{v} = s(\mathbf{\nabla}\cdot\mathbf{v}) \qquad \text{(A1.3-27)}$$

$$\mathbf{\nabla}\cdot(s\mathbf{v}) = \mathbf{\nabla}s\cdot\mathbf{v} + s(\mathbf{\nabla}\cdot\mathbf{v}) \qquad \text{(A1.3-28)}$$

$$\mathbf{\nabla}\cdot(s\mathbf{T}) = \mathbf{\nabla}s\cdot\mathbf{T} + s(\mathbf{\nabla}\cdot\mathbf{T}) \qquad \text{(A1.3-29)}$$

$$\mathbf{\nabla}\cdot(\mathbf{T}\cdot\mathbf{v}) = \mathbf{T}:\mathbf{\nabla}\mathbf{v} + \mathbf{v}\cdot(\mathbf{\nabla}\cdot\mathbf{T}) \qquad \text{(A1.3-30)}$$

for all scalar fields s, vector fields $\mathbf{v}$, $\mathbf{v}_1$, and $\mathbf{v}_2$, and tensor fields $\mathbf{T}$.

We conclude by noting that, for a volume V enclosed by an orientable (i.e., a two-sided) bounding surface S,

$$\int_V \mathbf{\nabla} \cdot \mathbf{v} \, dV = \int_S \mathbf{n} \cdot \mathbf{v} \, dS \qquad\qquad \text{(A1.3-31)}$$

$$\int_V \mathbf{\nabla} \cdot \mathbf{T} \, dV = \int_S \mathbf{n} \cdot \mathbf{T} \, dS \qquad\qquad \text{(A1.3-32)}$$

for all vector fields $\mathbf{v}$ and tensor fields $\mathbf{T}$, where $\mathbf{n}$ denotes the unit ($|\mathbf{n}| = 1$) outer normal to an element dS of the surface S: this is Gauss' divergence theorem. Note that the volume V may be multiply connected and S may be disconnected. Thus, for example, a water droplet in air comprises a singly (or simply) connected volume V (the water) with a connected surface S (where the water and air meet), water flowing through a pipe with its inlet connected to its outlet comprises a multiply connected (in fact, a doubly connected) volume V (the water) with a connected surface S (where the water meets the pipe wall), while a water droplet in air with an air bubble enclosed in the droplet comprises a singly connected volume V (the water) with a disconnected surface S (one part where the water and surrounding air meet, the other where the water and the enclosed air bubble meet).

BASIC EQUATIONS OF FLUID MECHANICS

A2.1 RATE OF STRAIN AND VORTICITY COMPONENTS

We give here the components of the rate of strain tensor $\mathbf{e}$ and vorticity vector $\boldsymbol{\omega}$ in rectangular, cylindrical polar, and spherical polar coordinates. In rectangular coordinates, with velocity components u_x, u_y, and u_z in the x, y, and z directions, respectively,

$$
\begin{aligned}
\mathbf{e} = \; & e_{xx}\mathbf{i}_x\mathbf{i}_x + e_{xy}\mathbf{i}_x\mathbf{i}_y + e_{xz}\mathbf{i}_x\mathbf{i}_z \\
& + e_{yx}\mathbf{i}_y\mathbf{i}_x + e_{yy}\mathbf{i}_y\mathbf{i}_y + e_{yz}\mathbf{i}_y\mathbf{i}_z \\
& + e_{zx}\mathbf{i}_z\mathbf{i}_x + e_{zy}\mathbf{i}_z\mathbf{i}_y + e_{zz}\mathbf{i}_z\mathbf{i}_z
\end{aligned}
\tag{A2.1-1}
$$

$$
e_{xx} = 2\frac{\partial u_x}{\partial x} \qquad e_{yy} = 2\frac{\partial u_y}{\partial y} \qquad e_{zz} = 2\frac{\partial u_z}{\partial z}
$$

$$
e_{xy} = e_{yx} = \frac{\partial u_x}{\partial y} + \frac{\partial u_y}{\partial x} \qquad e_{xz} = e_{zx} = \frac{\partial u_x}{\partial z} + \frac{\partial u_z}{\partial x} \tag{A2.1-2}
$$

$$
e_{yz} = e_{zy} = \frac{\partial u_z}{\partial y} + \frac{\partial u_y}{\partial z}
$$

$$
\boldsymbol{\omega} = \omega_x\mathbf{i}_x + \omega_y\mathbf{i}_y + \omega_z\mathbf{i}_z \tag{A2.1-3}
$$

$$
\omega_x = \frac{\partial u_z}{\partial y} - \frac{\partial u_y}{\partial z} \qquad \omega_y = \frac{\partial u_x}{\partial z} - \frac{\partial u_z}{\partial x} \qquad \omega_z = \frac{\partial u_y}{\partial x} - \frac{\partial u_x}{\partial y} \tag{A2.1-4}
$$

In cylindrical polar coordinates with velocity components u_r, u_θ, and u_z in the r, θ, and z directions, respectively,

$$\mathbf{e} = e_{rr}\mathbf{i}_r\mathbf{i}_r + e_{r\theta}\mathbf{i}_r\mathbf{i}_\theta + e_{rz}\mathbf{i}_r\mathbf{i}_z + e_{\theta r}\mathbf{i}_\theta\mathbf{i}_r + e_{\theta\theta}\mathbf{i}_\theta\mathbf{i}_\theta$$
$$+ e_{\theta z}\mathbf{i}_\theta\mathbf{i}_z + e_{zr}\mathbf{i}_z\mathbf{i}_r + e_{z\theta}\mathbf{i}_z\mathbf{i}_\theta + e_{zz}\mathbf{i}_z\mathbf{i}_z \tag{A2.1-5}$$

$$e_{rr} = 2\frac{\partial u_r}{\partial r} \qquad e_{\theta\theta} = \frac{2}{r}\frac{\partial u_\theta}{\partial\theta} + \frac{2}{r}u_r \qquad e_{zz} = 2\frac{\partial u_z}{\partial z}$$

$$e_{r\theta} = e_{\theta r} = r\frac{\partial}{\partial r}\left(\frac{1}{r}u_\theta\right) + \frac{1}{r}\frac{\partial u_r}{\partial\theta} \qquad e_{rz} = e_{zr} = \frac{\partial u_r}{\partial z} + \frac{\partial u_z}{\partial r} \tag{A2.1-6}$$

$$e_{z\theta} = e_{\theta z} = \frac{1}{r}\frac{\partial u_z}{\partial\theta} + \frac{\partial u_\theta}{\partial z}$$

$$\boldsymbol{\omega} = \omega_r\mathbf{i}_r + \omega_\theta\mathbf{i}_\theta + \omega_z\mathbf{i}_z \tag{A2.1-7}$$

$$\omega_r = \frac{1}{r}\frac{\partial u_z}{\partial\theta} - \frac{\partial u_\theta}{\partial z} \qquad \omega_\theta = \frac{\partial u_r}{\partial z} - \frac{\partial u_z}{\partial r} \qquad \omega_z = \frac{1}{r}\frac{\partial}{\partial r}(ru_\theta) - \frac{1}{r}\frac{\partial u_r}{\partial\theta} \tag{A2.1-8}$$

In spherical polar coordinates with velocity components u_r, u_θ, and u_α in the r, θ, and α directions, respectively,

$$\mathbf{e} = e_{rr}\mathbf{i}_r\mathbf{i}_r + e_{r\theta}\mathbf{i}_r\mathbf{i}_\theta + e_{r\alpha}\mathbf{i}_r\mathbf{i}_\alpha + e_{\theta r}\mathbf{i}_\theta\mathbf{i}_r + e_{\theta\theta}\mathbf{i}_\theta\mathbf{i}_\theta$$
$$+ e_{\theta\alpha}\mathbf{i}_\theta\mathbf{i}_\alpha + e_{\alpha r}\mathbf{i}_\alpha\mathbf{i}_r + e_{\alpha\theta}\mathbf{i}_\alpha\mathbf{i}_\theta + e_{\alpha\alpha}\mathbf{i}_\alpha\mathbf{i}_\alpha \tag{A2.1-9}$$

$$e_{rr} = 2\frac{\partial u_r}{\partial r} \qquad e_{\theta\theta} = \frac{2}{r}\frac{\partial u_\theta}{\partial\theta} + \frac{2}{r}u_r$$

$$e_{\alpha\alpha} = \frac{2}{r\sin\theta}\frac{\partial u_\alpha}{\partial\alpha} + \frac{2}{r}u_r + \frac{2\cot\theta}{r}u_\theta$$

$$e_{r\theta} = e_{\theta r} = \frac{1}{r}\frac{\partial u_r}{\partial\theta} - \frac{1}{r}u_\theta + \frac{\partial u_\theta}{\partial r} \tag{A2.1-10}$$

$$e_{r\alpha} = e_{\alpha r} = \frac{1}{r\sin\theta}\frac{\partial u_r}{\partial\alpha} - \frac{1}{r}u_\alpha + \frac{\partial u_\alpha}{\partial r}$$

$$e_{\theta\alpha} = e_{\alpha\theta} = \frac{1}{r}\frac{\partial u_\alpha}{\partial\theta} + \frac{1}{r\sin\theta}\frac{\partial u_\theta}{\partial\alpha} - \frac{\cot\theta}{r}u_\alpha$$

$$\boldsymbol{\omega} = \omega_r\mathbf{i}_r + \omega_\theta\mathbf{i}_\theta + \omega_\alpha\mathbf{i}_\alpha \tag{A2.1-11}$$

$$\omega_r = \frac{1}{r\sin\theta}\left(\frac{\partial}{\partial\theta}(\sin\theta u_\alpha) - \frac{\partial u_\theta}{\partial\alpha}\right) \qquad \omega_\theta = \frac{1}{r\sin\theta}\frac{\partial u_r}{\partial\alpha} - \frac{1}{r}\frac{\partial}{\partial r}(ru_\alpha)$$

$$\omega_\alpha = \frac{1}{r}\frac{\partial}{\partial r}(ru_\theta) - \frac{1}{r}\frac{\partial u_r}{\partial\theta} \tag{A2.1-12}$$

A2.2 MASS CONSERVATION EQUATION FOR FLUID OF CONSTANT DENSITY

We give here the forms of the mass conservation equation for a fluid of constant density ρ, that is, for an incompressible fluid

$$\nabla \cdot \mathbf{u} = 0 \qquad (A2.2\text{-}1)$$

in rectangular, cylindrical polar, and spherical polar coordinates.

In rectangular coordinates, with velocity components u_x, u_y, and u_z in the x, y, and z directions, respectively,

$$\frac{\partial u_x}{\partial x} + \frac{\partial u_y}{\partial y} + \frac{\partial u_z}{\partial z} = 0 \qquad (A2.2\text{-}2)$$

In cylindrical polar coordinates with velocity components u_r, u_θ, and u_z, in the r, θ, and z directions, respectively,

$$\frac{1}{r}\frac{\partial}{\partial r}(ru_r) + \frac{1}{r}\frac{\partial u_\theta}{\partial \theta} + \frac{\partial u_z}{\partial z} = 0 \qquad (A2.2\text{-}3)$$

In spherical polar coordinates with velocity components u_r, u_θ, and u_α in the r, θ, and α directions, respectively,

$$\frac{1}{r^2}\frac{\partial}{\partial r}(r^2 u_r) + \frac{1}{r\sin\theta}\frac{\partial}{\partial \theta}(\sin\theta u_\theta) + \frac{1}{r\sin\theta}\frac{\partial u_\alpha}{\partial \alpha} = 0 \qquad (A2.2\text{-}4)$$

A2.3 LINEAR MOMENTUM CONSERVATION EQUATION FOR FLUID OF CONSTANT DENSITY AND CONSTANT VISCOSITY

We give here the forms of the linear momentum conservation equation for a fluid of constant density ρ and constant viscosity μ, that is, the Navier-Stokes equations

$$\rho\frac{\partial \mathbf{u}}{\partial t} + \rho\mathbf{u}\cdot\nabla\mathbf{u} = \rho\mathbf{g} - \nabla p + \mu\Delta\mathbf{u} \qquad (A2.3\text{-}1)$$

in rectangular, cylindrical polar, and spherical polar coordinates.

In rectangular coordinates, with velocity components u_x, u_y, and u_z in the x, y, and z directions, respectively,

$$\rho\left(\frac{\partial u_x}{\partial t} + u_x\frac{\partial u_x}{\partial x} + u_y\frac{\partial u_x}{\partial y} + u_z\frac{\partial u_x}{\partial z}\right)$$

$$= \rho g_x - \frac{\partial p}{\partial x} + \mu\left(\frac{\partial^2 u_x}{\partial x^2} + \frac{\partial^2 u_x}{\partial y^2} + \frac{\partial^2 u_x}{\partial z^2}\right) \qquad (x \text{ component})$$

$$\rho\left(\frac{\partial u_y}{\partial t} + u_x\frac{\partial u_y}{\partial x} + u_y\frac{\partial u_y}{\partial y} + u_z\frac{\partial u_y}{\partial z}\right) \qquad (A2.3\text{-}2)$$

$$= \rho g_y - \frac{\partial p}{\partial y} + \mu\left(\frac{\partial^2 u_y}{\partial x^2} + \frac{\partial^2 u_y}{\partial y^2} + \frac{\partial^2 u_y}{\partial z^2}\right) \qquad (y \text{ component})$$

$$\rho\left(\frac{\partial u_z}{\partial t} + u_x\frac{\partial u_z}{\partial x} + u_y\frac{\partial u_z}{\partial y} + u_z\frac{\partial u_z}{\partial z}\right)$$

$$= \rho g_z - \frac{\partial p}{\partial z} + \mu\left(\frac{\partial^2 u_z}{\partial x^2} + \frac{\partial^2 u_z}{\partial y^2} + \frac{\partial^2 u_z}{\partial z^2}\right) \qquad (z\ component)$$

(A2.3-2)
(Cont.)

In cylindrical polar coordinates with velocity components u_r, u_θ, and u_z in the r, θ, and z directions, respectively,

$$\rho\left(\frac{\partial u_r}{\partial t} + u_r\frac{\partial u_r}{\partial r} + \frac{1}{r}u_\theta\frac{\partial u_r}{\partial \theta} - \frac{1}{r}u_\theta^2 + u_z\frac{\partial u_r}{\partial z}\right)$$

$$= \rho g_r - \frac{\partial p}{\partial r} + \mu\left\{\frac{\partial}{\partial r}\left[\frac{1}{r}\frac{\partial}{\partial r}(ru_r)\right] + \frac{1}{r^2}\frac{\partial^2 u_r}{\partial \theta^2} - \frac{2}{r^2}\frac{\partial u_\theta}{\partial \theta} + \frac{\partial^2 u_r}{\partial z^2}\right\} \qquad (r\ component)$$

$$\rho\left(\frac{\partial u_\theta}{\partial t} + u_r\frac{\partial u_\theta}{\partial r} + \frac{1}{r}u_\theta\frac{\partial u_\theta}{\partial \theta} + \frac{1}{r}u_r u_\theta + u_z\frac{\partial u_\theta}{\partial z}\right)$$

$$= \rho g_\theta - \frac{1}{r}\frac{\partial p}{\partial \theta} + \mu\left\{\frac{\partial}{\partial r}\left[\frac{1}{r}\frac{\partial}{\partial r}(ru_\theta)\right] + \frac{1}{r^2}\frac{\partial^2 u_\theta}{\partial \theta^2} + \frac{2}{r^2}\frac{\partial u_r}{\partial \theta} + \frac{\partial^2 u_\theta}{\partial z^2}\right\}$$

(A2.3-3)

(θ component)

$$\rho\left(\frac{\partial u_z}{\partial t} + u_r\frac{\partial u_z}{\partial r} + \frac{1}{r}u_\theta\frac{\partial u_z}{\partial \theta} + u_z\frac{\partial u_z}{\partial z}\right)$$

$$= \rho g_z - \frac{\partial p}{\partial z} + \mu\left\{\frac{1}{r}\frac{\partial}{\partial r}\left(r\frac{\partial u_z}{\partial r}\right) + \frac{1}{r^2}\frac{\partial^2 u_z}{\partial \theta^2} + \frac{\partial^2 u_z}{\partial z^2}\right\} \qquad (z\ component)$$

In spherical polar coordinates with velocity components u_r, u_θ, and u_α in the r, θ, and α directions, respectively,

$$\rho\left(\frac{\partial u_r}{\partial t} + u_r\frac{\partial u_r}{\partial r} + \frac{1}{r}u_\theta\frac{\partial u_r}{\partial \theta} + \frac{1}{r\sin\theta}u_\alpha\frac{\partial u_r}{\partial \alpha} - \frac{1}{r}(u_\theta^2 + u_\alpha^2)\right)$$

$$= \rho g_r - \frac{\partial p}{\partial r} + \mu\left[\frac{1}{r^2}\frac{\partial}{\partial r}\left(r^2\frac{\partial u_r}{\partial r}\right) + \frac{1}{r^2\sin\theta}\frac{\partial}{\partial \theta}\left(\sin\theta\frac{\partial u_r}{\partial \theta}\right)\right.$$

$$+ \frac{1}{r^2\sin^2\theta}\frac{\partial^2 u_r}{\partial \alpha^2} - \frac{2}{r^2}u_r - \frac{2}{r^2}\frac{\partial u_\theta}{\partial \theta} - \frac{2\cot\theta}{r^2}u_\theta$$

(A2.3-4)

$$\left. - \frac{2}{r^2\sin\theta}\frac{\partial u_\alpha}{\partial \alpha}\right] \qquad (r\ component)$$

$$\rho\left(\frac{\partial u_\theta}{\partial t} + u_r\frac{\partial u_\theta}{\partial r} + \frac{1}{r}u_\theta\frac{\partial u_\theta}{\partial \theta} + \frac{1}{r\sin\theta}u_\alpha\frac{\partial u_\theta}{\partial \alpha} + \frac{1}{r}u_r u_\theta - \frac{\cot\theta}{r}u_\alpha^2\right)$$

$$= \rho g_\theta - \frac{1}{r}\frac{\partial p}{\partial \theta} + \mu\left[\frac{1}{r^2}\frac{\partial}{\partial r}\left(r^2\frac{\partial u_\theta}{\partial r}\right) + \frac{1}{r^2\sin\theta}\frac{\partial}{\partial \theta}\left(\sin\theta\frac{\partial u_\theta}{\partial \theta}\right)\right.$$

$$+ \frac{1}{r^2 \sin^2 \theta} \frac{\partial^2 u_\theta}{\partial \alpha^2} + \frac{2}{r^2} \frac{\partial u_r}{\partial \theta} - \frac{1}{r^2 \sin^2 \theta} u_\theta$$

$$- \frac{2 \cos \theta}{r^2 \sin^2 \theta} \frac{\partial u_\alpha}{\partial \alpha} \bigg) \qquad (\theta \text{ component})$$

$$\rho \left(\frac{\partial u_\alpha}{\partial t} + u_r \frac{\partial u_\alpha}{\partial r} + \frac{1}{r} u_\theta \frac{\partial u_\alpha}{\partial \theta} + \frac{1}{r \sin \theta} u_\alpha \frac{\partial u_\alpha}{\partial \alpha} + \frac{1}{r} u_\alpha u_r + \frac{\cot \theta}{r} u_\theta u_\alpha \right) \qquad \text{(A2.3-4)}$$
$$\text{(Cont.)}$$

$$= \rho g_\alpha - \frac{1}{r \sin \theta} \frac{\partial p}{\partial \alpha} + \mu \left[\frac{1}{r^2} \frac{\partial}{\partial r} \left(r^2 \frac{\partial u_\alpha}{\partial r} \right) + \frac{1}{r^2 \sin \theta} \frac{\partial}{\partial \theta} \left(\sin \theta \frac{\partial u_\alpha}{\partial \theta} \right) \right.$$

$$+ \frac{1}{r^2 \sin^2 \theta} \frac{\partial^2 u_\alpha}{\partial \alpha^2} - \frac{1}{r^2 \sin^2 \theta} u_\alpha + \frac{2}{r^2 \sin \theta} \frac{\partial u_r}{\partial \alpha}$$

$$+ \frac{2 \cos \theta}{r^2 \sin^2 \theta} \frac{\partial u_\theta}{\partial \alpha} \bigg] \qquad (\alpha \text{ component})$$

A2.4 LINEAR MOMENTUM CONSERVATION EQUATION FOR FLUID OF CONSTANT DENSITY AND VARIABLE VISCOSITY

We give here the forms of the linear momentum conservation equation for a fluid of constant density ρ and variable viscosity μ

$$\rho \frac{\partial \mathbf{u}}{\partial t} + \rho \mathbf{u} \cdot \nabla \mathbf{u} = \rho \mathbf{g} - \nabla p + \nabla \cdot (\mu \mathbf{e}) \qquad (A2.4-1)$$

in rectangular, cylindrical polar, and spherical polar coordinates.

In rectangular coordinates, with velocity components u_x, u_y, and u_z in the x, y, and z directions, respectively,

$$\rho \left(\frac{\partial u_x}{\partial t} + u_x \frac{\partial u_x}{\partial x} + u_y \frac{\partial u_x}{\partial y} + u_z \frac{\partial u_x}{\partial z} \right)$$

$$= \rho g_x - \frac{\partial p}{\partial x} + \frac{\partial}{\partial x} (\mu e_{xx}) + \frac{\partial}{\partial y} (\mu e_{yx}) + \frac{\partial}{\partial z} (\mu e_{zx}) \qquad (x \text{ component})$$

$$\rho \left(\frac{\partial u_y}{\partial t} + u_x \frac{\partial u_y}{\partial x} + u_y \frac{\partial u_y}{\partial y} + u_z \frac{\partial u_y}{\partial z} \right) \qquad (A2.4-2)$$

$$= \rho g_y - \frac{\partial p}{\partial y} + \frac{\partial}{\partial x} (\mu e_{xy}) + \frac{\partial}{\partial y} (\mu e_{yy}) + \frac{\partial}{\partial z} (\mu e_{zy}) \qquad (y \text{ component})$$

$$\rho \left(\frac{\partial u_z}{\partial t} + u_x \frac{\partial u_z}{\partial x} + u_y \frac{\partial u_z}{\partial y} + u_z \frac{\partial u_z}{\partial z} \right)$$

$$= \rho g_z - \frac{\partial p}{\partial z} + \frac{\partial}{\partial x}(\mu e_{xz}) + \frac{\partial}{\partial y}(\mu e_{yz}) + \frac{\partial}{\partial z}(\mu e_{zz}) \qquad (z \text{ component})$$

where the components of **e** are given by Eqs. (A2.1-2).

In cylindrical polar coordinates with velocity components u_r, u_θ, and u_z in the r, θ, and z directions, respectively,

$$\rho\left(\frac{\partial u_r}{\partial t} + u_r \frac{\partial u_r}{\partial r} + \frac{1}{r} u_\theta \frac{\partial u_r}{\partial \theta} - \frac{1}{r} u_\theta^2 + u_z \frac{\partial u_r}{\partial z}\right)$$

$$= \rho g_r - \frac{\partial p}{\partial r} + \frac{1}{r}\frac{\partial}{\partial r}(r\mu e_{rr}) + \frac{1}{r}\frac{\partial}{\partial \theta}(\mu e_{\theta r}) + \frac{\partial}{\partial z}(\mu e_{zr}) - \frac{1}{r}\mu e_{\theta\theta} \qquad (r \text{ component})$$

$$\rho\left(\frac{\partial u_\theta}{\partial t} + u_r \frac{\partial u_\theta}{\partial r} + \frac{1}{r} u_\theta \frac{\partial u_\theta}{\partial \theta} + \frac{1}{r} u_r u_\theta + u_z \frac{\partial u_\theta}{\partial z}\right)$$

$$\qquad\qquad\qquad\qquad\qquad\qquad\qquad\qquad\qquad\qquad (A2.4\text{-}3)$$

$$= \rho g_\theta - \frac{1}{r}\frac{\partial p}{\partial \theta} + \frac{1}{r^2}\frac{\partial}{\partial r}(r^2 \mu e_{r\theta}) + \frac{1}{r}\frac{\partial}{\partial \theta}(\mu e_{\theta\theta}) + \frac{\partial}{\partial z}(\mu e_{z\theta}) \qquad (\theta \text{ component})$$

$$\rho\left(\frac{\partial u_z}{\partial t} + u_r \frac{\partial u_z}{\partial r} + \frac{1}{r} u_\theta \frac{\partial u_z}{\partial \theta} + u_z \frac{\partial u_z}{\partial z}\right)$$

$$= \rho g_z - \frac{\partial p}{\partial z} + \frac{1}{r}\frac{\partial}{\partial r}(r\mu e_{rz}) + \frac{1}{r}\frac{\partial}{\partial \theta}(\mu e_{\theta z}) + \frac{\partial}{\partial z}(\mu e_{zz}) \qquad (z \text{ component})$$

where the components of **e** are given by Eqs. (A2.1-6).

In spherical polar coordinates with velocity components u_r, u_θ, and u_α in the r, θ, and α directions, respectively,

$$\rho\left(\frac{\partial u_r}{\partial t} + u_r \frac{\partial u_r}{\partial r} + \frac{1}{r} u_\theta \frac{\partial u_r}{\partial \theta} + \frac{1}{r \sin \theta} u_\alpha \frac{\partial u_r}{\partial \alpha} - \frac{1}{r}(u_\theta^2 + u_\alpha^2)\right)$$

$$= \rho g_r - \frac{\partial p}{\partial r} + \frac{1}{r^2}\frac{\partial}{\partial r}(r^2 \mu e_{rr}) + \frac{1}{r \sin \theta}\frac{\partial}{\partial \theta}(\sin \theta \mu e_{\theta r})$$

$$+ \frac{1}{r \sin \theta}\frac{\partial}{\partial \alpha}(\mu e_{\alpha r}) - \frac{1}{r}(\mu e_{\theta\theta} + \mu e_{\alpha\alpha}) \qquad (r \text{ component})$$

$$\rho\left(\frac{\partial u_\theta}{\partial t} + u_r \frac{\partial u_\theta}{\partial r} + \frac{1}{r} u_\theta \frac{\partial u_\theta}{\partial \theta} + \frac{1}{r \sin \theta} u_\alpha \frac{\partial u_\theta}{\partial \alpha} + \frac{1}{r} u_r u_\theta - \frac{\cot \theta}{r} u_\alpha^2\right) \qquad (A2.4\text{-}4)$$

$$= \rho g_\theta - \frac{1}{r}\frac{\partial p}{\partial \theta} + \frac{1}{r^3}\frac{\partial}{\partial r}(r^3 \mu e_{r\theta}) + \frac{1}{r \sin \theta}\frac{\partial}{\partial \theta}(\sin \theta \mu e_{\theta\theta})$$

$$+ \frac{1}{r \sin \theta}\frac{\partial}{\partial \alpha}(\mu e_{\alpha\theta}) - \frac{\cot \theta}{r}\mu e_{\alpha\alpha} \qquad (\theta \text{ component})$$

$$\rho\left(\frac{\partial u_\alpha}{\partial t} + u_r \frac{\partial u_\alpha}{\partial r} + \frac{1}{r} u_\theta \frac{\partial u_\alpha}{\partial \theta} + \frac{1}{r \sin \theta} u_\alpha \frac{\partial u_\alpha}{\partial \alpha} + \frac{1}{r} u_\alpha u_r + \frac{\cot \theta}{r} u_\theta u_\alpha\right)$$

$$= \rho g_\alpha - \frac{1}{r \sin \theta}\frac{\partial p}{\partial \alpha} + \frac{1}{r^3}\frac{\partial}{\partial r}(r^3 \mu e_{r\alpha}) + \frac{1}{r \sin \theta}\frac{\partial}{\partial \theta}(\sin \theta \mu e_{\theta\alpha})$$

$$+ \frac{1}{r \sin \theta} \frac{\partial}{\partial \alpha} (\mu e_{\alpha\alpha}) + \frac{\cot \theta}{r} \mu e_{\alpha\theta} \qquad (\alpha \text{ component})$$

where the components of $\mathbf{e}$ are given by Eqs. (A2.1-10).

A2.5 ENERGY CONSERVATION EQUATION FOR FLUID OF CONSTANT DENSITY, CONSTANT VISCOSITY, AND CONSTANT THERMAL CONDUCTIVITY

We give here the forms of the energy conservation equation for a fluid of constant density ρ, constant viscosity μ, and constant thermal conductivity k

$$\rho c \frac{\partial T}{\partial t} + \rho c \mathbf{u} \cdot \nabla T = k \nabla^2 T + \frac{1}{2} \mu \mathbf{e}{:}\mathbf{e} \qquad (A2.5\text{-}1)$$

in rectangular, cylindrical polar, and spherical polar coordinates. Note, incidentally, that the forms of the energy conservation equation are unchanged if in fact the fluid is of variable viscosity.

In rectangular coordinates, with velocity components u_x, u_y, and u_z in the x, y, and z directions, respectively,

$$\rho c \left(\frac{\partial T}{\partial t} + u_x \frac{\partial T}{\partial x} + u_y \frac{\partial T}{\partial y} + u_z \frac{\partial T}{\partial z} \right)$$
$$= k \left(\frac{\partial^2 T}{\partial x^2} + \frac{\partial^2 T}{\partial y^2} + \frac{\partial^2 T}{\partial z^2} \right) + 2\mu \left[\left(\frac{\partial u_x}{\partial x} \right)^2 + \left(\frac{\partial u_y}{\partial y} \right)^2 + \left(\frac{\partial u_z}{\partial z} \right)^2 \right]$$
$$+ \mu \left[\left(\frac{\partial u_x}{\partial y} + \frac{\partial u_y}{\partial x} \right)^2 + \left(\frac{\partial u_x}{\partial z} + \frac{\partial u_z}{\partial x} \right)^2 + \left(\frac{\partial u_y}{\partial z} + \frac{\partial u_z}{\partial y} \right)^2 \right] \qquad (A2.5\text{-}2)$$

In cylindrical polar coordinates with velocity components u_r, u_θ, and u_z in the r, θ, and z directions, respectively,

$$\rho c \left(\frac{\partial T}{\partial t} + u_r \frac{\partial T}{\partial r} + \frac{1}{r} u_\theta \frac{\partial T}{\partial \theta} + u_z \frac{\partial T}{\partial z} \right)$$
$$= k \left[\frac{1}{r} \frac{\partial}{\partial r} \left(r \frac{\partial T}{\partial r} \right) + \frac{1}{r^2} \frac{\partial^2 T}{\partial \theta^2} + \frac{\partial^2 T}{\partial z^2} \right] + 2\mu \left[\left(\frac{\partial u_r}{\partial r} \right)^2 + \left(\frac{1}{r} \frac{\partial u_\theta}{\partial \theta} + \frac{1}{r} u_r \right)^2 + \left(\frac{\partial u_z}{\partial z} \right)^2 \right]$$
$$+ \mu \left[\left(\frac{\partial u_r}{\partial z} + \frac{\partial u_z}{\partial r} \right)^2 + \left(\frac{\partial u_\theta}{\partial z} + \frac{1}{r} \frac{\partial u_z}{\partial \theta} \right)^2 + \left(\frac{1}{r} \frac{\partial u_r}{\partial \theta} + r \frac{\partial}{\partial r} \left(\frac{1}{r} u_\theta \right) \right)^2 \right] \qquad (A2.5\text{-}3)$$

In spherical polar coordinates with velocity components u_r, u_θ, and u_α in the r, θ, and α directions, respectively,

$$\rho c \left(\frac{\partial T}{\partial t} + u_r \frac{\partial T}{\partial r} + \frac{1}{r} u_\theta \frac{\partial T}{\partial \theta} + \frac{1}{r \sin \theta} u_\alpha \frac{\partial T}{\partial \alpha} \right)$$

$$= k \left[\frac{1}{r^2} \frac{\partial}{\partial r} \left(r^2 \frac{\partial T}{\partial r} \right) + \frac{1}{r^2 \sin \theta} \frac{\partial}{\partial \theta} \left(\sin \theta \frac{\partial T}{\partial \theta} \right) \right.$$

$$+ \left. \frac{1}{r^2 \sin^2 \theta} \frac{\partial^2 T}{\partial \alpha^2} \right] + 2\mu \left[\left(\frac{\partial u_r}{\partial r} \right)^2 + \left(\frac{1}{r} \frac{\partial u_\theta}{\partial \theta} + \frac{1}{r} u_r \right)^2 \right.$$

$$+ \left. \left(\frac{1}{r \sin \theta} \frac{\partial u_\alpha}{\partial \alpha} + \frac{1}{r} u_r + \frac{\cot \theta}{r} u_\theta \right)^2 \right]$$

$$+ \mu \left\{ \left[r \frac{\partial}{\partial r} \left(\frac{1}{r} u_\theta \right) + \frac{1}{r} \frac{\partial u_r}{\partial \theta} \right]^2 + \left[\frac{1}{r \sin \theta} \frac{\partial u_r}{\partial \alpha} + r \frac{\partial}{\partial r} \left(\frac{1}{r} u_\alpha \right) \right]^2 \right.$$

$$+ \left. \left[\frac{\sin \theta}{r} \frac{\partial}{\partial \theta} \left(\frac{1}{\sin \theta} u_\alpha \right) + \frac{1}{r \sin \theta} \frac{\partial u_\theta}{\partial \alpha} \right]^2 \right\} \qquad \text{(A2.5-4)}$$

A2.6 AVERAGE MASS AND LINEAR MOMENTUM CONSERVATION EQUATIONS FOR FLUID OF CONSTANT DENSITY AND CONSTANT VISCOSITY IN A TURBULENT FLOW

We give here the forms of the ensemble average mass and linear momentum conservation equations for a fluid of constant density ρ and constant viscosity μ, that is, Reynolds' equations

$$\nabla \cdot \bar{\mathbf{u}} = 0 \qquad \text{(A2.6-1)}$$

$$\rho \frac{\partial \bar{\mathbf{u}}}{\partial t} + \nabla \cdot (\rho \bar{\mathbf{u}}\bar{\mathbf{u}}) = \rho \mathbf{g} - \nabla \bar{p} + \mu \Delta \bar{\mathbf{u}} + \nabla \cdot (-\rho \overline{\mathbf{u}'\mathbf{u}'}) \qquad \text{(A2.6-2)}$$

in rectangular, cylindrical polar, and spherical polar coordinates.

In rectangular coordinates, with velocity components u_x, u_y, and u_z in the x, y, and z directions, respectively,

$$\frac{\partial \bar{u}_x}{\partial x} + \frac{\partial \bar{u}_y}{\partial y} + \frac{\partial \bar{u}_z}{\partial z} = 0 \qquad \text{(A2.6-3)}$$

$$\rho \left(\frac{\partial \bar{u}_x}{\partial t} + \bar{u}_x \frac{\partial \bar{u}_x}{\partial x} + \bar{u}_y \frac{\partial \bar{u}_x}{\partial y} + \bar{u}_z \frac{\partial \bar{u}_x}{\partial z} \right) = \rho g_x - \frac{\partial \bar{p}}{\partial x} + \mu \left(\frac{\partial^2 \bar{u}_x}{\partial x^2} + \frac{\partial^2 \bar{u}_x}{\partial y^2} + \frac{\partial^2 \bar{u}_x}{\partial z^2} \right)$$

$$+ \frac{\partial}{\partial x} (-\rho \overline{u_x' u_x'}) + \frac{\partial}{\partial y} (-\rho \overline{u_y' u_x'}) + \frac{\partial}{\partial z} (-\rho \overline{u_z' u_x'}) \qquad (x \text{ component})$$

$$\rho \left(\frac{\partial \bar{u}_y}{\partial t} + \bar{u}_x \frac{\partial \bar{u}_y}{\partial x} + \bar{u}_y \frac{\partial \bar{u}_y}{\partial y} + \bar{u}_z \frac{\partial \bar{u}_y}{\partial z} \right) = \rho g_y - \frac{\partial \bar{p}}{\partial y} + \mu \left(\frac{\partial^2 \bar{u}_y}{\partial x^2} + \frac{\partial^2 \bar{u}_y}{\partial y^2} + \frac{\partial^2 \bar{u}_y}{\partial z^2} \right)$$

$$\qquad \qquad \qquad \qquad \qquad \qquad \qquad \qquad \qquad \qquad \qquad \qquad \text{(A2.6-4)}$$

$$+ \frac{\partial}{\partial x} (-\rho \overline{u_x' u_y'}) + \frac{\partial}{\partial y} (-\rho \overline{u_y' u_y'}) + \frac{\partial}{\partial z} (-\rho \overline{u_z' u_y'}) \qquad (y \text{ component})$$

$$\rho\left(\frac{\partial \overline{u}_z}{\partial t} + \overline{u}_x \frac{\partial \overline{u}_z}{\partial x} + \overline{u}_y \frac{\partial \overline{u}_z}{\partial y} + \overline{u}_z \frac{\partial \overline{u}_z}{\partial z}\right)$$

$$= \rho g_z - \frac{\partial \overline{p}}{\partial z} + \mu\left(\frac{\partial^2 \overline{u}_z}{\partial x^2} + \frac{\partial^2 \overline{u}_z}{\partial y^2} + \frac{\partial^2 \overline{u}_z}{\partial z^2}\right) + \frac{\partial}{\partial x}(-\rho\overline{u_x'u_z'})$$

(A2.6-4)
(*Cont.*)

$$+ \frac{\partial}{\partial y}(-\rho\overline{u_y'u_z'}) + \frac{\partial}{\partial z}(-\rho\overline{u_z'u_z'}) \qquad (z \text{ component})$$

In cylindrical polar coordinates with velocity components u_r, u_θ, and u_z in the r, θ, and z directions, respectively,

$$\frac{1}{r}\frac{\partial}{\partial r}(r\overline{u}_r) + \frac{1}{r}\frac{\partial \overline{u}_\theta}{\partial \theta} + \frac{\partial \overline{u}_z}{\partial z} = 0 \qquad\qquad (A2.6-5)$$

$$\rho\left(\frac{\partial \overline{u}_r}{\partial t} + \overline{u}_r \frac{\partial \overline{u}_r}{\partial r} + \frac{1}{r}\overline{u}_\theta \frac{\partial \overline{u}_r}{\partial \theta} - \frac{1}{r}\overline{u}_\theta^2 + \overline{u}_z \frac{\partial \overline{u}_r}{\partial z}\right)$$

$$= \rho g_r - \frac{\partial \overline{p}}{\partial r} + \mu\left(\frac{\partial}{\partial r}\left(\frac{1}{r}\frac{\partial}{\partial r}(r\overline{u}_r)\right)\right) + \frac{1}{r^2}\frac{\partial^2 \overline{u}_r}{\partial \theta^2} - \frac{2}{r^2}\frac{\partial \overline{u}_\theta}{\partial \theta} + \frac{\partial^2 \overline{u}_r}{\partial z^2}\right)$$

$$+ \frac{1}{r}\frac{\partial}{\partial r}(-r\rho\overline{u_r'u_r'}) + \frac{1}{r}\frac{\partial}{\partial \theta}(-\rho\overline{u_\theta'u_r'}) + \frac{\partial}{\partial z}(-\rho\overline{u_z'u_r'})$$

$$+ \frac{1}{r}\rho\overline{u_\theta'u_\theta'} \qquad (r \text{ component})$$

$$\rho\left(\frac{\partial \overline{u}_\theta}{\partial t} + \overline{u}_r \frac{\partial \overline{u}_\theta}{\partial r} + \frac{1}{r}\overline{u}_\theta \frac{\partial \overline{u}_\theta}{\partial \theta} + \frac{1}{r}\overline{u}_r\overline{u}_\theta + \overline{u}_z \frac{\partial \overline{u}_\theta}{\partial z}\right)$$

(A2.6-6)

$$= \rho g_\theta - \frac{1}{r}\frac{\partial \overline{p}}{\partial \theta} + \mu\left(\frac{\partial}{\partial r}\left(\frac{1}{r}\frac{\partial}{\partial r}(r\overline{u}_\theta)\right) + \frac{1}{r^2}\frac{\partial^2 \overline{u}_\theta}{\partial \theta^2} + \frac{2}{r^2}\frac{\partial \overline{u}_r}{\partial \theta} + \frac{\partial^2 \overline{u}_\theta}{\partial z^2}\right)$$

$$+ \frac{\partial}{\partial r}(-\rho\overline{u_r'u_\theta'}) + \frac{1}{r}\frac{\partial}{\partial \theta}(-\rho\overline{u_\theta'u_\theta'}) + \frac{\partial}{\partial z}(-\rho\overline{u_z'u_\theta'}) - \frac{2}{r}\rho\overline{u_r'u_\theta'} \qquad (\theta \text{ component})$$

$$\rho\left(\frac{\partial \overline{u}_z}{\partial t} + \overline{u}_r \frac{\partial \overline{u}_z}{\partial r} + \frac{1}{r}\overline{u}_\theta \frac{\partial \overline{u}_z}{\partial \theta} + \overline{u}_z \frac{\partial \overline{u}_z}{\partial z}\right)$$

$$= \rho g_z - \frac{\partial \overline{p}}{\partial z} + \mu\left(\frac{1}{r}\frac{\partial}{\partial r}\left(r\frac{\partial \overline{u}_z}{\partial r}\right) + \frac{1}{r^2}\frac{\partial^2 \overline{u}_z}{\partial \theta^2} + \frac{\partial^2 \overline{u}_z}{\partial z^2}\right)$$

$$+ \frac{1}{r}\frac{\partial}{\partial r}(-r\rho\overline{u_r'u_z'}) + \frac{1}{r}\frac{\partial}{\partial \theta}(-\rho\overline{u_\theta'u_z'}) + \frac{\partial}{\partial z}(-\rho\overline{u_z'u_z'}) \qquad (z \text{ component})$$

In spherical polar coordinates with velocity components u_r, u_θ, and u_α in the r, θ, and α directions, respectively,

$$\frac{1}{r^2}\frac{\partial}{\partial r}(r^2\overline{u}_r) + \frac{1}{r \sin \theta}\frac{\partial}{\partial \theta}(\sin \theta\, \overline{u}_\theta) + \frac{1}{r \sin \theta}\frac{\partial \overline{u}_\alpha}{\partial \alpha} = 0 \qquad (A2.6-7)$$

$$\rho\left[\frac{\partial \overline{u}_r}{\partial t} + \overline{u}_r \frac{\partial \overline{u}_r}{\partial r} + \frac{1}{r}\overline{u}_\theta \frac{\partial \overline{u}_r}{\partial \theta} + \frac{1}{r \sin \theta}\overline{u}_\alpha \frac{\partial \overline{u}_r}{\partial \alpha} - \frac{1}{r}(\overline{u}_\theta^2 + \overline{u}_\alpha^2)\right] \qquad (A2.6-8)$$

$$
= \rho g_r - \frac{\partial \bar{p}}{\partial r} + \mu \left(\frac{1}{r^2} \frac{\partial}{\partial r} \left(r^2 \frac{\partial \bar{u}_r}{\partial r} \right) + \frac{1}{r^2 \sin \theta} \frac{\partial}{\partial \theta} \left(\sin \theta \frac{\partial \bar{u}_r}{\partial \theta} \right) \right.
$$

$$
\left. + \frac{1}{r^2 \sin^2 \theta} \frac{\partial^2 \bar{u}_r}{\partial \alpha^2} - \frac{2}{r^2} \bar{u}_r - \frac{2}{r^2} \frac{\partial \bar{u}_\theta}{\partial \theta} - \frac{2 \cot \theta}{r^2} \bar{u}_\theta - \frac{2}{r^2 \sin \theta} \frac{\partial \bar{u}_\alpha}{\partial \alpha} \right)
$$

$$
+ \frac{1}{r^2} \frac{\partial}{\partial r} (-r^2 \rho \overline{u_r' u_r'}) + \frac{1}{r \sin \theta} \frac{\partial}{\partial \theta} (-\sin \theta \, \rho \overline{u_\theta' u_r'})
$$

$$
+ \frac{1}{r \sin \theta} \frac{\partial}{\partial \alpha} (-\rho \overline{u_\alpha' u_r'}) + \frac{1}{r} \rho \overline{u_\theta' u_\theta} + \frac{1}{r} \rho \overline{u_\alpha' u_\alpha} \qquad (r \text{ component})
$$

$$
\rho \left(\frac{\partial \bar{u}_\theta}{\partial t} + \bar{u}_r \frac{\partial \bar{u}_\theta}{\partial r} + \frac{1}{r} \bar{u}_\theta \frac{\partial \bar{u}_\theta}{\partial \theta} + \frac{1}{r \sin \theta} \bar{u}_\alpha \frac{\partial \bar{u}_\theta}{\partial \alpha} + \frac{1}{r} \bar{u}_r \bar{u}_\theta - \frac{\cot \theta}{r} \bar{u}_\alpha^2 \right)
$$

$$
= \rho g_\theta - \frac{1}{r} \frac{\partial \bar{p}}{\partial \theta} + \mu \left(\frac{1}{r^2} \frac{\partial}{\partial r} \left(r^2 \frac{\partial \bar{u}_\theta}{\partial r} \right) + \frac{1}{r^2 \sin \theta} \frac{\partial}{\partial \theta} \left(\sin \theta \frac{\partial \bar{u}_\theta}{\partial \theta} \right) \right.
$$

$$
\left. + \frac{1}{r^2 \sin^2 \theta} \frac{\partial^2 \bar{u}_\theta}{\partial \alpha^2} + \frac{2}{r^2} \frac{\partial \bar{u}_r}{\partial \theta} - \frac{1}{r^2 \sin^2 \theta} \bar{u}_\theta - \frac{2 \cos \theta}{r^2 \sin^2 \theta} \frac{\partial \bar{u}_\alpha}{\partial \alpha} \right)
$$

$$
+ \frac{1}{r^3} \frac{\partial}{\partial r} (-r^3 \rho \overline{u_r' u_\theta'}) + \frac{1}{r \sin \theta} \frac{\partial}{\partial \theta} (-\sin \theta \, \rho \overline{u_\theta' u_\theta'})
$$

$$
+ \frac{1}{r \sin \theta} \frac{\partial}{\partial \alpha} (-\rho \overline{u_\alpha' u_\theta'}) + \frac{\cot \theta}{r} \rho \overline{u_\alpha' u_\alpha'} \qquad (\theta \text{ component})
$$

(A2.6-8)
(*Cont.*)

$$
\rho \left(\frac{\partial \bar{u}_\alpha}{\partial t} + \bar{u}_r \frac{\partial \bar{u}_\alpha}{\partial r} + \frac{1}{r} \bar{u}_\theta \frac{\partial \bar{u}_\alpha}{\partial \theta} + \frac{1}{r \sin \theta} \bar{u}_\alpha \frac{\partial \bar{u}_\alpha}{\partial \alpha} + \frac{1}{r} \bar{u}_\alpha \bar{u}_r + \frac{\cot \theta}{r} \bar{u}_\theta \bar{u}_\alpha \right)
$$

$$
= \rho g_\alpha - \frac{1}{r \sin \theta} \frac{\partial \bar{p}}{\partial \alpha} + \mu \left(\frac{1}{r^2} \frac{\partial}{\partial r} \left(r^2 \frac{\partial \bar{u}_\alpha}{\partial r} \right) + \frac{1}{r^2 \sin \theta} \frac{\partial}{\partial \theta} \left(\sin \theta \frac{\partial \bar{u}_\alpha}{\partial \theta} \right) \right.
$$

$$
\left. + \frac{1}{r^2 \sin^2 \theta} \frac{\partial^2 \bar{u}_\alpha}{\partial \alpha^2} - \frac{1}{r^2 \sin^2 \theta} \bar{u}_\alpha + \frac{2}{r^2 \sin \theta} \frac{\partial \bar{u}_r}{\partial \alpha} + \frac{2 \cos \theta}{r^2 \sin^2 \theta} \frac{\partial \bar{u}_\theta}{\partial \alpha} \right)
$$

$$
+ \frac{1}{r^3} \frac{\partial}{\partial r} (-r^3 \rho \overline{u_r' u_\alpha'}) + \frac{1}{r \sin \theta} \frac{\partial}{\partial \theta} (-\sin \theta \, \rho \overline{u_\theta' u_\alpha'})
$$

$$
+ \frac{1}{r \sin \theta} \frac{\partial}{\partial \alpha} (-\rho \overline{u_\alpha' u_\alpha'}) - \frac{\cot \theta}{r} \rho \overline{u_\alpha' u_\theta'} \qquad (\alpha \text{ component})
$$

PROBLEMS

The superscript that follows most problem numbers indicates the number of the chapter on which the problem is set. Where there is no superscript, the problem is set on material in more than one chapter.

1^1 Because the earth rotates, a reference frame attached to it is not inertial. Under what circumstances do allowances need to be made for this in the interpretation of experiments conducted on earth?

2^1 Use elementary trigonometry to relate the components of a vector $\mathbf{v}$ in cylindrical and spherical polar coordinates to its components in rectangular coordinates.

3^1 In rectangular coordinates (x, y, z), the divergence of a vector field $\mathbf{v}$ is given by

$$\mathbf{\nabla} \cdot \mathbf{v} = \frac{\partial v_x}{\partial x} + \frac{\partial v_y}{\partial y} + \frac{\partial v_z}{\partial z}$$

Use the chain rule of partial differentiation to obtain expressions for $\mathbf{\nabla} \cdot \mathbf{v}$ in cylindrical and spherical polar coordinates.

4^2 The Lagrangian specifications of two velocity fields are

$$\hat{\mathbf{u}}(\mathbf{x}_0(X, t_0), t - t_0) = \epsilon x_0 e^{\epsilon(t-t_0)} \mathbf{i}_x - \epsilon y_0 e^{\epsilon(t-t_0)} \mathbf{i}_y$$

and

$$\hat{\mathbf{u}}(\mathbf{x}_0(X, t_0), t - t_0) = \gamma y_0 \mathbf{i}_x$$

Here X denotes a material point, t time, and $\mathbf{x}_0$ the position of X at time t_0

$$\mathbf{x}_0 = x_0 \mathbf{i}_x + y_0 \mathbf{i}_y + z_0 \mathbf{i}_z$$

in rectangular coordinates. The parameters ϵ and γ are constants. Obtain Eulerian specifications $\mathbf{u}(x, t)$ for these two velocity fields.

5^2 The velocity field $\mathbf{u}$ of an incompressible fluid has no θ component and its r and z components are independent of θ in a cylindrical polar coordinate system (r, θ, z). Use mass conservation arguments to show that $\mathbf{u}$ can be represented in terms of a Stokes stream function ψ_s as follows

$$\mathbf{u} = -\frac{1}{r}\frac{\partial \psi_s}{\partial z}\mathbf{i}_r + \frac{1}{r}\frac{\partial \psi_s}{\partial r}\mathbf{i}_z$$

6^2 If a velocity field $\mathbf{u}$ can be represented in terms of a Stokes stream function ψ_s, show that ψ_s is constant along a streamline.

7^2 The velocity field $\mathbf{u}$ of a fluid flowing in a long pipe of radius R is found experimentally to be given over most of the length of the pipe by

$$\mathbf{u} = U\left(1 - \frac{r^2}{R^2}\right)\mathbf{i}_z$$

provided the flow is laminar. Here r denotes distance from the centerline of the pipe and z denotes distance along the centerline; U is a constant.

(a) Is the flow solenoidal? If so, what is a suitable vector potential for it?

(b) Is the flow irrotational? If so, what is a suitable scalar potential for it?

8^3 Verify Cauchy's fundamental theorem for stress as follows. Consider an elementary tetrahedron, three faces of which are mutually perpendicular isosceles triangles. A rectangular coordinate system (x, y, z) is oriented so that these three faces are normal to the x, y, and z directions, respectively.

(a) Let $\mathbf{n}_i$ ($i = 1, 2,$ or 3) denote the unit outer normal to each of these three mutually perpendicular faces: $\mathbf{n}_1 = -\mathbf{i}_x$, $\mathbf{n}_2 = -\mathbf{i}_y$, $\mathbf{n}_3 = -\mathbf{i}_z$. Let $\mathbf{n}$ denote the unit outer normal to the remaining fourth face of the tetrahedron which is an equilateral triangle. Show that

$$\mathbf{n} = \frac{1}{\sqrt{3}}\mathbf{i}_x + \frac{1}{\sqrt{3}}\mathbf{i}_y + \frac{1}{\sqrt{3}}\mathbf{i}_z$$

(b) Let $\mathbf{t}_i$ ($i = 1, 2,$ or 3) denote the stress vector on each of the three mutually perpendicular faces and $\mathbf{t}$ denote the stress vector on the remaining face. Perform a force balance to show that

$$\mathbf{t} \rightarrow -\frac{1}{\sqrt{3}}(\mathbf{t}_1 + \mathbf{t}_2 + \mathbf{t}_3) \qquad \text{as } \Delta \rightarrow 0$$

where Δ denotes the length of the common sides of the three mutually perpendicular faces.

(c) Hence show that the stress tensor $\boldsymbol{\tau}$ defined by $\mathbf{t} = \mathbf{n} \cdot \boldsymbol{\tau}$ is given by

$$\boldsymbol{\tau} = -t_{1x}\mathbf{i}_x\mathbf{i}_x - t_{1y}\mathbf{i}_x\mathbf{i}_y - t_{1z}\mathbf{i}_x\mathbf{i}_z$$
$$-t_{2x}\mathbf{i}_y\mathbf{i}_x - t_{2y}\mathbf{i}_y\mathbf{i}_y - t_{2z}\mathbf{i}_y\mathbf{i}_z$$
$$-t_{3x}\mathbf{i}_z\mathbf{i}_x - t_{3y}\mathbf{i}_z\mathbf{i}_y - t_{3z}\mathbf{i}_z\mathbf{i}_z$$

where
$$\mathbf{t}_i = t_{ix}\mathbf{i}_x + t_{iy}\mathbf{i}_y + t_{iz}\mathbf{i}_z.$$

9^4 The mass conservation equation in rectangular coordinates (x, y, z) is

$$\frac{\partial \rho}{\partial t} + \frac{\partial}{\partial x}(\rho u_x) + \frac{\partial}{\partial y}(\rho u_y) + \frac{\partial}{\partial z}(\rho u_z) = 0$$

where ρ denotes density, $\mathbf{u}$ velocity, and t time. Derive this equation by performing a mass balance on an elementary rectangular parallelepiped, the faces of which are of length Δx in the x direction, Δy in the y direction, and Δz in the z direction. Does it matter whether or not the parallelepiped is fixed in space?

10^4 A discontinuity surface such as a shock wave moves through a material at local velocity $\mathbf{u}_s$. Derive the mass conservation equation relating the local density ρ_A and velocity $\mathbf{u}_A$ on one side of the surface to the local density ρ_B and velocity $\mathbf{u}_B$ on the other.

11^5 A Newtonian fluid is sometimes defined as one for which the shear stress τ is related to the shear rate γ by $\tau = \mu\gamma$, where μ denotes the viscosity. In what way or ways does this definition fail to meet the criteria for an acceptable constitutive equation for stress?

12^6 A sphere of radius R rotates with angular velocity Ω in an unbounded viscous fluid which is stationary at infinity. What boundary conditions can be imposed on the velocity field **u**? What symmetry conditions, if any, can be imposed?

13[7] A fluid of constant density ρ and constant viscosity μ flows steadily in a pipe of radius R and length L at a volumetric flow rate Q. Assume that the flow is symmetrical about the centerline of the pipe.

(a) Obtain mass and linear momentum conservation equations for the flow in an appropriate coordinate system. Incorporate gravitational effects in the pressure. Obtain boundary conditions for the flow, including those at the pipe entrance and exit.

(b) Make the flow equations and boundary conditions dimensionless by scaling radial distances with R, axial distances with L, and other variables in any suitable way. Show that two dimensionless groups arise naturally, one of which might be the Reynolds number $\text{Re} = \rho Q / \mu R$ and the other of which might be the length-to-radius ratio L/R.

(c) What simplifications can be made to the flow equations and boundary conditions if

- $\text{Re} \ll 1$?
- $\text{Re} \gg 1$?
- $L/R \gg 1$?
- $L/R \gg 1$ and $\text{Re} \ll 1$?
- $L/R \gg 1$ and $\text{Re} \gg 1$?

(d) Suppose that, instead of a constant viscosity, the fluid has a variable viscosity μ with a power-law dependence on shear rate γ:

$$\mu = \mu_0 \gamma^{(1/m)-1}$$

where μ_0 denotes the consistency and m the shear-rate exponent. How might a Reynolds number now be defined?

14[8] A sphere of radius R rotates with angular velocity Ω in an unbounded, incompressible, Newtonian fluid of density ρ and viscosity μ. The fluid at infinity is stationary. Assume that the flow is slow, that is, $\text{Re} = \rho |\Omega| R^2 / \mu \ll 1$. Suppose that $\Omega = \Omega \mathbf{i}_z$, so that the sphere rotates about the z axis. Define spherical polar coordinates (r, θ, α) with origin at the center of the sphere and r direction aligned with the z direction when $\theta = 0$. Assume that the flow is symmetric and swirling about the z axis, so that the velocity field $\mathbf{u} = u_\alpha(r, \theta) \mathbf{i}_\alpha$.

(a) Show that the assumed form for **u** automatically satisfies the mass conservation equation.

(b) Use the linear momentum conservation equation to obtain an equation for u_α.

(c) Obtain appropriate boundary conditions on u_α. Motivated by the form of these conditions, solve for u_α.

(d) Because the sphere is rotating, a moment **M** is exerted on the fluid by the sphere. Show that

$$\mathbf{M} = 8\pi\mu R^3 \Omega.$$

15[8] A solid spherical particle of density $\rho^\dagger$ and radius R is projected vertically upwards at an initial speed U_0 into an incompressible Newtonian gas of density ρ and viscosity μ. Show that the maximum altitude s attained by the particle above the point of projection is given approximately by

$$s = \frac{U_0^2}{\alpha g} \left[1 - \frac{1}{\alpha} \ln(1 + \alpha) \right]$$

assuming that the flow is slow. Here $\alpha = 9\mu U_0 / 2\rho^\dagger R^2 g$; g denotes the magnitude of the gravitational acceleration. Why is this expression for s approximate? Under what condition is the flow slow?

16[8] A cylindrical filament is extended along its axis. The velocity field **u** in the filament is given in component form by $\mathbf{u} = u_r \mathbf{i}_r + u_\theta \mathbf{i}_\theta + u_z \mathbf{i}_z$, where r denotes distance from the centerline of the filament, θ denotes angular position about the centerline, and z denotes distance along the centerline. Provided that inertial and gravitational effects are negligible, symmetry means that the radius of the filament is independent of axial position. Show that (a) u_r is a function of r and t only; (b) u_θ vanishes; and (c) u_z is a function of z and t only (t denotes time).

17⁹ Show that the inertial added mass m' per unit length of an infinitely long cylinder of radius R moving normal to its axis in a fluid of density ρ is given by $m' = \pi R^2 \rho$.

18⁹ Generalize d'Alembert's paradox by using Bernoulli's equation to show that there is no drag force on any body with fore and aft symmetry moving through an inviscid fluid in the direction normal to the symmetry plane of the body.

19⁹ A gas bubble rises through a liquid in a vertical pipe of radius R. The bubble is so large that it almost completely fills the cross section of the pipe. Its shape is virtually constant, any increase in volume merely leading to an increase in the length of the bubble. It is observed experimentally that, under steady conditions, the speed of rise U of the bubble is given by $U \simeq 0.48\sqrt{gR}$, where g denotes the magnitude of the gravitational acceleration. Assume that viscous and interfacial tension effects are negligible.

Show that the thickness d of the annular liquid film that drains down the wall of the pipe past the rising bubble is given by

$$d \simeq 0.17\sqrt{R^3/x}$$

where x denotes the vertical distance from the top of the bubble.

20⁹ A gas stream flows vertically upward at a constant volumetric flow rate Q through a thin-walled nozzle protruding into a large reservoir of otherwise stationary liquid. The gas stream breaks up there: assume that it forms spherical bubbles of uniform diameter D.

(a) Show that mass conservation yields $D = (6Qt/\pi)^{1/3}$, where t denotes the time that has elapsed since the formation of the given bubble started.

(b) Then show that a balance of inertial added mass and buoyancy forces yields $s = \frac{1}{2}gt^2$, where s denotes the distance that the center of the forming bubble has risen above the upper end of the nozzle and g denotes the magnitude of the gravitational acceleration.

(c) Hence show that, if bubble detachment occurs when $s = \frac{1}{2}D$,

$$D = \frac{(6Q/\pi)^{2/5}}{g^{1/5}}$$

(d) In practice, this expression can give a good estimate of the bubble volume V given by $V = \pi D^3/6$, but not of D because the bubble tends not to be spherical. Why not?

21⁹ A cylindrical tank of radius R and height H is full of liquid. At time $t = 0$, the tank is instantaneously completely ruptured. The liquid then surges over the ground, which is flat and horizontal, toward a circular retaining wall a distance L from the initial position of the tank wall. Assume that viscous and interfacial tension effects are negligible. Show that the liquid first reaches the wall at time $t = L/2\sqrt{gH}$, where g denotes the magnitude of the gravitational acceleration. What happens near the retaining wall when $t > L/2\sqrt{gH}$?

22¹⁰ The viscosity μ of a fluid usually increases with an increase in pressure p. When pressure variations become large, variations in viscosity can become significant. Adequate agreement with experimental results can often be achieved by assuming that $\mu = \mu_1 e^{\alpha(p-p_1)}$, where μ_1 denotes the viscosity at pressure p_1, and α is a positive constant. A fluid with such a pressure-dependent viscosity flows in a pipe of length L and radius R ($<<L$). Let p_0 denote the entrance pressure and p_L the exit pressure in the pipe and take p_1 to be p_L. Show that the pressure drop $(p_0 - p_L)$ is given approximately by

$$p_0 - p_L = -\frac{1}{\alpha}\ln\left(1 - \frac{8\mu_1 QL\alpha}{\pi R^4}\right)$$

where Q denotes the volumetric flow rate. Why is this expression for $(p_0 - p_L)$ only approximate? What happens when $8\mu_1 QL\alpha/\pi R^4 \to 1$?

23¹⁰ An incompressible Newtonian fluid of density ρ and viscosity μ flows as a result of a pressure difference $(p_0 - p_L)$ along a rectangular channel of length L, width W, and height H. The channel is thin (i.e., $H << W$ and $H << L$). Show that the volumetric flow rate Q is given approximately by $Q = WH^3(p_0 - p_L)/12\mu L$.

24[10] A fluid of constant density ρ and constant viscosity μ flows in a very long pipe of length L. The radius R of the pipe is not constant but varies linearly with axial position: $R = R_0$ at the entrance of the pipe; $R = R_L$ at its exit. Show that the volumetric flow rate Q through the pipe is given approximately by

$$Q = \frac{3\pi R_0^3 R_L^3 (p_0 - p_L)}{8\mu L (R_0^2 + R_0 R_L + R_L^2)}$$

where p_0 denotes the pressure at the pipe entrance and p_L the pressure at its exit.

25[10] The surface of a large, flat, vertical wall is to be protected from a gaseous atmosphere by being covered with a thin film of paint of uniform thickness H. When it is wet, the paint is a non-Newtonian fluid and can be regarded as a Bingham plastic, for which the shear stress τ is related to the shear rate γ by

$$\tau = \mu\gamma + \tau_0 \qquad \text{if } \tau > \tau_0$$
$$\gamma = 0 \qquad \text{if } \tau \leq \tau_0$$

where μ denotes viscosity and τ_0 yield stress. Because of gravity, there is a volumetric flow rate Q' of paint down the wall per unit width of wall. Let g denote the magnitude of the gravitational acceleration and ρ the density of the paint. Show that

$$Q = \begin{cases} 0 & \text{when } \alpha \geq 1 \\ \dfrac{\rho g H^3}{3\mu}(1 - \alpha)^2 \left(1 + \dfrac{1}{2}\alpha\right) & \text{when } \alpha < 1 \end{cases}$$

where $\alpha = \tau_0/\rho g H$. Note that there is no tendency for paint to flow down the wall when $H \leq \tau_0/\rho g$.

26[10] An incompressible Newtonian fluid of density ρ and viscosity μ flows between two parallel flat plates of length L, separation $2H$ ($\ll L$), and infinite width. The flow is induced by a pressure difference $(p_0 - p_L)$ over the length of the plates. The two plates are porous: by injecting more of the same fluid through one of the plates and removing it through the other, a uniform crossflow is generated. Use rectangular coordinates (x, y, z) with origin on the centerplane of the entrance to the plates, x direction aligned along the length of the plates, and y direction aligned normal to the plates. Then uniform crossflow means that the transverse velocity component u_y is constant, U say. Show that the axial velocity component u_x is given by

$$u_x = \frac{(p_0 - p_L)H^2}{\mu L} \frac{1}{\text{Re}} \left[\frac{y}{H} - 1 + \frac{e^{\text{Re}} - e^{\text{Re}y/H}}{\sinh \text{Re}} \right]$$

where the crossflow Reynolds number $\text{Re} = \rho U H/\mu$, assuming that there is no slip at the porous plates. Sketch the axial velocity profile for $\text{Re} \ll 1$, $\text{Re} \sim 1$, and $\text{Re} \gg 1$.

27[11] An incompressible Newtonian fluid of density ρ and viscosity μ flows toward and past a semi-infinite flat plate. The flow far from the plate is uniform and parallel to the plate with speed U_∞. A laminar boundary layer forms on the plate: its thickness at a distance x from the leading edge of the plate is δ. Show that the $\text{Re}_\delta = \rho U_\infty \delta/\mu$ is related to $\text{Re}_x = \rho U_\infty x/\mu$ as follows

$$\text{Re}_\delta \sim \sqrt{\text{Re}_x}.$$

28[11] A laminar boundary layer forms on a flat plate. The local shear stress on the wall can be expressed in dimensionless form in terms of a local drag coefficient C_f. The mean drag coefficient $\overline{C_f}$ on the plate is related to C_f thus

$$\overline{C_f} = \frac{1}{x} \int_0^x C_f \, dx$$

where x denotes distance from the leading edge of the plate. Why is this an appropriate mean?

29[11] A fluid of constant density ρ and constant viscosity μ flows steadily at volumetric flow rate Q from a very large reservoir into a very long channel of height H and width W ($>>H$). The entrance length L_e is the distance along the channel from its entrance beyond which the flow is essentially fully developed. Argue that $L_e \sim H\mathrm{Re}$, where the Reynolds number $\mathrm{Re} = \rho Q/W\mu$.

30[11] An incompressible, Newtonian fluid of density ρ and viscosity μ issues as a jet from a circular orifice into a very large reservoir of the same fluid which is otherwise stationary. Let (r, θ, z) be cylindrical polar coordinates with origin at the center of the orifice and z direction aligned with the flow out of the orifice. Assume that the jet spreads radially only slowly, so that the total flux J of axial momentum is approximately constant.

(a) Obtain appropriate equations and boundary conditions for the velocity components u_z and u_r.

(b) Express u_z and u_r in terms of a Stokes stream function ψ_s. Assume that

$$\psi_s = \frac{\mu z}{\rho} f(\eta)$$

where the similarity variable $\eta = r/z$. Show that

$$\frac{d}{d\eta}\left(\frac{d^2 f}{d\eta^2} - \frac{1}{\eta}\frac{df}{d\eta}\right) + \frac{d}{d\eta}\left(\frac{f}{\eta}\frac{df}{d\eta}\right) = 0$$

and

$$f = 0 \qquad \frac{df}{d\eta} = 0 \qquad \text{at } \eta = 0$$

$$\frac{df}{d\eta} \to 0 \qquad \text{as } \eta \to \infty$$

(c) The solution of this equation subject to these conditions is

$$f = \frac{3J\rho\eta^2/(16\pi\mu^2)}{1 + 3J\rho\eta^2/(64\pi\mu^2)}.$$

Verify that J is independent of z. Show that the mass flow rate $M = 8\pi\mu z$. What is the physical significance of the independence of M from J and ρ?

31[11] An incompressible Newtonian fluid of density ρ and viscosity μ flows at volumetric flow rate Q in a pipe of radius R and length L. The pipe is so long that

$$\frac{2R}{L}\mathrm{Re} << 1$$

where $\mathrm{Re} = 2\rho Q/\pi\mu R$. As a result, the velocity field $\mathbf{u}$ is fully developed

$$\mathbf{u} = \frac{2Q}{\pi R^2}\left(1 - \frac{r^2}{R^2}\right)\mathbf{i}_z$$

where r denotes distance from the centerline of the pipe and z denotes distance along the pipe from its entrance. The fluid enters the pipe at a uniform temperature T_e which is different from the uniform pipe wall temperature T_w. The Graetz number Gz is defined by

$$\mathrm{Gz} = \frac{2R}{L}\mathrm{Pe}$$

where the Péclet number $\mathrm{Pe} = 2\rho c Q/\pi k R$; c denotes the specific heat of the fluid and k its thermal conductivity. Because $\mathrm{Gz} >> 1$, the temperature field T is not fully developed.

(a) Show that the energy conservation equation reduces to

$$\rho c u_z \frac{\partial T}{\partial z} = \frac{k}{r}\frac{\partial}{\partial r}\left(r\frac{\partial T}{\partial r}\right)$$

assuming that axial conduction and heat generation by viscous dissipation are negligible, and that the boundary conditions are

$$T = T_e \quad \text{at } z = 0 \qquad \frac{\partial T}{\partial r} = 0 \quad \text{at } r = 0 \qquad T = T_w \quad \text{at } r = R$$

(b) Because Gz is large, argue that a thin thermal boundary layer forms by the pipe wall. In this layer, argue that

$$\frac{1}{r}\frac{\partial}{\partial r}\left(r\frac{\partial T}{\partial r}\right) \simeq \frac{\partial^2 T}{\partial r^2}$$

and

$$u_z \simeq (R - r)\frac{4Q}{\pi R^3}$$

(c) Argue that a similarity solution exists.
(d) A suitable similarity variable η is given by

$$\eta = \frac{1 - r/R}{(9z/\text{Gz}L)^{1/3}}$$

Thus the energy conservation equation becomes

$$\frac{d^2 T}{d\eta^2} + 3\eta^2 \frac{dT}{d\eta} = 0$$

with boundary conditions $T = T_w$ at $\eta = 0$ and $T \to T_e$ as $\eta \to \infty$.
Show that

$$\frac{T - T_w}{T_e - T_w} = \frac{\displaystyle\int_0^\eta e^{-\eta^{\#3}}d\eta^\#}{\displaystyle\int_0^\infty e^{-\eta^{\#3}}d\eta^\#}$$

This is Lévêque's solution.

32[12] Obtain the ensemble average form of the energy conservation equation for an incompressible Newtonian, Fourier fluid in a turbulent flow. What is the physical significance of each of the terms in the averaged equation?

33[12] The linear sublayer in a turbulent boundary layer on a flat plate is the region where the axial velocity component $\bar{u}_x$ varies linearly with the distance y from the plate. Show that the transverse velocity component $\bar{u}_y$ varies as the square of y and that $\overline{u'_y u'_x}$ varies as the cube of y. Here $\bar{f}$ and f' denote the average and fluctuating components of a flow variable f, respectively.

34[12] An incompressible Newtonian fluid of kinematic viscosity ν flows through a long, smooth pipe of radius R at volumetric flow rate Q. The Reynolds number Re $= 2Q/\pi\nu R$. The flow is laminar if Re $<$ Re$_c$, transitional if Re$_c$ $<$ Re $<$ Re$_C$ and turbulent if Re$_C$ $<$ Re. What is the expected effect on Re$_c$ and Re$_C$ of varying the relative turbulent intensity I_r of the flow into the pipe?

35 An incompressible Newtonian fluid of density ρ and viscosity μ flows at volumetric flow rate Q' per unit height between two large, vertical, flat plates inclined at an angle $2\theta_w$ to each other.
(a) Assume that the velocity field **u** is purely radial

$$\mathbf{u} = u_r(r, \theta)\mathbf{i}_r$$

where r denotes radial distance from the virtual apex of the plates and θ denotes the angle from the centerplane of the plates. Show that

$$u_r = \frac{1}{r}\, f(\theta)$$

and that the function f satisfies the equation

$$\frac{d^3 f}{d\theta^3} + \frac{2\rho f}{\mu}\frac{df}{d\theta} + 4\frac{df}{d\theta} = 0$$

and the conditions

$$f = 0 \quad \text{at } \theta = \pm\theta_w \quad \int_{-\theta_w}^{+\theta_w} f d\theta = Q'$$

(b) Show that

$$\bullet u_r = \frac{3Q'}{4r\theta_w}\left(1 - \frac{\theta^2}{\theta_w^2}\right) \quad \text{when } \theta_w \ll 1 \text{ and } \frac{\rho Q'}{\mu\theta_w} \not\gg 1$$

$$\bullet u_r = \frac{Q'}{r}\frac{\cos(2\theta) - \cos(2\theta_w)}{\sin(2\theta_w) - 2\theta_w \cos(2\theta_w)} \quad \text{when } \frac{\rho Q'}{\mu\theta_w} \ll 1$$

$$\bullet u_r = \frac{Q'}{2r\theta_w} \quad \text{when } \frac{\rho Q'}{\mu\theta_w} \gg 1$$

What is the physical significance of these three asymptotic cases?

36 A liquid of constant density ρ and constant viscosity μ is in laminar flow in a long pipe of radius R_1. It issues from the pipe into an otherwise stationary gas and forms a jet of circular cross section. The radius of the jet tends to a constant value R_2 downstream. Assume that the flow in the pipe is fully developed right up to the exit of the pipe and that interfacial tension and gravitational effects are negligible. Show that $R_2/R_1 = \sqrt{3}/2$.

37 Why does air in the atmosphere tend to flow along isobars, that is, along lines of constant pressure, rather than normal to them? Is this tendency stronger near the poles or near the equator?

38 It can be argued that if the parameters which describe a physical process can be combined to give a single dimensionless group, that group is a constant of unit order of magnitude. Why does this argument fail if the parameters combine to give more than one dimensionless group?

39 Is there a practical case when the length scale of the smallest fluctuations in a turbulent flow is so small that the flow cannot be regarded as a continuum?

BIBLIOGRAPHY

There are many general texts on fluid mechanics. Two excellent ones are

Batchelor, G. K. *An Introduction to Fluid Dynamics,* Cambridge University Press, Cambridge, 1967.
Denn, M. M. *Process Fluid Mechanics,* Prentice-Hall, Englewood Cliffs, N.J. 1980.

A good understanding of fluid mechanics is greatly enhanced by observing real flows. An outstanding collection of photographs, many of which illustrate flows discussed here, is to be found in

Van Dyke, M. *An Album of Fluid Motion,* Parabolic Press, Stanford, 1982.

Many more are to be found in

Tritton, D. J. *Physical Fluid Dynamics,* Van Nostrand Reinhold, New York, 1977.

Good introductions to the mathematics of scalar, vector, and tensor fields and to continuum mechanics are

Aris, R. *Vectors, Tensors, and the Basic Equations of Fluid Mechanics,* Prentice-Hall, Englewood Cliffs, N.J., 1962.
Astarita, G., and Marrucci, G. *Principles of Non-Newtonian Fluid Mechanics,* McGraw-Hill, London, 1974.
Bird, R. B., Armstrong, R. C., and Hassager, O. *Dynamics of Polymeric Liquids,* Vol. 1, Wiley, New York, 1977.

More advanced texts are

Leigh, D. C. *Nonlinear Continuum Mechanics,* McGraw-Hill, New York, 1968.
Truesdell, C. A. *A First Course in Rational Continuum Mechanics,* Vol. 1, Academic Press, New York, 1977.

Tables and properties of the mathematical functions used here are given in

Abramowitz, M., and Stegun, I. A. (Eds.). *Handbook of Mathematical Functions,* Dover, New York, 1965.

It is as yet impossible to prove whether or not the flow equations for an incompressible Newtonian fluid have a solution and, if so, whether the solution is unique for arbitrary initial and boundary conditions. At present, all that can be proved is that a unique weak solution exists subject to certain conditions that are less restrictive if the flow is two dimensional, steady, or slow. A classical solution of a differential equation is, roughly speaking, one that is smooth enough for all the terms in the differential equation to be defined. A weak solution, in contrast, is one that satisfies an integral equation derived from the differential equation but is not usually smooth enough for all the terms in the differential equation to be defined. A classical solution is thus a weak one but a weak solution is not usually a classical one. The standard text on this subject is

Ladyzhenskaya, O. A. *Mathematical Theory of Viscous Incompressible Flow*, Gordon & Breach, New York, 1969.

A more recent one is

Témam, R. *Navier-Stokes Equations*, North-Holland, Amsterdam, 1977.

A rather more readable account is in

Shinbrot, M. *Lectures on Fluid Mechanics*, Gordon & Breach, New York, 1973.

The existence and uniqueness of a solution of the flow equations do not guarantee that the flow corresponding to the solution can actually occur in practice: the flow must also be stable. The stability of flows is discussed in

Drazin, P. G. and Reid, W. H. *Hydrodynamic Stability*, Cambridge University Press, Cambridge, 1981.

A comprehensive treatment of slow flow is to be found in

Happel, J. and Brenner, H. *Low Reynolds Number Hydrodynamics*, Noordhoff, Leyden, 1973.

An excellent text, with particularly good coverage of the paradoxes which arise in slow and inviscid flow is

Van Dyke, M. *Perturbation Methods in Fluid Mechanics*, Academic Press, New York, 1964.

A standard text on inviscid flow despite its age is

Lamb, H. *Hydrodynamics*, Cambridge University Press, Cambridge, 1932.

A very good more recent one is

Lighthill, M. J. *An Informal Introduction to Theoretical Fluid Mechanics*, Clarendon Press, Oxford, 1986.

A masterly discussion of wave motions in fluids is to be found in

Lighthill, M. J. *Waves in Fluids*, Cambridge University Press, Cambridge, 1978.

A very readable account of (nearly) nonaccelerating flow and boundary layer flow is given in

White, F. M. *Viscous Fluid Flow*, McGraw-Hill, New York, 1974.

The standard text on boundary layer flow is

Schlichting, H. *Boundary-Layer Theory*, McGraw-Hill, New York, 1968.

An extremely comprehensive one is

Rosenhead, L. (Ed.). *Laminar Boundary Layers*, Clarendon Press, Oxford, 1963.

An excellent introductory text on turbulence is

Tennekes, H., and Lumley, J. L. *A First Course in Turbulence,* M.I.T. Press, Cambridge, 1972.

A more advanced one is

Bradshaw, P. (Ed.). *Turbulence,* Springer-Verlag, Berlin, 1976.

Turbulence models are discussed in

Launder, B. E. and Spalding, D. B. *Mathematical Models of Turbulence,* Academic Press, London, 1972.

A thorough discussion of the role of similarity solutions, with particular applications to turbulent flows, is given in

Barenblatt, G. I. *Similarity, Self-similarity, and Intermediate Asymptotics,* Consultants Bureau, New York, 1979.

Many texts are devoted to the use of computers for numerical solution of the flow equations. A particularly good introduction is given in

Roache, P. J. *Computational Fluid Dynamics,* Hermosa, Albuquerque, 1972.

A very readable text on fluid mechanics, heat and mass transfer is

Bird, R. B., Stewart, W. E., and Lightfoot, E. N. *Transport Phenomena,* Wiley, New York, 1960.

The role of thermodynamics in fluid mechanics is discussed in

Astarita, G. *An Introduction to Non-Linear Continuum Thermodynamics,* Società Editrice di Chimica, Milan, 1975.

An introductory text on inelastic non-Newtonian fluid mechanics is

Skelland, A. H. P. *Non-Newtonian Flow and Heat Transfer,* Wiley, New York, 1967.

A good text on the mechanics of isolated bubbles, drops, and particles is

Clift, R., Grace, J. R., and Weber, M. E. *Bubbles, Drops and Particles,* Academic Press, New York, 1978.

A comprehensive text on multiphase flow is

G. Hetsroni (Ed.): *Handbook of Multiphase Systems,* Hemisphere, Washington, D.C., 1982.

NOMENCLATURE

We give here a list of the symbols which are most commonly used, together with their most common meanings. Note that the symbol for a vector or tensor quantity is always written in boldface.

a	acceleration
A, B, C	empirically determined constants in a turbulent flow
Br	Brinkman number
c	constant or specific heat or phase velocity
$c_1, c_2, \ldots$	constants
C_D, C_f	drag coefficients
e	rate of strain
Eo	Eotvos number
f	friction factor
F	force
Fr	Froude number
g	gravitational acceleration
H	height or separation or thickness or enthalpy per unit mass
i	unit vector
I	unit tensor
k	thermal conductivity or average kinetic energy of fluctuating flow per unit mass or length scale of roughness
K	permeability
L	length
m, n	shear-rate exponents
n	unit normal

Na	Nahme number
p	pressure
$\tilde{p}, \bar{\tilde{p}}$	modified pressures
Pe	Péclet number
$\mathbf{q}$	heat flux
Q	volumetric flow rate
r, θ, z	cylindrical polar coordinates
r, θ, α	spherical polar coordinates
R	radius
Re	Reynolds number
$R_{\mathrm{I}}, R_{\mathrm{II}}$	principal radii of curvature
s	scalar field
S	area or entropy per unit mass
t	time
$\mathbf{t}$	stress (vector) or unit tangent
T	temperature
$\mathbf{T}$	tensor field
$\mathbf{u}$	velocity
u_τ	friction velocity
U	speed or internal energy per unit mass
$\mathbf{v}$	vector field
V	volume
$\mathbf{w}$	vorticity (tensor)
W	width
We	Weber number
x, y, z	rectangular coordinates
$\mathbf{x}$	position
X	material point
γ	shear rate
δ	boundary layer thickness or shear layer thickness
ϵ	extension rate or porosity or dissipation
ζ	temperature exponent
η	similarity variable
κ	wave number
λ	microscale or wavelength
Λ	macroscale
μ	viscosity
μ_0	consistency
ν	kinematic viscosity
ν_t	eddy viscosity
ξ_1, ξ_2, ξ_3	right-handed orthogonal coordinates
ρ	density
σ	interfacial tension
τ	stress (tensor)
$\boldsymbol{\tau}_D$	deviatoric stress

$\boldsymbol{\tau}_E$	extra stress
ϕ	scalar potential
ψ	stream function
ψ_s	Stokes stream function
$\boldsymbol{\Psi}$	vector potential
ω	frequency
$\boldsymbol{\omega}$	vorticity (vector)
$\boldsymbol{\Omega}$	angular velocity

Superscripts

$\char`\^$	Lagrangian specification
#	dummy variable
*	dimensionless variable
†	property of dispersed phase
'	per unit length or per unit width or fluctuating component
—	average value

Subscripts

a	ambient value
b	value associated with bubble or value associated with backward characteristic
c	characteristic value or critical value
f	value associated with forward characteristic
I	value in inertial reference frame
L	exit value
n	normal value
R	value in rotating reference frame
t	tangential value or terminal value
x	value based on axial position
w	wall value
0	entrance value or initial value
δ	value based on thickness of boundary layer
∞	value at infinity

INDEX